LIVING ON THE WIND

LIVING ON

Across the Hemisphere

SCOTT

THE WIND

with Migratory Birds

WEIDENSAUL

North Point Press

A division of Farrar, Straus and Giroux

New York

North Point Press
A division of Farrar, Straus and Giroux
19 Union Square West, New York 10003

Copyright © 1999 by Scott Weidensaul
All rights reserved
Distributed in Canada by Douglas & McIntyre Ltd.
Printed in the United States of America
Designed by Jonathan D. Lippincott
Maps designed by Jeffrey L. Ward
First edition, 1999

Library of Congress Cataloging-in-Publication Data
Weidensaul, Scott.
    Living on the wind : across the hemisphere with migratory birds /
Scott Weidensaul.
        p.   cm.
    Includes bibliographical references and index.
    ISBN 0-86547-543-1 (alk. paper)
    1. Birds—Migration.   I. Title.
QL698.9.W45   1999
598.156'8—dc21   99-11693

For Samuel C. Gundy

A fine ornithologist and an even finer teacher

Contents

Preface ix

Part One

SOUTHBOUND

1: Beringia 5

2: A Far-flung Tapestry 29

3: The Way South 57

4: Riding the Sea Wind 81

5: Rivers of Hawks 105

Part Two

HIATUS

6: La Selva Maya 129

7: Hopping Dick and Betsy Kick-up 153

8: Aguilucheros 173

9: When Anywhere Is Better than Home 197

10: Uneasy Neighbors 221

Part Three

NORTHBOUND

11: The Gulf Express   *249*

12: Heartland   *275*

13: Hopscotch   *303*

14: Catching the Wave   *323*

15: Trouble in the Woods   *345*

Afterword   *371*

Notes and Bibliography   *375*

Acknowledgments   *399*

Index   *403*

Preface

At whatever moment you read these words, day or night, there are birds aloft in the skies of the Western Hemisphere, migrating.

If it is spring or fall, the great pivot points of the year, then the continents are swarming with billions of traveling birds—a flood so great that even the most ignorant and unobservant notice, if nothing else, the skeins of geese and the flocks of robins.

But the migration's breadth goes far beyond those obvious watersheds, shifting endlessly across distance and season. In the middle of July, Hudsonian godwits lift off from the iceberg-choked shores of the Beaufort Sea, heading southeast along the northern rim of Canada to Labrador, then vaulting south in a nonstop flight to Venezuela. In the snow squalls of December, goshawks and golden eagles fly south along the ridges of the Appalachians, over oak trees that rattle their last stiff, dead leaves at the wind. Even within the tropics, a land where migration would seem unnecessary, birds move with the seasonal rains and droughts across hundreds of miles, following the blossoming of flowers or the ripening of fruit.

They don't all follow the expected course, nor do they always travel by wing. With the first winter snows, blue grouse leave the more temperate foothills of the Western mountains and migrate—on foot, no less—into the bitter, wind-drifted high country, searching for conifer needles to eat.

A very few travel even beyond the bounds of the hemisphere. Tiny songbirds from Alaska leap west across the Bering Sea to the Philippines, and others from eastern Canada cross the North Atlantic to Europe and Central Africa. Short-tailed albatrosses from Japan glide down the coast of Washington in summer on wings as fragile as a whisper; in those same waters the albatrosses pass shearwaters from New Zealand and storm-petrels from Antarctica and the Galápagos.

Even the darkness moves with the passage of birds. On soft spring midnights, the air is alive with the flight notes of unseen warblers and vireos, thrushes and orioles, sparrows and tanagers, filtering down through the moonlight like the voices of stars.

Bird migration is the one truly unifying natural phenomenon in the world, stitching the continents together in a way that even the great weather systems, which roar out from the poles but fizzle at the equator, fail to do. It is an enormously complex subject, perhaps the most compelling drama in all of natural history.

That such delicate creatures undertake these epic journeys defies belief. Only recently have scientists discovered that some shorebirds apparently fly nonstop from the southern tip of South America to the coast of New Jersey, a journey of ten days—240 hours of uninterrupted flight. Even more remarkable are the four-ounce Arctic terns that leave the northern fringe of the continent each autumn, flying east across the Atlantic to Europe. They push south along the bulge of Africa, *recross* the Atlantic to the edge of South America, and spend the winter months moving east off Antarctica. In spring they reverse course, moving up southern Africa and lancing back to Canada— a figure eight inscribed on half the globe, a track that returns them, often as not, to precisely the same sheltered nook where they nested the summer before.

Even scientists have little grasp of the numbers of birds involved in this seasonal ebb and flow. In spring, hordes of warblers, tanagers, vireos, and other tropical migrants cross the Gulf of Mexico each night, arriving on the U.S. coast at a rate that may exceed a hundred thousand songbirds per mile of shoreline, with tens of millions making landfall each day. On a single autumn night several years ago, radar on Cape Cod indicated that 12 million songbirds passed overhead. And on the narrow coastal plain of Veracruz, Mexico, biologists discovered only recently one of the greatest raptor migrations in the world, where nearly a million hawks have been counted in a single day. In all, scientists guess, more than 5 billion birds annually weave this incredible tapestry across the hemisphere.

Because they travel such extraordinary distances, often with differing requirements for food and shelter along the way, migratory birds pose one of the stickiest conservation challenges in the world. In the past, preservation programs focused on saving breeding areas, but experts now realize they must also save wintering grounds and migratory stopovers if this global web isn't to unravel.

There are serious signs of fraying. Birders have warned for years that songbird numbers were dropping, and now scientists have hard evidence that some beloved species, like wood thrushes* and cerulean warblers, have declined by more than 75 percent in the past quarter century. Entire communities of birds, like those that nest in open grassland, are in freefall. In the past, the dangers to birds were mostly direct persecution, and the remedy was legal. The Migratory Bird Treaty Act, passed in 1918 and amended several times since, is the cornerstone of federal protection, levying harsh fines and jail time for killing all but a relatively few game birds. The law grew out of a time when songbirds and shorebirds were still routinely shot for the dinner table and terns and egrets were slaughtered so their feathers could decorate hats.

Today, however, the biggest threats to migratory birds do not come from the barrel of a gun, nor are they easily cured by passing laws. They arise from habitat loss and the wholesale environmental changes we have imposed on the natural world. Laws like the federal Endangered Species Act provide last-ditch support for almost-extinct birds, but conservationists now realize the smartest approach is to recognize the trouble early and try to stabilize populations while they are still relatively common. This recognition, coupled with an impending sense of crisis, has sparked an unprecedented international conservation effort, probably the largest and most ambitious in history, involving virtually every country in North, Central, and South America in research, education, and habitat protection. By way of example, one such multinational approach is the Western Hemisphere Shorebird Reserve Network, which has identified (and in some cases prompted the preservation of) 4 million acres of wetlands between Argentina and Alaska, upon which more than 30 million migratory shorebirds depend.

---

* A note about bird names: While it isn't one of the crucial issues of the day, there is a testy little debate in ornithological circles about whether or not to capitalize the complete, proper names of birds, like Wood Thrush and Lesser Goldfinch (but not the more general thrush or goldfinch, when the specific name does not appear). Generally, those writing for scientific works tend to capitalize, arguing that it reduces confusion (which it does), while those pursuing a more literary style of writing do not, considering the practice pretentious and visually awkward (which it is). Bird names are not capitalized in this book, but they are in many of the writings from which I quote, which can make for some bumpy reading; likewise, the word "neotropical," referring to the New World tropics and the birds that inhabit them, is subject to similar ups and downs, depending on who is being quoted.

In writing *Living on the Wind*, I have tried to convey the sweep and drama of bird migration in the Western Hemisphere, which forms a largely self-contained system—the problems that migrants face and the outlook for their future. It is also the story of the people—researchers, amateur birders, land managers, and others—who are trying to understand the mysteries of migration and make the world a safer place for the birds. Propelled by an ancient faith deep within their genes, billions of birds hurdle the globe each season, a grand passage across the heavens that we can only dimly comprehend and are just coming to fully appreciate. They are not residents of any single place but of the whole, and their continued survival rests almost entirely in our hands.

The book opens on the Alaska Peninsula, a crossroad for migrants that fan out toward winter homes as far-flung as Australia, Africa, China, Fiji, and Tierra del Fuego. In subsequent chapters of the first section, "Southbound," the narrative follows the main flow of the migration across Canada and the United States—down the mountain ranges, along the gale-raked coasts, through the Plains, and into Mexico. The second portion of the book, "Hiatus," focuses on the birds and their wintering grounds, from the frozen tundra alongside Hudson Bay to the steamy rain forests of Central America and the grassy pampas of Argentina. In the final section, "Northbound," the book reverses course, keeping pace with the surging return of the spring migrants that flood the Gulf Coast plain and the Southwestern deserts, sweep across the prairie heartland of the continent, and bedeck the woodlands of North America in time for the nesting season, closing the yearly cycle once more.

This book covers a lot of ground. Over the course of more than six years, I traveled virtually the length of the hemisphere, logging nearly seventy thousand miles by jet, car, bush plane, sailing ketch, tundra buggy, dugout canoe, horseback, and on foot—yet traveling fewer miles than a single small sandpiper would in its short lifetime, propelled only by muscle and the instinct to migrate.

LIVING ON THE WIND

Part One

SOUTHBOUND

I

Beringia

Mist clouded my glasses, and a breeze lashed the long, coarse grass on the steep bluff where I sat. The knife-edge headland jutted into the gray Bering Sea, its slopes spangled with the late-summer blue of monkshood flowers. Fifty or sixty feet below me, waves washed on a cobbled beach of round black rocks, each surge making the smallest pebbles snicker lightly against one another.

The tide was half out, and looking through gaps in the fog across the half-mile width of Izembek Lagoon, I could see a delicate ruffle to the surface of the water—the tips of great eelgrass beds were just dimpling the top, and smooth channels of open water branched through them like roads. In one of these channels, a harbor seal's blocky head emerged from the waves for a brief moment, blinked its eyes, then disappeared. Nearby, two sea otters lay sprawled on their backs, wrapped in the sinuous eelgrass to keep from floating away; one was working at dinner, using its paws and teeth to wrench apart a crab.

Through the fog came a sound, a creaky, three-noted honk—*Claa-ha-aa! Claa-ha-aa!*—like someone blowing on a clarinet reed. A small flock of emperor geese broke into clear air, eight of them rowing against the wind—dark, purplish-black bodies and white heads stained with rusty orange from feeding in iron-rich tundra ponds. As the mist lifted, I saw birds in almost every direction: skeins of small, dark geese called brant, in hurried flight; squadrons of eiders and scoters and other sea ducks riding the swells, pudgy as footballs; swirls of kittiwakes, long-winged gulls with balletic grace; and smokelike tendrils against the horizon that were thousands of shorebirds, rising and falling on distant mudflats exposed by the dropping tide.

At the very edge of the continent, I was looking off across the Bering Sea toward hidden Siberia and the Old World. Izembek National Wildlife Refuge lies at the tip of the Alaska Peninsula, where this crescent of rugged land dissolves into the thousand-mile arc of the Aleutian Islands. It is a land of fire and ice, of alpine snowfields and active volcanoes, like one of the rugged cones that flank nearby Cold Bay.

I came to Izembek in early September, after weeks of wandering across western Alaska, because this tendril of land is a global crossroads. Fifteen or twenty thousand years ago, during the height of the last ice age, continental glaciers captured much of the world's water and sea levels were hundreds of feet lower. What is now the shallow Bering Sea was then a marshy land bridge, a port of entry between Asia and America. Across this bridge, named Beringia by geologists, humans wandered from Siberia, hunting mammoths, giant bison, and other ice age game, colonizing a new world.

Beringia was inundated roughly eleven thousand years ago, as the glaciers melted and the seas rose. But Asia and America still nearly touch here, a brushing kiss across the 50-mile-wide Bering Strait, and Beringia is still a way station of international significance. Some of the travelers come by sea: gray whales from Baja, salmon returning to their natal streams from the black waters off Japan and Korea, northern fur seals hauling out on the rocky islands of the Pribilofs. But far more journey by air. Many birds whose travels span the globe breed in western Alaska, and now, as summer faded to autumn, they were taking to the wind once more.

They were not leaving because the weather would soon turn cold—although it would, the raw, bone-deep cold of coastal Alaska, with its sea ice and wet snow and howling winds. Migration is, fundamentally, about food, not temperature; those birds that can continue to find enough to eat during the winter rarely migrate—why bother?—while those whose food supplies are seasonal must flee. Almost all of the more than five hundred North American species that migrate depend on weather-sensitive food supplies—the ducks and wading birds whose marshes are sealed with ice, for instance, or the insectivorous songbirds that can't find bugs in a December snowstorm. Seed-eaters are less likely to migrate than insect-eaters and tend not to go as far when they do; they can find plenty of weed seeds in North Dakota in January, but for flying insects a bird must travel at least to the Gulf States, or all the way to the tropics.

Behind me, a small cove lay in the protective lee of the bluff. Windrows of dead eelgrass formed thick, snaky ropes at the high-water mark, black against the oily gray of the mudflats. Flocks of shorebirds were feeding there with restless energy—scurrying, probing, poking into the rich tidal muck for small invertebrates. Most of the birds were dunlin, small sandpipers with drooping bills, many still in breeding plumage, with reddish backs and black bellies, as though they'd squatted in soot. Every few minutes, responding to some silent signal of alarm, they would leap into the air, wings flashing white, then twist and circle to earth again to resume foraging.

Migration is not the simple, north-to-south-and-back-again affair that most of us assume, and the shorebirds feeding on that mudflat were a perfect example. Dunlin breed in much of Alaska and the Canadian Arctic, but they form three distinct populations, each with radically different migration routes. Those that I was watching were probably of the subspecies *pacifica*, which nests in this part of southwestern Alaska and travels relatively short distances along the coast, stopping anywhere from the southeastern Alaskan panhandle to Baja California. Dunlin that nest in the central Canadian Arctic, on the other hand, migrate overland to the southern Atlantic and Gulf Coasts of the United States. And those that breed in the Northwest Territories and northern Alaska cross the Bering Strait to Siberia, where they join Russian dunlin and migrate on to eastern China, Japan, and Korea.

Among the masses of dunlin were a number of other species of shorebirds, each laying on fat before departing for far-flung destinations. Rock sandpipers, matching the gray volcanic stone, would barely budge; although some would travel to California, many would pass the wet, gloomy winter right here. Least sandpipers no bigger than sparrows, like little, buffy windup toys, would skirt the coast to South America. Among them were two Pacific golden-plovers, the color of hand-worn brass, which might take one of two routes from Alaska. Some cross the Bering Strait and follow the Asian coast, finally veering southeast to Australia or the Pacific islands. Others fly southwest across the open ocean to Hawaii, and then on to the islands of the South Pacific; some apparently make the flight to the Marshall and Line Islands—a thousand miles south of Hawaii, and a trip of nearly 4,000 miles—in a single, nonstop flight. Others on the beach and flats that day would make similar migrations—the wandering tattler, a whimsically named sandpiper, which winters across much of the South Pacific, and the squat, piebald ruddy turnstone, which travels to Southeast Asia, Australia, and the islands of Oceania, as well as to the California coast.

This single acre of mud held examples of a dozen different migration strategies. Not far from the Pacific golden-plovers stood five American golden-plovers, close cousins distinguished (with difficulty) by their duller plumage and somewhat longer wings. They also breed in western Alaska and make equally epic trips, but in the opposite direction. First they fly east across Arctic Canada, feasting on its wealth of berries and insects before gathering in the Atlantic Maritime provinces; then they swing south, cutting across the western Atlantic nearly 2,000 miles to South America, eventually ending their trip in the grasslands of Argentina. Joining them on the pampas would be greater yellowlegs, which I had seen earlier in the day along freshwater streams flowing into Izembek Lagoon—graceful sandpipers with steely-gray plumage and colorful, ripe-lemon legs. From the muskeg forests of Alaska and Canada, the yellowlegs spill south each autumn across the hemisphere, but they take an overland route, some stopping as far north as the Pacific Northwest and mid-Atlantic coasts, but others pushing on clear to Patagonia.

Simply cataloging the avian wanderers that pass through western Alaska would be a lengthy chore. Virtually every black brant on earth stops at Izembek in autumn, feeding on the world's most expansive eelgrass beds, before moving out in November for destinations as far away as Mexico. Most of the world's emperor geese and Steller's eiders, the latter a threatened species, congregate at Izembek in the fall; the geese move on to the Aleutians for the winter, but the ducks stay put in the sheltered lagoons.

In Beringia, a naturalist may find Hudsonian godwits bound for Tierra del Fuego and bar-tailed godwits headed for New Zealand; the small greenish songbird known as the Arctic warbler, which migrates to the Philippines, and Wilson's warbler, yellow with a black cap, which flies to Central America. There are fox sparrows and golden-crowned sparrows that winter in Pacific coastal woodlands, and gray-cheeked thrushes that travel to the Amazon. All morning, I had been watching wheatears, black-and-white songbirds that look like slim thrushes. They'll join their Siberian brethren and fly to China and India, then continue on to eastern Africa together, while wheatears from the eastern Arctic swing across Greenland and Iceland to reach western Africa by a European route, embracing the world in a wishbone of movement.

Out on the ocean around Izembek were more migratory wonders. The day before, an Aleutian gale had savaged the region, ripping even protected harbors to foam and fury. But from a sheltered nook overlooking Cold Bay, I had watched dark shapes skimming the waves, gliding on stiff, narrow wings that barely cleared the tops of violent whitecaps, to all appearances blithely unconcerned by the storm. These were shearwaters, fittingly named seabirds that spend virtually their entire lives on the open sea and that are among the world's most accomplished migrants.

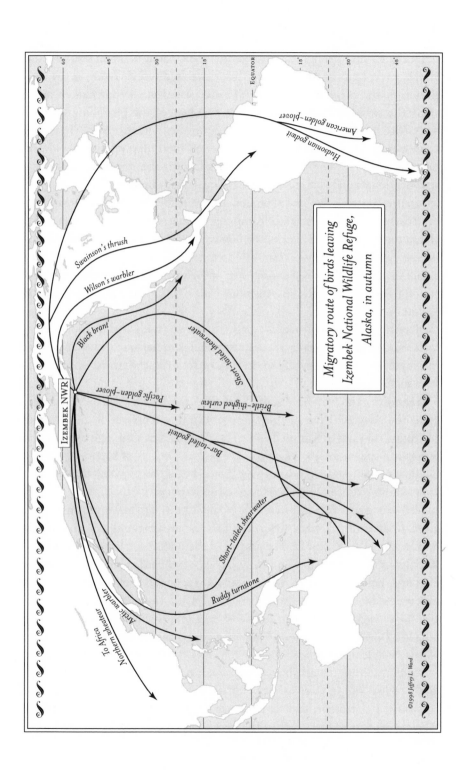

Migratory route of birds leaving
Izembek National Wildlife Refuge,
Alaska, in autumn

©1998 Jeffrey L. Ward

This particular species, the short-tailed shearwater, nests on small islands off the south coast of Australia and Tasmania. In April and May, at the conclusion of the breeding season, millions of shearwaters pour rapidly north along the western Pacific rim, taking advantage of the prevailing winds—across Oceania, past Japan, and into the Bering Sea a month later, where they enjoy perpetual subarctic daylight and a rich food supply. They remain off Beringia until September, when they head down the American coast and across the central Pacific, again aided by local wind currents. The shearwaters return to Australia by late October or November, when observers along the coast of New South Wales have seen as many as 60,000 *an hour* going past, a torrent of birds fueled by the bounty of the Bering Sea. A short-tailed shearwater may cover more than 18,000 miles in a single year, carving a vast circuit on the Pacific, aided at each step of the way by the prevailing breeze—and yet they arrive back in Australia within the same eleven-day period each year.

Migrations like this leave us staggered; we are such stodgy, rooted creatures. To think of crossing thousands of miles under our own power is as incomprehensible as jumping to the moon. Yet even the tiniest of birds perform such miracles.

Some days before, four hundred miles to the east, I had been camping along the Alagnak River, a crystalline stream that roars out of the Aleutian Mountains of Katmai National Park. The Alagnak flows through open tundra and sparse spruce forests, its banks wrapped in thickets of alder and stunted birch; several times each day, I saw brown bears lunging into the river for forty-pound king salmon as large and red as fireplugs.

The streamside thickets were full of small birds, streaky and tinged with green—blackpoll warblers, five and a half inches long and weighing barely half an ounce each; you could mail two of them for a single first-class stamp. Warblers are a largely tropical family, either as permanent residents or winter migrants, but the blackpoll is the most northerly of the clan, breeding from western Alaska across the midsection of Canada to Hudson Bay, Labrador, and New England. It is also the most southerly in its wintering grounds, migrating as far as the western Amazon, giving it the longest migration of any North American songbird.

That would be remarkable enough even if the blackpoll took the most direct course. But it doesn't. Instead, like the American golden-plovers, they cross first a continent, and then an ocean. As August wanes, most of the Alaskan birds travel east, across the boreal forests of Canada, all the way to the Maritime provinces and the coast of New England. For a blackpoll born on the banks of the Alagnak, that alone is a journey of roughly 3,000 miles. While some of them then hug the shore toward Florida, it appears that many blackpolls strike out south over the open ocean, departing the Northeast coast at dusk, ordinarily picking a night with a brisk, northerly tailwind after the passage of a cold front.

They will need the help. For the next forty or fifty hours, the tiny songbirds will fly over the western Atlantic, wings buzzing at twenty flaps a second, climbing to altitudes of more than 5,000 feet. They will show up on weather radar as they pass Bermuda and the Greater Antilles, glowing green specks that form diffuse blobs on the monitors, like ghosts beneath the moon.

The warblers follow a curving track, steered and abetted by the wind. At first, the northwesterlies carry them out to sea, the wind's push adding to the 20 miles per hour that the warblers can fly on their own. Midway, somewhere around Bermuda, the northwesterlies fail and the migrants come under the influence of the subtropical trade winds, which blow from the northeast. The tiny birds are shepherded back to the southwest, toward South America, finally making landfall along the coast of Venezuela or Guyana, an overwater trip of about 2,000 miles—a passage with no rest, no refueling, no water, during which each will have flapped its wings nearly 3 million times. "If a Black-poll Warbler were burning gasoline instead of its reserves of body fat, it could boast of getting 720,000 miles to the gallon," note two researchers.

Nor are the birds finished; although some blackpolls do winter in the rain forests of northern South America, others continue south as far as northern Bolivia and western Brazil, another 1,500 miles or so. Then, in April and May, they reverse course, making a less-spectacular but still daunting traverse of the Gulf of Mexico or the western Caribbean and returning to their breeding grounds via the interior of North America. In all, the elliptical round trip for an Alaskan blackpoll warbler may cover eleven or twelve thousand miles.

I have seen blackpolls crowding the spruce woodlands of coastal Maine in late September, the gathered multitudes of the northern woods, and heard their slightly buzzy call notes in the dark as they set off over the sea, staking their lives on an exhausting journey through storm-raked skies. Knowing that some have already come from as far as the Alaska Peninsula is humbling. Yet, to my thinking, the most astonishing of all the migrants that leave Beringia each fall are two kinds of long-legged shorebirds, the bristle-thighed curlew and the bar-tailed godwit, some of which cross not just part of the Atlantic but the entire width of the Pacific Ocean.

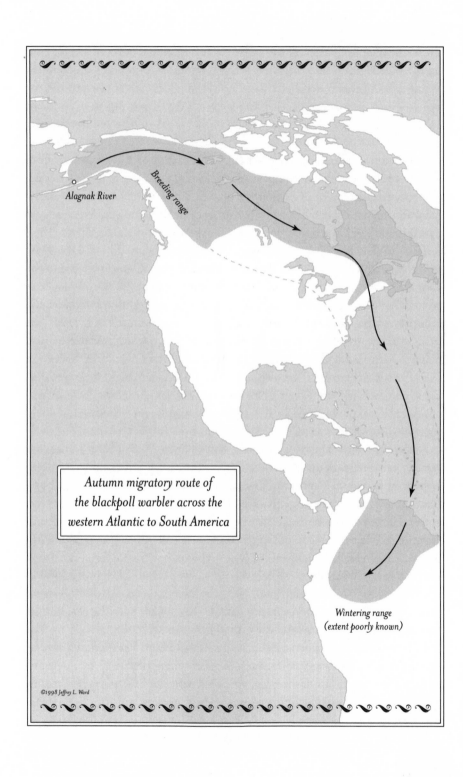

Alagnak River

Breeding range

Autumn migratory route of
the blackpoll warbler across the
western Atlantic to South America

Wintering range
(extent poorly known)

©1998 Jeffrey L. Ward

The bristlethigh has always been something of an enigma. About seven-teen inches long, it looks very similar to a common, widespread shorebird called the whimbrel—the same mottled brown-and-buff plumage, the same strongly down-curved bill and striped head. Differences between the two are subtle enough to give most birders fits: a pale base to the curlew's beak and a more cinnamon wash to its body feathers. Only at very close range can one see the unique feature that gives the bristlethigh its name—tufts of long, shiny, hairlike feathers that grow around the upper legs and whose function remains unknown.

Bristle-thighed curlews nest only in a small part of western Alaska, in the heart of Beringia from the Seward Peninsula south through the Yukon River delta. It took a long time to discover that fact. The species was recognized by scientists as early as the 1760s, when it was collected by Captain Cook's expe-dition, but the first nest was not discovered by ornithologists until 1948, in the remote tundra near the mouth of the Yukon. For another forty years its breeding ecology was essentially unknown, until fieldwork by biologists began in the 1980s. Although a sketchy picture of the curlew's life history has emerged, it remains one of the least-understood shorebirds in the world.

The curlews, biologists have found, arrive on the breeding grounds in May, when pairs set up their territories in areas with shrubby cover, placing their small, bowl-like nests beneath the tangled, protective canopy of dwarf willows. The eggs—lovely things of warm ocher, blotched with chestnut—are laid in clutches of four and incubated for nearly a month, during which time the defensive curlews may actually strike intruding predators like eagles and foxes.

Just a week or two after hatching, the parents lead their small, downy chicks overland for miles, joining other flightless broods on windswept hill-tops to form creches, or nurseries, with just a few adults watching over them. The reason for such gatherings, which may include up to thirty curlews as well as golden-plovers, whimbrels, and godwits, is obscure, but is probably related to protection against predators, especially because sharing guard duty allows the adults to feed more often, gaining back the weight they lost during the arduous nesting season.

Not long after the chicks can fly—about four weeks of age—the adult females desert them entirely, moving hundreds of miles to staging areas in the central Yukon Delta, where they will feed, gather in flocks, and prepare to migrate. The males follow shortly thereafter, and the adults gain a precious week or two in which, freed of all family responsibilities, they can focus solely on eating before migration.

In late August and early September, the entire species, first the adults, then some weeks later the juveniles, leaves Alaska and flies southwest, across the Aleutians and over the Pacific. Some will land for the winter 3,000 miles later on Laysan, Lisianski, and other tiny islands in the northwestern Hawaiian chain. Others will push on for thousands of miles more, eventually coming down on a crescent of tiny South Sea islets, from the Caroline Islands in Micronesia through Fiji, Tonga, and French Polynesia. (It was on Tahiti that Captain Cook's crew first encountered the curlew in 1769, and for the next century scientists assumed they were permanent residents of the islands.) Lacking waterproof plumage, they cannot rest on the sea, and there is no evidence that most of them stop off in Hawaii along the way. Some of the curlews may make a journey of more than 5,000 miles in a single flight, spreading out across the Pacific archipelagos.

It is extraordinary to consider what a young bristle-thighed curlew must accomplish, what obstacles it must overcome. Only five weeks out of the egg, it sits with a flock of its peers on the treeless, marshy land of Beringia. Its parents have abandoned it, fleeing to the tropics as the nights turned frosty. The curlew feeds voraciously, gobbling down purple-black crowberries plucked from the tundra and insects that swarm in the shallow ponds and across the hummocky ground. Fat builds under its skin, a thick, yellowish layer along the sides of its body, in the hollow of its neck, and under its wings; each gram of the stuff contains eight times the energy of an equal amount of protein, essential fuel for the trial ahead. The rapidly diminishing amount of daylight triggers changes within its brain—an instinctive orientation toward the south-southwest and a growing agitation that sends the flock into frequent short flights, like premigratory drills, during which they may test the winds aloft, searching for the right combination of speed and direction.

Then, one day, it is a drill no longer; the birds vault into the sky, find the right winds, and stream inexorably toward their ancestral wintering grounds. There are no adults to lead them; their parents are long gone, already in the South Sea. It is a comfortable myth that experienced elders lead the flocks of migrating birds south—for curlews, as for most species, the novices are on their own. Instead, they rely on genetically programmed cues. They orient themselves using the sun by day and the stars by night, innately compensating for those markers' changing positions in the sky; they also apparently sense the earth's magnetic field, and may use other clues as well—polarized light at sunrise and sunset that is invisible to human eyes, perhaps even very low-frequency sound waves generated by trade winds and ocean surf.

For most birds, migration is a leap of blind faith, an instinctive urge over which they have no real control. The curlew does not "know," in a conscious sense, that coconut palms and placid coral atolls await it in Tonga or Fiji—it can sense only an urgency to fly in a certain direction for a certain length of time, following a path graven in its genes and marked by the stars.

Some will not make it. They have been born with a faulty navigational sense, their internal compass skewed a few degrees; these most often die lonely deaths in the vastness of the Pacific. Others will not survive because they couldn't lay on enough fat for the long, nonstop flight—a fault of their physiology, perhaps, or a failing of their environment, a drought or late frost that reduced the berry crop. Storms will claim some, predators others, weakness and exhaustion and disease and all the other implacable agents of natural selection taking still more.

Yet most *will* survive, to descend on quivering wings to coral sand beaches and shimmering mudflats beside cobalt waters. Banding studies have shown that bristle-thighed curlews are extremely long-lived, with some surviving more than twenty-three years, and they have exceptionally low adult mortality from year to year. The system, impossible as it sounds to us, works very well for the birds. The South Pacific is as different a place from the clammy, permafrost-girded tundra as it is possible to imagine, yet the curlews slip neatly into the local ecosystem. In the islands, bristle-thighed curlews become opportunistic hunters, using their long, probing bills to capture bugs and spiders, crabs, even scorpions; they are not above scavenging carrion or stealing eggs from under the breasts of incubating seabirds. There are even reports of tool use by the curlews, which may throw small stones at especially obdurate eggs.

The Pacific islands were once an idyllic place for a bird; most of the islands had no naturally occurring mammalian predators, and the curlews—alone among the world's shorebirds—evolved the tendency to molt most of their large wing feathers at once after reaching their wintering grounds, rendering them flightless. That became a disadvantage once the Polynesians arrived, a thousand or more years ago; humans started trapping the curlews for food, and introduced animals like pigs and dogs (and later, rats and cats from European ships) took a toll. The curlews are no longer found in any significant numbers on many of the larger islands, like the main Hawaiian chain, and today's population of roughly 10,000 is undoubtedly smaller than in centuries past, though as far as anyone can tell, the species is fairly stable.

But stunning as the curlew's migration is, the bar-tailed godwit's exceeds it. The same size as the curlew, but with a long, straight, ice pick of a bill, the bartail nests in eastern Siberia and western Alaska. In mid-July, the Alaskan birds leave their tundra breeding grounds and congregate by the tens of thousands along the Alaska Peninsula, in places like Nelson Lagoon and the mouth of the Egegik River, where they gorge on tiny clams pulled from the intertidal mud. They eat so much that the fat forms in thick rolls—more, by proportion of body weight, than any other wild bird, up to 55 percent of their total mass. And then, when they can eat no more, the godwits undergo a remarkable internal change. Their kidneys, liver, and intestines atrophy, shrinking to a fraction of their usual size, a trait scientists suspect many long-distance migrants share. Obese with fuel, freed of the baggage of heavy guts, the godwits are ready for the air.

They wait for wind, for one of the big autumn storms that rip through southwest Alaska every few days in September. As soon as the front has passed and the pounding wind shifts out of the north, they take off by the hundreds or the thousands, aiming south over the open ocean.

The godwits will not stop again until they reach New Zealand, as much as 6,800 miles away, a trip that will last at least four or five days, even with the wind at their backs and their flashing wings pushing them at 45 miles per hour. Robert E. Gill, Jr., of the Alaska Biological Sciences Center, who has been studying godwits and curlews for years (and who has documented much of this remarkable story), believes the godwits' transpacific flight is the longest nonstop bird migration in the world—for unlike seabirds like terns or shearwaters, which can rest and feed along the way, the godwits (like the curlews) will drown if they land on the ocean.

But wait a minute, the skeptical part of the mind protests, how could this be? How does a migration like this evolve? It's easy to imagine how migratory paths might lengthen on a continent—perhaps birds pushing farther and farther north each year as ice-age glaciers retreat, returning in winter to a traditional nonbreeding ground that lies farther away with each generation. And that is precisely how ornithologists have long believed some migratory routes did evolve. But how did a godwit first find New Zealand? And how on earth did a bristle-thighed curlew first find a flyspeck like tiny Rangiroa Atoll in the vastness of the Pacific?

Perhaps by accident. A pioneering bird, like a pioneering human seafarer, has no way of knowing what's beyond the horizon. Experiments have proven that migratory birds inherit a genetic urge to travel in particular directions at particular times of the year, thus retracing a path their ancestors successfully followed for generations. But whenever an animal blunders off in new directions, their biological compass awry, any discoveries they make are pure serendipity.

The bristle-thighed curlew's nearest relative is believed to be the whimbrel, which is found in much of the Northern Hemisphere, including Alaska and Siberia. Siberian whimbrels migrate along the western Pacific rim, down through Japan to Southeast Asia and the islands of Australasia. Alaskan whimbrels also take a coastal route to the Pacific shores of Central and South America. Whimbrels do not regularly migrate straight across the Pacific—but they *do* show up in places like Hawaii occasionally, the odd vagrant victimized by its own bad sense of direction.

Such misfits would ordinarily be culled out of the population—if you fly blindly into an ocean as wide as the Pacific, your chances of finding land are pretty slim. But every once in a while, one will stumble across an island, the dumb-luck payoff for having a mistuned guidance system. Once, before humans found the islands, this would have been like hitting the jackpot—a warm, safe place with no ground predators. Some of their progeny, inheriting the gene for the "defective" migratory route, would repeat the journey, although only those capable of making the arduous, nonstop flight would survive to breed. If this transoceanic trip offers them a survival advantage (and a lack of competition on the wintering grounds is a very big advantage indeed), then over time, a distinct population of long-distance migrants arises. If there is geographic or behavioral isolation on the breeding grounds, which keeps them from mingling with others of their kind and diluting the genetic changes, they may eventually become a separate species—the bristle-thighed curlew.

That's one possibility. The other, first proposed in the 1950s by ornithologist Dean Amadon, is that the bristle-thighed curlew is a "relict species," one whose range had once been much greater, like the tattlers and sanderlings that also migrate to the islands but winter on the mainland as well.

"It is natural to suggest that the Bristle-thighed Curlew is a relict species whose wintering range *once* included such continental shores, but has *now* been restricted to the central Pacific islands, presumably because it could not compete successfully with shorebirds that winter in continental areas," Amadon wrote. This theory also assumes that the curlews stumbled onto the islands by accident, but unlike the first hypothesis, they would already have evolved into a separate species. Because the curlew's breeding range is as restricted as its wintering grounds, Amadon theorized that conditions on the islands keep the species from growing in numbers and reclaiming larger areas of mainland. "Gradual restriction to an insular winter range as a species declines may, it is suggested, have produced certain remarkable migrations otherwise difficult to explain."

This is still just speculation; since Amadon's hypothesis was published in 1953, no one has made a serious attempt to decipher the origins of the curlew or its migratory pattern. One thing is certain, however—transpacific shorebird migration has been going on for a very long time. On the Hawaiian island of Oahu, Pacific golden-plover bones dating back more than 120,000 years provide graphic evidence that birds from the Arctic have been crossing the world's largest ocean for millennia.

As I sat on the bluff overlooking Izembek's mudflats, staring through my binoculars at the shorebird flocks below, it dawned on me that the misty rain had stopped and the sky was brightening. Within seconds, the low overcast shredded, revealing craggy Amak Island offshore, black against a suddenly golden sea, and snow-crowned Frosty Peak, an inactive volcano fifteen miles inland, rugged and dramatic with its wreath of clouds. I loosened the hood of my raincoat, welcoming a shaft of late-day sun on my face. The weather on the doorstep of the Aleutians is infamously bad, but from time to time it doles out these priceless moments, when the climate suspends its ill-tempered war just long enough to assure a visitor that sunshine still exists.

With my hood down and the blinders off, I saw that flocks of small birds were moving nearby. I "pished" a few times to bring them closer—that is, I made the kind of sibilant, hissing sound with my lips that one uses to attract a cat, *pishpishpishpish*, and which also draws curious songbirds. Most of the birds were Lapland longspurs, beefy, sparrowlike birds colored with fawn and russet. The males were already molting out of their breeding plumage, and lighter feathers were pushing through their once solidly black faces and chests. It gave them a frazzled, unfinished look, like men who have run out of the house unshaven. With the longspurs were a few snow buntings, whose wings flashed white whenever they fluttered, and two wheatears picking insects along the periphery of the flock.

The longspurs are what ornithologists call "complete migrants"—that is, they (like the godwits and the curlews) entirely vacate their breeding grounds at the end of the nesting season and migrate to a separate winter range. Although we tend to think of migratory birds as flying far to the south, the destination needn't be tropical, and the longspurs travel only to the prairies and beaches of the northern United States. Snow buntings also winter across southern Canada and the northern states; I often see them near home in

Pennsylvania, scratching with longspurs for seeds in the strips of manure spread by Amish farmers on snowy fields. But because a small portion of the population remains year-round in western Alaska, snow buntings are technically "partial migrants," the term for a species that leaves only a portion (even a substantial portion) of its breeding range.

The northern wheatear is a complete migrant, and one whose complex travel patterns still bear the imprint of the last ice age—the classic example of how migratory routes can evolve. They are elegant birds, pewter-gray on the back and head, white below, with black wings and a black robber's mask. Their curious name comes from the Anglo-Saxon term *hwit-oers*—"white-ass"—and sure enough, when one of the wheatears flew past me, skimming the ground, its rump feathers glowed in the sun, like a golf ball cutting a low trajectory.

There are nearly two dozen species of wheatears, almost all of them restricted to the Old World, especially Africa and the Middle East. The lone exception is the northern wheatear, which, in addition to breeding in Europe and northern Asia, is also found in Alaska and the Yukon and in the eastern Canadian Arctic south to Labrador. In autumn, the Eastern population crosses Baffin Bay to southern Greenland, then waits for northwest winds to make a single-stage flight over Iceland and Britain, down the Iberian Peninsula to Gibraltar. Finally, they cross the Sahara to the tropical savannas of West Africa's bulge, having flown roughly 5,000 miles.

These Alaskan and Yukon birds, on the other hand, migrate west across the Bering Sea, where they join Siberian wheatears in trekking across Russia before angling south through Turkey and Syria. They cross the Sinai and Red Sea into East Africa and fetch up for the winter in the grasslands of Tanzania, a trip of nearly 7,000 miles—one that trades caribou and brown bears as neighbors for elephants and lions.

Such long migrations entail not only behavioral adaptations but physical changes, too. If you take museum specimens of northern wheatears from eastern Canada, Europe, and northern Africa and compare the length of the outermost wing feathers—what ornithologists call "primary projection"—you'll see that the African birds have relatively short, stubby wings, just adequate for crossing the Sahara. European wheatears have somewhat longer primary feathers, while the Canadian birds have noticeably long, pointed wings—a clear aerodynamic advantage for birds that must negotiate thousands of miles of land and water.

Differences in primary feather length also crop up between closely related species of birds, even among subspecies, depending on their migratory habits.

The blackpoll warbler, the most highly migratory of its colorful family, has the longest primary projection; the wing lengths of cerulean and bay-breasted warblers, which also winter in South America, aren't far behind. Common yellowthroats and pine warblers, on the other hand, may travel only to the Gulf States, and their fairly short wings reflect this much less draconian migration.

As I said, the northern wheatear's split migration is considered the text-book example of how migratory routes may evolve. During the last ice age, wheatears were presumably restricted to ice-free zones south of Europe, but as the glaciers melted 12,000 years ago, they began expanding north. Each year, they moved farther—into Europe, and east across northern Asia, returning to Africa each winter. Colonization of even a large continent can take place very rapidly; in North America, birds like cardinals and mockingbirds, once southern in distribution, pushed quickly north over just the past century and are now found into southern Canada. Strong winds can help spread birds to islands like Britain, Iceland, and Greenland, and even farther; spring storms often deposit northern European songbirds like fieldfares in eastern Canada, and cattle egrets, an African heron blown across the Atlantic, spread completely across North America in just sixty years, developing their own migratory route back to the New World tropics.

Wheatears are still on the march westward, having colonized the eastern Arctic even as Siberian birds moved across Beringia into Alaska and western Canada. Nor were they alone; several other species of Asian birds have expanded into Alaska, including the yellow wagtail, which still migrates each year to Southeast Asia, and the Arctic warbler, which travels as far as the Philippines. Meanwhile, the exchange is working in the opposite direction, too. Pectoral sandpipers and long-billed dowitchers, two North American shorebirds, have colonized northern Siberia, but most of them recross the Bering Strait to winter in the Western Hemisphere—as far south as Argentina in the case of the pectoral sandpiper.

But why journey so far, when suitable wintering grounds could be found much closer? The short answer is that a bird has no way of knowing that if it flies southeast instead of southwest, it will wind up in a tropical forest in Guatemala, as opposed to its ancestral tropical forest in Thailand. It must rely on its genetic program, which often forces it to retrace the long, laborious route followed by preceding generations. That may be why blackpoll warblers from Alaska first fly east across the continent before turning south; they may be reversing the course their ancestors followed in colonizing western North

America. Such "crooked, illogical, out-of-the-way routes," states one ornithol-ogy text, "are not the paths of biological necessity, but are the fruits of tradi-tion, much as the crooked streets of Boston trace their lineage to ancestral cowpaths."

Migration isn't simply a way for birds to avoid the cold—in fact, some non-migratory birds tolerate the most frigid temperatures on the planet, in the Arctic and along the edges of Antarctica. We think of migration as a moving away from something unpleasant, when it is just as often a *moving toward* something beneficial—moving from the tropics toward a summer breeding ground full of food, for example. As ornithologist Paul Kerlinger has noted, "Migration evolved as a way for birds to exploit resources that are seasonally abundant and avoid times . . . or places where resources are scarce or weather is very harsh. Many species can tolerate cold temperatures if food is plentiful; when food is not available they must migrate."

It is a North American bias that we think of birds "going south for the winter"; the scope of migration in the Western Hemisphere is vastly more complex than that. In the cone of South America, places like Argentina and Chile with clear-cut winters like ours, many species of birds fly *north* to escape the cold; some of them arrive in the tropics just as boreal migrants are depart-ing for North America, an avian changing of the guard. Even within the trop-ics themselves—ecosystems that, to our uncritical eyes, would seem to have no discernible periods of scarcity—there are great migrations of parrots, tou-cans, hummingbirds, and other species, often timed to coincide with the ripening of fruit or bamboo seeds, or the blossoming of flowers.

As the short-tailed shearwaters demonstrate, the oceans of the world seethe with migrating birds, few of which are moving to avoid bad weather. Most often, they are traveling from their breeding ranges on tiny, isolated islands to rich feeding grounds that may be many thousands of miles away—and very close to the opposite pole, since warm equatorial waters tend to have less food. It is also a pursuit of the sun: the Arctic tern, which nests at high northern latitudes and winters in the extreme south, enjoys a greater percent-age of daylight in its life (and thus more hours in which to hunt) than any other animal on earth.

On land, migration may be partially east-to-west instead of north-to-south, as shown by the blackpoll warbler and the American golden-plover; red-heads, common diving ducks, may migrate from Utah to the Atlantic for the winter, while harlequin ducks that nest in the Rockies migrate west to the Pacific. Then, too, migration may be vertical rather than horizontal—up and down mountains, instead of north and south. This kind of altitudinal migration is common in North America among many Western mountain species—Clark's nutcrackers, white-tailed ptarmigan, gray-crowned rosy finches, and dark-eyed juncos among them—moving to lower elevations in the winter, then back up to alpine meadows and forests in summer.

But even *that* commonsense arrangement has an exception. The blue grouse spends the summer in deciduous woodlands of lower valleys, where there are plenty of insects, berries, and seeds for the chicks. With autumn, the birds begin to move—on foot—higher into the Rockies or the Sierras, passing the other altitudinal migrants heading down. By the time winter hits, the grouse are up near the timberline, feeding on the needles of conifers; one grouse may remain in a single tree for weeks, nibbling its needles, sheltering beneath its snow-draped limbs, and leaving enormous piles of droppings when the snow finally melts.

Wherever a migrant bird goes, it must fit itself to the environment. During migration, it may pass through dozens of different habitats and ecosystems, often requiring exquisite timing to arrive just as local food resources are near their peak, fueling the next leg of the trip. And like the wheatears that trade Alaska for Africa, the changes between summer and winter homes can be dramatic.

It was once thought that migrants were perennial misfits on their wintering grounds, existing in the odd corners, moving from place to place like vagabonds to avoid competition with resident birds—"the view that migrants are forever the 'new kids on the block,'" as ornithologists Eugene S. Morton and Russell Greenberg put it. Research in the past twenty years has shown how false that notion is, however. Migrant birds fit masterfully into their wintering habitat, requiring an almost chameleonlike ability to change their basic behavior, such as switching from a diet of insects in the north to one of fruit in the tropics. Others have co-evolved with tropical plants, the most remarkable example of which may be the orchard oriole, a lovely bird of the East and Plains with black-and-burnt-orange plumage. On its Central American wintering grounds, it feeds heavily on nectar, especially from a coral bean tree in the genus *Erythrina*. As the orioles open the flowers to reach the nectaries within, they expose dark orange petals that match the color of dominant adult males. After the flock has been feeding for a while, the tree looks as if it is full of male orioles, which seems to spook the original flock into moving to another *Erythrina*—in the process carrying the pollen, dusted on their faces by the flowers, to the next tree, ensuring cross-pollination. As far as Morton can determine, while other species of birds feed on its nectar, only male orchard orioles pollinate the *Erythrina* tree.

Humans have known about bird migration for thousands of years, even if they didn't understand the causes. Even the most primitive hunter-gatherers— *especially* hunter-gatherers, whose lives depended on such knowledge—recognized that birds were more common at certain times of the year and that their movements sometimes had purposeful direction. Where they thought the birds were going, and what they were doing, is lost to history, but the Old Testament observes that "the stork in the heaven knoweth her appointed times; and the turtle[dove] and the crane and the swallow observe the time of their coming."

The Greeks got some of it right. In the eighth century B.C., Homer wrote of "cranes which flee from the coming of winter and sudden rain and fly with clamor toward the streams of the ocean." Aristotle, several hundred years later, was a bit more specific: "Some creatures can make provision against change without stirring from their ordinary haunts; others migrate . . . as in the case of the crane; for these birds migrate from the steppes of Scythia to the marshlands south of Egypt where the Nile has its source . . . Pelicans also migrate, and fly from the Strymon to the Ister, and breed on the banks of this river." But he also made a comment that stained the study of bird migration for more than a thousand years—for straight through the Middle Ages, Aristotle's word was law among learned men. Aristotle declared that "certain birds (as the kite and the swallow) . . . decline the trouble of migration and simply hide themselves where they are." Storks, dippers, doves, and larks, Aristotle claimed, all hid themselves in holes in the ground. "Swallows, for instance, have often been found in holes, quite denuded of their feathers, and the kite on its first emergence from torpidity has been seen to fly from out some hiding-place." The Greeks also believed in transformations; Aristotle asserted that the European redstart, a common summer bird, simply changed into the Old World robin, in truth a different species that migrates south into Greece in the winter, when the redstarts have all gone to Africa. Educated medieval Europeans took the whole notion of metamorphosis a giant step further. They saw small, black-and-white geese appear each winter, but did not know they migrated from the Arctic—so they named them barnacle geese, assuming that they sprouted from the peculiar goose-necked barnacles that cling to floating driftwood.

The barnacle misunderstanding was cleared up in the 1600s, but the concept that swallows wintered beneath the water was a persistent one. In 1555,

Olaus Magnus, the Archbishop of Uppsala, threw his clerical weight behind the idea. "They cling beak to beak, wing to wing, foot to foot, having bound themselves together in the first days of autumn in order to hide among the canes and reeds," he wrote. Fishermen often hauled the torpid swallows from the water in their nets; inexperienced men who took them home, where the warmth of the fire roused the birds, soon had a house full of swallows, he said.

This belief in aquatic swallows lingered as late as the eighteenth century and counted a number of prominent biologists, including the great Carolus Linnaeus, as its supporters. Ironically, in 1555, the same year that the archbishop wrote his treatise, a Frenchman named Pierre Belon published *History of the Nature of Birds with their Descriptions and Naive Portraits from Nature* based on his travels in the Middle East. He had little patience with Aristotle's denuded doves or submerged swallows. "We shall say frankly that, although some people have believed that turtle-doves hide and lose their plumage in the winter, we have seen them at this season when they are away from us," Belon wrote, with refreshing common sense and the benefit of firsthand experience. "As swallows cannot spend the winter in Europe both because of the great cold and because they would not find food, they go to Africa, Egypt and Arabia, where, since winter resembles our summer, they have no lack of nourishment."

There is an ironic twist to this idea of hibernation, however, that somewhat vindicates Aristotle. In 1804, in what is now North Dakota, explorer Meriwether Lewis found a poorwill, a nightjar related to the whip-poor-will, in such a torpid state that it was easily captured and barely moved for several days. No one ascribed any importance to the incident until 1946, when an ornithologist in the California desert discovered a poorwill hidden in a rocky hollow, immobile in a deep torpor. Its body temperature was half the normal level, and its heartbeat was so slow as to be almost undetectable. Since then, a number of other hibernating poorwills have been found, to the amazement of scientists—but not the Hopi tribe in Arizona, which long ago dubbed the poorwill *Hölchoko*, "the sleeping one." Hummingbirds that live in high, cold elevations may drop into a profound torpor on chilly nights to conserve energy, as to a lesser extent do swifts and chickadees, the latter in harsh Northern winters. But aquatic hibernation still seems to be strictly the purview of turtles and frogs.

Scientific interest in migration—especially the mechanisms by which birds orient themselves and navigate across vast distances—increased dramatically in the twentieth century, first through the use of lightweight, numbered leg bands, and more recently through high-tech wizardry like radio telemetry. Still, it is surprising how many fundamental gaps exist in our knowledge of bird migration—even such basic questions as where certain species go. In the case of Beringia's spectacled eider, that was answered by sheer good fortune.

Eiders are large, burly sea ducks, and the male spectacled eider, found in western Alaska and eastern Siberia, is one of the most arrestingly plumaged birds in the world—black-and-white body, a pale, pastel-green head with large white "goggles" surrounding the eyes. Females are drab brown, but have the same pale goggles, giving them a wide-eyed, perpetually startled expression.

As recently as the mid-1990s, no one was sure where spectacled eiders went in the winter, after leaving the ponds and lakes of coastal Beringia. "Winter quarters remain unfound but presumably [are] in Bering Sea area, perhaps variable each year according to extension of pack ice," speculated one scholarly work on the waterfowl of the world. The location of the wintering grounds was of more than simply academic interest; the number of spectacled eiders breeding in Alaska had dropped drastically since 1986, and the species was listed as threatened by the U.S. Fish and Wildlife Service (USFWS) in 1993. The causes for the decline were unclear, and many scientists wondered if something on the undiscovered wintering grounds was killing the eiders.

Once, ornithologists had fairly limited tools for studying migratory birds, but the miniaturization revolution that swept electronics in recent decades opened new avenues for research. In the summer of 1994, federal biologists working on the Yukon–Kuskokwim Delta in western Alaska surgically implanted tiny transmitters in the body cavities of fourteen spectacled eiders. Unlike older styles of transmitters, which emit radio signals that are picked up by hand-held antennas, these high-tech devices turn themselves on and off at preset intervals and beam their coordinates directly to satellites orbiting far overhead. That way, for months on end, scientists sitting in a distant office can cheaply and easily monitor dozens of birds scattered across remote, almost inaccessible terrain.

At least, that's what should have happened. Instead, the eiders' body heat wore down the implanted battery packs more quickly than expected, and the transmitters fell silent long before the ducks moved to their mysterious winter quarters. But six months after the equipment failed, the transmitter implanted in one hen eider sparked back to life just long enough to broadcast a set of coordinates, far out in the Bering Sea.

Must be a mistake, biologists thought; the coordinates were deep inside the winter pack ice, far from the open water that ducks would need. But it was all they had to go on, so in March 1995, USFWS biologists Greg Balogh and Bill Larned climbed into a plane and roared off into one of the most hostile environments on earth. What they found astounded them—more than 155,000 eiders, jammed together in small openings and leads in the pack ice, all in a twenty-mile-wide area more than one hundred miles from St. Matthew Island, the nearest land. As far as Balogh and Larned could tell, the openings were maintained in the thirty-below cold solely by the body heat and movement of the eiders themselves, which were presumably feeding on mollusks and invertebrates in the waters below.

Besides solving a long-standing mystery, the discovery of the spectacled eider's wintering grounds had another, unexpected result—the realization that the bird is a good deal more plentiful than originally thought. Instead of barely 50,000 in the world, as believed when the eider was given Endangered Species Act protection, Larned estimated the 1996 wintering population at a quarter million. It now appears that, while the western Alaskan population did decline drastically, spectacled eiders breeding on the North Slope and in Siberia were more common than anyone suspected.

For millions of years, the birds of the Western Hemisphere have adapted to their environment, running an ever-shifting gauntlet of hazards and roadblocks. Glaciers came and went, the climate warmed and cooled, continental land bridges linking North America with Asia and South America formed and breached and formed again. Yet nothing in geologic history quite compares to the overwhelming changes that mankind has imposed on the global environment, or the rapidity with which these changes are taking place.

Migratory birds are facing what may be their greatest test since the earth entered the ice ages, nearly 3 million years ago, and maybe their greatest challenge ever. Migration depends upon links—food, safe havens, quiet roost sites, clean water, and a host of other resources, strung out in due measure and regular occurrence along routes that may cross thousands of miles. But we are breaking those links with abandon. Whereas most natural changes in climate or habitat are incremental, spread out over many hundreds, even thousands of bird generations, we are altering the landscape of migration in a heartbeat.

A blackpoll warbler from Alaska must face the same dangers its ancestors did, transcontinental and transoceanic flights. But even if it reaches South America safely, there is no guarantee that its wintering habitat—lowland evergreen forest—will be there; although Amazonia still has vast, untouched areas of rain forest, they are shrinking rapidly, and what was virgin forest last year may be a clear-cut or a pasture this season. If it finds a place for the winter, the warbler must still survive a spring flight that takes it up through Central America to the Yucatán Peninsula, through one of the most altered landscapes in the tropics, through countries like El Salvador that have virtually no natural habitat left.

And before we North Americans point self-righteous fingers at the south, consider how badly we have tattered our piece of the migration tapestry. From the Yucatán, the warbler must fly 600 miles across the Gulf of Mexico—a piece of cake if the bird catches a southerly tailwind, but a nightmare if it meets a cold front halfway, as happens frequently in spring. Then its survival may depend on finding suitable forests as soon as it reaches the Gulf Coast, insect-rich woods where it can quickly regain its lost fat reserves. But here again, human development has all but destroyed the maritime forests on which these travelers depend, turning them into vacation resorts, cattle pastures, casinos, and commercial pine plantations.

Also crossing the Gulf in spring are millions of other songbirds comprising dozens of species, most of which breed farther south than the blackpolls' boreal range. They must contend with the radical alteration of the continent's middle latitudes—the fragmentation of once-contiguous hardwood and conifer forests into increasingly isolated islands, laced by roads, powerlines, and housing tracts, awash in nest parasites like brown-headed cowbirds that hijack their nests and predators like house cats, blue jays, and raccoons that eat their eggs and chicks. Grassland birds like bobolinks and dickcissels face an even tougher hurdle; in many parts of the Great Plains, more than 98 percent of the original prairie has been converted to crops.

Yet the picture is not all dismal. We have, at last, begun to recognize the aesthetic and ecological value of the migratory whole and to work to preserve—in some cases, even restore—the land and resources upon which hundreds of species of migrant birds depend. After decades of piecemeal, provincial efforts, conservationists are taking an international approach to migrant protection, through cooperative research, education, and land preservation on a hemispheric scale. For a few species it is already too late, but for others, there is still hope.

Most important of all, we are finally coming to more fully appreciate the drama, the sweep, of bird migration in the Western Hemisphere. It is by turns surprising, uplifting, morbid, exhilarating—but it is always humbling. Earthbound, we watch creatures whose lives are spent on the wind. It leaves us awestruck, wishing that we, too, had wings.

2

A Far-flung Tapestry

Our ancestors saw the legions of birds come and go with the seasons and wondered why; the only tools they had for finding an answer were their eyes. Today, we can use miniature electronics, radar, remote imaging, and computers to study the phenomenon of migration, but sometimes eyes are still best.

It was the twenty-third of September, and the moon was a night past full, casting a lovely, pearly light through rainclouds that thinned and fragmented with the passing cold front. I sat on the front steps of my old farmhouse, adjusting the legs of a tripod and focusing my spotting scope on the face of the moon. Checking my watch, I made a note on the clipboard by my side, squinted through the eyepiece, and began waiting.

At first there was nothing. The sky was still a little hazy from the rain, and the moon glowed through a veil of thin, high overcast across which, every few moments, drifted the tattered remnants of a thicker, murkier cloud.

Then, flashing across the moon in less time than it took to catch my breath, was the tiny, wing-beating form of a songbird. A long pause, then two more in tandem, and a third. Another pause, and more silhouettes passed. For the next hour I sat and squinted and counted the birds, which rode the storm front south through the moonlit night.

Most people are surprised to find out that birds migrate after dark—not just nocturnal species like owls, but hundreds of otherwise diurnal, or day-active, species—songbirds like warblers, flycatchers, vireos, tanagers, orioles, and sparrows; sandpipers, plovers, ducks, geese, and swans; egrets, herons, and other wading birds. In fact, the vast majority of migration goes on at night, out of sight and, for most humans, out of mind. In autumn, the heaviest flights tend to be after the passage of cold fronts, when the northerly winds provide a boost for the travelers, as much as doubling their ground speed and saving them a great deal of energy they would otherwise squander from their carefully hoarded fat deposits.

That so many diurnal birds should migrate at night has always been puzzling. Nocturnal migration may allow them to avoid predators like hawks, or free up daylight hours for feeding and rest. But the most powerful reason may be that the night sky is a more forgiving medium through which to fly. A warbler, flapping its wings at a constant twenty beats per second, generates tremendous amounts of muscular heat, raising its body temperature from about 100 degrees Fahrenheit at rest to more than 108 degrees; this the chilly night air wicks away. Likewise, night air tends to be damp, especially in the humid aftermath of rain, reducing the amount of water a small bird loses with each of the nearly one hundred breaths it takes every minute. Nor is the atmosphere as turbulent as in the day; the winds tend to be less chaotic, with fewer jarring updrafts.

Every rainless autumn night, half an hour after sunset, the land sighs a great, upward breath of birds, which climb to one or two thousand feet before leveling out, their beaks pointed to the south. More and more join the exodus as the night wears on, reaching a peak around ten o'clock; as midnight comes on, the birds begin to drop out, although some continue to fly all night.

One way that scientists know this is by doing what I was doing—moonwatching, one of the oldest techniques for studying nocturnal migration. It was not a major flight that evening; my counts averaged four birds per tenminute block of time. Of course, the moon's disk is only a fractional part of the sky, but it represents that cornerstone of science, a random sample. If you were to draw a line from horizon to horizon, the full moon covers about half a degree of the night sky along it—$\frac{1}{695}$ of a circle, or $\frac{1}{347.45}$ of the visible sky, to be more precise. To estimate the number of birds passing during my count session, I simply multiplied my average of four birds by six for the hour, then multiplied those twenty-four songbirds by 347.45. According to my observations, about 8,340 birds should have passed within view of my twenty-power scope—a span about a mile and a half wide—during that one hour. That sounds like a lot, but along the Gulf Coast, the launching point for many species that winter in Central America, researchers have documented as many as a million songbirds per mile passing during a five-hour period at the peak of the autumn migration. And on one exceptional night off Cape Cod some years ago, radar showed an estimated 12 million songbirds pouring south.

Staggering as those numbers are, they are only half the story—much less than half of it, I think. What was more important to me, as the north wind began to whinny through the limbs of the shade trees, was the knowledge that the night was alive with secret movement, unknown and almost unknowable. And alive with secret sounds; for even after I folded my tripod and put away my clipboard, the multitudes overhead spilled music across the land— not the complex songs of spring, but monosyllabic flight calls, lovely in their simplicity, like Shaker hymns. I heard the chips and squeaks of several species of warblers, some notes pure and sweet, others hoarse and fuzzy, interrupted by the explosive *eek!* of a rose-breasted grosbeak, which sounded as though someone had just goosed it. A flock of Swainson's thrushes passed above me, their reedy, piping voices like the spring peepers that sing in my marsh each April. Perhaps this is what the ancients meant when they spoke of the "music of the spheres."

Birds migrate because the earth is tilted. The planet's axis is tipped off the vertical by an average of 23.5 degrees, and as it swings through its orbit each year, first the Northern Hemisphere and then the Southern are pointed toward the sun, producing the seasons. The seasons, in turn, produce alternating times of plenty and times of hardship, which spur the great hegiras of birds to and from the high latitudes.

Some movements are subtle, like those of dark-eyed juncos—sooty, somber little sparrows that filter back to my weedy meadow each autumn, usually arriving within a day or two of October 12. Other migrations shake the sky with their clamor. On November 2, 1995, following an unusually balmy autumn, a powerful blizzard exploded out of the Canadian prairies, pushing ducks and geese down the midsection of the continent like a meteorological broom, in numbers so great and flocks so compressed that they overwhelmed the air traffic control radar at the Des Moines, Kansas City, and Omaha airports in quick succession. Waterfowl biologists estimated that at least 50 million, and perhaps as many as 80 million, ducks and geese were involved in the mass migration.

When I say that these birds were migrating, everyone knows what I mean. But what is migration? The question is far more complicated than it appears. Migration, in fact, means different things to different people. To entomologists, it often refers to significant, single-direction movements of insects, a one-way flight with no return. The study of insects and oceanic plankton "has revealed movement patterns which, although they do not involve round trips, have the same function as so-called true migrations," writes entomologist Hugh Dingle. "That is, they allow exploitation of different habitats as life history requirements alter or as environments change seasonally or succes-sionally." Infectious-disease specialists talk about the potential for northward migration of malaria or dengue fever if the global climate warms. That is also what "migration" means when we talk about humans—an expansion of cul-tures or ethnic groups, like the black migration from the South following World War II, or a flood of refugees fleeing war.

However, to ornithologists, and to most non-scientists, migration means a seasonal to-and-from movement—what is often called "return migration," because the animal eventually winds up where it began. The problem is that, even among birds, there is no sharp dividing line between what is migration and what is not. A hermit warbler that moves twice each year between a grove of Douglas-fir trees in western Oregon and a stand of pines in the high-lands of Oaxaca, Mexico, is clearly migrating. A green-winged teal that swims from one side of a pond to the other, most of us would agree, is clearly not. But what about a red-tailed hawk that shifts its activity range a couple of miles each winter, away from snow-covered uplands and toward a more shel-tered river bottom? Or a snowy owl that spends years at a stretch in the Arc-tic, but comes south once in its life when the lemming population crashes? In their own ways, these birds are migrating, too.

Robin Baker, a British zoologist who has written extensively about migra-tion, points out that there is a smooth continuum between epic, seasonal journeys, single-direction travel like the dispersal of young after the breeding season, and day-in, day-out movements like foraging for food. Casting a wide net, Baker therefore defines animal migration as "the act of moving from one spatial unit to another," a definition that even he acknowledges is so broad as to be nearly useless—although he spends another thousand pages in his doorstopper of a book, *The Evolutionary Ecology of Animal Migration*, dissect-ing the possibilities. Dingle narrows the field somewhat by drawing a distinc-tion based on behavior: "Migration is specialized behavior especially evolved for the displacement of the individual in space," a definition that, while more specific than Baker's, still includes baby spiders "ballooning" on windborne silk threads and fluffy milkweed seeds drifting on the breeze. On the other hand, Peter Berthold, a leading German migration researcher, restricts bird migration to "regular seasonal movement from breeding areas to resting grounds (often winter quarters)," a definition that jibes with most people's commonsense understanding of the phenomenon.

Simply agreeing on what to call different kinds of migratory birds remains a thorny issue among ornithologists. For generations, a sort of boreal chauvinism reigned in ornithology; scientists from the United States or Canada talked blithely about "North American birds," overlooking the fact that many species spend a greater portion of their lives south of the Mexican border than they do in the north. Smithsonian ornithologist Neal Smith, pointing out that some birds spend as little as four months a year in North America, has written: "The phrases 'to winter' and 'wintering grounds,' when applied to migratory birds, are parochial anachronisms. They have represented until recently a kind of biological thinking that has acted as a barrier to understanding the evolution of migratory systems and the adaptations to a migratory life. . . . Once freed of this mental inhibition, a biologist then might consider that many aspects of the appearance and behavior of a species are the result of selective forces acting during the non-reproductive season outside the Temperate Zones."

By common use, birds that breed in North America and migrate to the New World tropics are termed "neotropical migrants." While this is an improvement over the days when the tropical connection was overlooked entirely, it is not completely accurate. For instance, many so-called neotrops have populations that winter well north of the tropics, such as Baltimore orioles, which can be found in north-central Mexico, well above the Tropic of Cancer. Likewise, many Arctic-nesting shorebirds are lumped with the neotropical migrants, even though they travel to Argentina and Chile, well south of the tropics. And as Smith suggests, such birds never experience winter on their "wintering" grounds; an upland sandpiper that travels from South Dakota to Uruguay simply flies from northern summer to southern summer.

In its increasingly common usage, "neotropical migrant" also leaves out an entire class of birds that migrate to the tropics—those that nest in the southern cone of South America and migrate north, a group usually known as austral migrants. Ornithologists are also discovering that many birds migrate within the tropics; these are perhaps the most thoroughly "neotropical" of the lot, but instead, they are usually termed intratropical migrants.

Floyd Hayes, a North American biologist working in the Caribbean, has proposed an entirely new terminology. Hayes suggests that birds that nest across the entire Northern Hemisphere, including North America, Europe, and Asia, and migrate to the tropics, be called boreal migrants; those found widely in the Southern Hemisphere (of which there are only a handful), which migrate north, would retain the title austral migrants. Birds that breed only in North America and migrate to the tropics would be called Nearctic or temperate migrants instead of neotropical, a term he applies to species breeding in southern South America and migrating north toward the equator—

although why Hayes thinks these are more "tropical" than their northern counterparts is unclear. Confused? Me too.

If nothing else, Hayes's suggestions show that the urge to tilt at windmills is still alive and well. However valid his arguments, the weight of burgeoning common use is against him—and against Smith's complaints about the term "wintering ground." Let's face it; no one is going to toss phrases like "non-reproductive resting area" into casual conversation. You'll see "neotropical migrant" and "wintering ground" often in this book; I gave up charging at windmills years ago.

The same north breeze that spurred the songbird flight across the face of the moon would, I knew, prompt another group of migrants come daybreak. So, well before dawn, when I could see little but the pale trunks of the tulip-trees and the silvered trail stretching uphill, I shouldered a heavy pack and started climbing the Kittatinny Ridge, a long, low mountain not far from my home. The Kittatinny runs more than 200 miles from northern New Jersey, across Pennsylvania, and almost to the Maryland line, part of the Appalachian ridge-and-valley system that brackets much of the eastern United States. One of the world's most famous migration corridors, the Kittatinny had a notorious reputation in the early years of the century, when rocky outcroppings along its spine became shooting galleries, with migrating hawks as the targets. Conservationists, appalled by the slaughter, bought the peak where the worst shooting occurred and created Hawk Mountain Sanctuary in 1934. It was the world's first refuge for raptors, but it took almost forty years more to completely silence the guns elsewhere on the ridge.

The craggy boulder field where I finally sank down for a rest was along the same ridge, some miles from Hawk Mountain. The sun was just cracking the horizon, but the wind was already starting to gust, making my sweaty shirt feel cold and clammy against my skin, and making the stunted red maples and mountain ashes that grew from clefts in the rocks strain against the breeze. I shrugged off my pack, changed into a dry shirt and a jacket, and began to open the mist nets that crowned the summit of the rock field.

Imagine an old-fashioned lady's hair net that is forty feet long and seven feet high, and you'll have a pretty good picture of a mist net, an essential tool for studying bird migration. Strung between metal poles, a mist net has four or five baggy, horizontal panels of nearly invisible black cotton mesh running its length; when a bird hits the net, it falls into the bag, where it is harmlessly restrained. My mist nets were arranged in a triangle, and as they opened, the wind made the bags stand out like sails on a ghostly galleon.

As I worked, many of the nocturnal migrants were still moving, but not with the purposeful, southbound flight of the previous night. Flocks of song-birds flitted along the summit of the ridge, moving from treetop to treetop, foraging for insects as they went; others were above the trees, but flying directly north instead of south. This seemingly odd, contrary behavior is known as morning flight, and it appears to be a result of night-flying songbirds overshooting appropriate habitat while traveling in the dark. Once daylight comes, the birds ricochet across the sky, sometimes backtracking completely to find the right kind of cover.

The small songbirds were the night shift, and as the sun rose higher, they dissolved away into the forest to feed and rest. Now the day shift took cen-ter stage—swarms of tree swallows weaving complicated patterns in the air; chimney swifts, stiff-armed and gray, twittering like crickets; blue jays, cedar waxwings, goldfinches, and robins in small bunches. High overhead, in the torn-tissue clouds, I could just see (and hear) flocks of Canada geese. Flickers passed down the ridge in swooping flight, bright yellow flashing beneath their wings like sunbursts.

But what drew me to the mountaintop that day, as on most rainless days each fall, were the raptors that made the Kittatinny famous. For more than a decade, I have captured migrating hawks, falcons, and even eagles at my bird-banding station, one of dozens along the migratory route in the Appalachi-ans. The smaller species are lured into the mist nets, while we catch the larger, stronger birds, like red-tailed hawks and goshawks, in a spring-loaded bownet the size of a kitchen table.

Before I had finished opening the nets, a small sharp-shinned hawk, not much bigger than a dove, coursed down the windward side of the ridge at treetop height. Its wings pulled back in a half-tuck, the hawk rode the updraft like a surfer riding a wave, scattering alarmed warblers in its wake like leaves in the slipstream of a car. The air was warmer now; I shed my jacket, and as I was folding it, I noticed another hawk wheeling up from the sunny south slope of the mountain, in the lee of the breeze. Another rose to join it, then three more, until within a minute, fifteen hawks were circling in close com-pany at eye level, rising majestically in a bubble of warm, buoyant air.

They were broad-winged hawks, a species common in the forested East— the size of crows, a bit on the pudgy side, but with fan-shaped tails and wide wings, perfect for catching the lift generated by rising air. In the peculiar lingo of hawk-watchers, these broadwings were "kettling," soaring in tight circles to stay within a thermal air current, which carried them effortlessly upward. The irony is that, in relation to the air mass, the hawks were going down, but doing so inside a column of heated air that was ascending faster than the birds were descending—as naturalist Edwin Way Teale neatly described it, "like a man walking slowly down the steps of a rapidly rising escalator." The result was that, without moving a muscle, the hawks were fast gaining altitude.

No matter how many thousands of times I see it, I am always struck by how quickly a thermaling hawk climbs; within thirty seconds, the kettle was more than 500 feet over my head, and the winds were shredding the invisible bubble. One by one, the hawks locked their wings flat and glided out of the disintegrating thermal, streaming single file down the ridge, steadily losing height as they did. When they were very nearly at the limit of my vision, and well below the top of the mountain again, they caught another thermal and, in close-order drill, began once more to rise.

If nocturnal migration is best for many birds, why not for all? For small songbirds that must flap constantly to stay aloft, the night's calm air and cool temperatures are a clear advantage. For hawks it is daylight, with its winds and thermals, that brings the biggest energy savings. This is especially crucial for the broad-winged hawk, which migrates as far south as Colombia, and which needs to save every possible gram of fat for a journey that will take it down the spine of the Appalachians, along the Gulf Coast to eastern Mexico, and through the narrow waist of Central America. What's more, some experts believe that broadwings fast during much of their migration; if true, then any energy savings becomes all the more critical.

Other diurnal migrants garner different advantages from daytime travel. Swallows and swifts, which catch flying insects, can eat while migrating. Cranes navigate by using visual landmarks, and like the broad-winged hawks they soar much of the time, so they benefit from sun-warmed thermals. Geese and ducks also use landmarks to find their way, but they will sometimes travel at night, especially if there is a bit of moonlight. Shorebirds also travel both by day and by night—many take such long, nonstop journeys that they have no choice. Most of the thrushes migrate at night, but the most common thrush, the American robin, does almost all its travel during the day, as do hummingbirds, migratory woodpeckers like flickers, and members of the jay family, including crows and ravens. Generally speaking, at least for small birds, long-distance migrants are more likely to travel by night, while those that stay within the borders of North America are more apt to do so by day.

The nets were ready and more hawks speckled the sky, so I slipped into a plywood blind, from which ran a spiderweb of lines and cords. From inside, I could manipulate the lures and trigger the bownet, much as a fisherman works a fly—teasing a migrating hawk from the sky and into my nets. Peering from the small windows, I spotted another sharpie coming down the ridge and gently pulled on one of the lines radiating from the blind. Inside the triangle of mist nets was a creamy domestic dove, harnessed in a thick leather jacket that was clipped to the cord; rocked off balance by my tug, the dove fluttered its wings for a moment.

That's all the hawk needed. Birds of prey have phenomenal eyesight; they are able to discriminate detail many times finer than humans and are especially sensitive to movement. The pale wings of the dove were like a beacon, and its unbalanced motion, different from the usual fluid wing beat of a wild bird, probably appeared sick or injured to the hawk—easy prey. The small hawk shifted its flight path, pulling in its wings until it looked like a teardrop, and plummeted down toward its meal. Focused on the dove, the sharp-shinned never saw the net, which it hit squarely in the middle, dropping into a pocket. The dove, who's been doing this for years, never even spared the hawk a glance.

Sprinting to the net, I caught the sharpie's yellow, soda-straw legs so it couldn't grab me with its sharp talons, and slipped it out of the mesh. The hawk glowered at me with lemon eyes, its hackles raised in anger and fright. It was a young Pringles bird—that is, it wore the brownish plumage of an immature sharp-shinned on its first migration, and it was a female that slid neatly into an empty Pringles potato-chip can, with one end removed and air holes cut in the other. Hawk banders use tubes of many sizes to confine the birds harmlessly while they are weighed, measured, and banded. Male raptors tend to be smaller than females; male sharp-shinned hawks, which are half the size of their sisters, fit perfectly into little cardboard salad crouton cans. Big red-tailed hawks, on the other hand, require a pair of three-pound coffee cans, taped end-to-end. Banders, always searching for the perfect fit, become inveterate scroungers: "Say, you aren't going to throw that can away, are you?"

Back inside the blind, I opened a tackle box and removed a string of jingling metal bands, each a quarter-inch in diameter and split along one side. I used a pair of custom-made pliers to open one of the lightweight rings, slid it over the hawk's right leg, and squeezed it shut to form a bracelet. Stamped on the band was the number 1493-26443, which I noted on my data sheet, and the cryptic message "AVISE BIRD BAND WRITE WASH DC USA." ("Avise" is close enough to "advise" in both English and Spanish to make the bands reasonably bilingual.)

I jotted some notes in shorthand—SSHA, F, HY, 742 (sharp-shinned hawk, female, age: hatching year, captured 7:42 A.M.). Then I weighed the hawk, still in its tube, took a dozen quick measurements, and released her just ten minutes after luring her into the net. Only a few months old, this hawk was marked for life, part of the largest ongoing study of birds in the world.

Besides simple observation, banding (known as ringing in Europe) is the oldest, and by far the most important, means of studying bird migration. The practice started, in an almost accidental way, with medieval falconers. There is an anecdote about Henry IV of France losing one of his trained peregrines around 1595, and the marked bird showing up the following day on the Mediterranean island of Malta. Gray herons, the European equivalent of the great blue heron, were marked with anklets (for reasons unknown) in the late seventeenth and early eighteenth centuries, including one caught by a falconer's bird in 1710 in Germany, which was found to be carrying a leg ring from Turkey.

In North America, John James Audubon is credited with being the first bander. In either 1803 or 1804 (the various sources disagree), he wondered if the phoebes nesting on his father's estate in Pennsylvania were the same individuals each year. Audubon removed the chicks from a nest in a local cave, and "I fixed a light silver thread on the leg of each, loose enough not to hurt the part, but so fastened that no exertions of theirs could remove it." To his delight, the marked birds returned the next year.

As Audubon knew, one wild bird looks pretty much the same as all the others of its species. To study the movements, life spans, or habits of a wild animal, you must be able to mark it in a simple, readily identifiable way. With birds, the best and safest way is with a very light aluminum-alloy band, stamped with numbers or letters. Scientists began to capitalize on this technique a century ago; starting in 1899, a Danish schoolteacher named Hans Mortensen made his own metal bands to study a variety of waterfowl, waders, and raptors, and within a few years the practice was all the rage on both sides of the Atlantic. In 1909 the private American Bird Banding Association was formed, and by 1920, banding had become so common that responsibility for coordinating it was passed to the federal government, which under international treaty has the job of protecting and managing migratory birds.

Today, several thousand federally licensed banders in the United States ring an average of 1.2 million birds every year, submitting their data to a central computer at the federal Bird Banding Laboratory (BBL) in Laurel, Maryland; in Canada banders work with the Canadian Wildlife Service, which has been a partner with the BBL since 1923. Each band carries a unique sequence of numbers and the abbreviated address. If anyone finds a banded bird—if it is captured by another bander, hit by a car, killed from flying into a window, or caught by a cat—and if the number is reported to the banding lab, another tiny chink in our knowledge of birds will be filled.

The odds of that happening are slim; the recovery rate, as it is called, ranges from just 1 or 2 percent for small birds to 10 or 15 percent for waterfowl. Since 1955, almost 300,000 sharp-shinned hawks have been banded, but only about 3,500 have been recovered. That's typical—out of 432,185 American robins banded during the same period, only 14,322 were recovered. Because of their popularity as game birds (and because hunters are generally good about reporting band numbers) mallards have an excellent recovery rate—773,276 out of 5.3 million banded, the most of any North American bird. Scientists studying fork-tailed storm-petrels, a Pacific coast seabird, haven't been as lucky; out of nearly 11,000 banded, just 173 have been caught or found again. And even that beats the rate for ruby-throated hummingbirds, which take a band so small you need a jeweler's loupe to read it. Almost 35,000 have been banded in the past forty years, but just 35 have been recovered—literally a one-in-a-thousand chance. In all, North American banders have marked 56 million birds this century, of which only 3 million were encountered again. Obviously, banders must play a numbers game, ringing as many birds as possible in order to get this meager, vital return on their investment.

What does banding tell us about migration? Much of the knowledge comes in bits, eventually forming a mosaic of where and when birds of a given species travel. As with Audubon's phoebes, banding proves that birds often return to precisely the same place to nest each year, and more recently, it has shown that some exhibit the same fidelity to their wintering sites in the tropics. A lesser yellowlegs banded in the 1930s on Cape Cod was found less than a week later on the island of Martinique, providing the first evidence of long-distance migration between North America and the Lesser Antilles—a route amply confirmed by later banding results. Recoveries of Arctic terns banded in the 1920s in Canada finally revealed this species' globe-circling migration. Large-scale banding of waterfowl in the first half of this century initially suggested that ducks and geese from different parts of the north follow discrete routes south, dubbed flyways—the Atlantic flyway along the coast, the Mississippi flyway down that river basin, and so on. Later, even more banding data showed that the flyway concept was flawed; while landforms like mountains and river valleys do seem to channel waterfowl migration, the birds actually make broad, sweeping movements that ignore the flyway boundaries.

Occasionally, banding solves long-standing mysteries in one fell swoop. For a century and a half, the wintering grounds of the chimney swift, one of the most familiar American birds, remained completely unknown. "There was great excitement in the bird banding office," one history of banding relates, "when a report, dated May 23, 1944, was received from the American Embassy in Lima, Peru, giving the serial numbers of 13 bands turned in by a group of Indians who had killed the birds that bore them at a point on the River Yanayaco, near the boundary between Peru and Colombia . . . Service files showed that these bands had been placed on the swifts in Ontario, Connecticut, Illinois, Tennessee, Alabama and Georgia."

In his book *Season at the Point,* about the birds and birders of Cape May, Jack Connor asserts, "Almost everything we know with certainty and precision about bird migration has been discovered or proven through banding, and it is not an exaggeration to say that the numbered leg band has been as important a tool in the science of avian migration as the telescope has been in planetary astronomy."

Over the years, untold numbers of letters addressed with a variant of BIRD BAND WASH DC USA have been delivered to the banding lab—a testament to the U.S. Postal Service, whatever its faults. (There is no need to send the band, let alone the dead bird, but not everyone figures that out, and according to banding folklore, some of the packages are, well, redolent by the time they reach Laurel.) They come from unexpected places; one letter from China, quoted in the late Erma Fisk's delightful booklet *The Bird with the Silver Bracelet,* began, "I feel very glad to wrote you. I did not know you and you did not know me. Who introduced I to you? It is your pigeon. She Flew To China. What a far way she flew! It is marvelous." The "pigeon" was actually a pintail banded in Louisiana, one of a surprising number of North American waterfowl that have turned up in Asia. Chinese or otherwise, the person reporting a banded bird receives a certificate of appreciation from the BBL that lists the bird's species, age, and sex, when and where it was banded, and by whom; a less flowery acknowledgment with the details of the recovery also goes to the bander.

Newer bands now carry a toll-free telephone number, which may avoid the kind of misunderstandings made in the past. Back in the 1940s, when the old federal Biological Survey ran the program, a series of bands went out with the misspelled abbreviation "Boil. Surv. Wash. D.C." The story may be apocryphal, but a Midwestern farmer who shot a crow wearing one of the bands is said to have written a tart letter to the Biological Survey: "I should report that I followed the instructions on the band but am badly disappointed with the result. I Washed, Boiled and Surved, but the durned thing still wasn't fit to eat."

Banding remains the most widely used technique for marking birds, but there are many variations. In addition to numbered bands, scientists often use colored plastic bracelets, which have the advantage of being visible at some distance without recapture; by juggling the combinations on both legs, many individuals can be marked for recognition at a glance. "Hey, it's Red-Silver/Yellow-Blue," a friend of mine once exclaimed, looking through his binoculars at a thrush in his study area, exactly as if saying, "Hey, it's George." Shorebirds, with their proportionately longer legs, are perfect for such tags. A red knot caught in Massachusetts may fly off with six bangles—a numbered metal band above the "knee" (actually analogous to our ankle joint) of one leg; a green band with a protruding tab, known as a flag, above the knee of the other, signifying it was banded in the United States; and a combination of four colored bands below the knees. Although this may sound cumbersome, the bands do not seem to interfere with a bird's movements.

Large species like eagles and storks are sometimes fitted with numbered wing tags, flexible plastic strips that fold over the skin webbing at the front of the wing, held in place with a harmless rivet; this is the only way to mark vultures and condors, which defecate down their legs, clogging metal bands and eventually producing gangrene. (This surprises you?) The ingenuity of ornithologists marking wild birds is amazing. Ducks may receive nasal tags that clip to the bill. Geese are fitted with plastic collars or neck rings, marked with large alphanumeric codes that can be read through binoculars; because birds attain their full size by the time they leave the nest, the collars and leg bands do not have to allow for growth. One autumn, as part of a migration study, I helped catch and mark more than 200 red-tailed hawks with tail streamers made of nothing more sophisticated than colored electrician's tape, half an inch wide and four inches long, designed to fall off after a few weeks.

The examples are endless. One hummingbird bander I know puts a dab of colored typewriter correction fluid on the birds' foreheads; another tints their bellies blue or green with food dye. Shorebird biologists often paint the feathers of their subjects with dilute picric acid, which turns the plumage bright yellow; this technique helped Brian Harrington and his colleagues at the Manomet Bird Observatory track migrant red knots from the New England coast to southern Argentina. Birdwatchers are often puzzled by such Technicolor surprises—canary sandpipers, purple terns, green plovers, blue swans. Sometimes, however, nature plays a trick of her own. California birders, hoping to report their sightings, tried without success to learn who was dyeing gulls there pink, only to eventually discover it was a natural stain, apparently picked up from something the gulls were eating.

There have been steady advances in the techniques and technology used to study migration. Moon-watching came into wide use in the early 1950s, culminating with a four-night moon watch by observers at 265 sites across the United States and Canada—an exercise that provided the first continental snapshot of nocturnal migration. (Scientists still watch the moon, but more often now they rely on ceilometers, which project a strong beam of light straight up into the sky, illuminating the birds that pass through it.) About the same time that moon-watching became popular, it was realized that radar could detect flocks of migrant birds—the "radar angels" that had puzzled, worried, and even awed the military during World War II. In fact, archived weather radar images from the 1960s onward provided some of the earliest and most compelling evidence of recent declines in songbird populations.

The latest Doppler radar system adds layers of detail to the picture, showing speed, altitude, and flight direction, and amateurs can access the same data as scientists, since images from the NEXRAD radar system are available over the Internet. Newer experiments include the use of infrared imaging devices that detect passing birds by the heat of their bodies, and a system that records, identifies, and tabulates the flight notes of nocturnal migrants.

Banding and marking are ways of determining where a bird goes; now scientists are turning to genetics to learn where a bird came from. Researchers are trying to map out genetic markers, such as parts of the mitochondrial DNA in a bird's cells, that are unique to particular populations or subspecies; the idea is to someday take a quick blood sample or feather snippet from a bird on its wintering grounds and quickly tell that it came from, say, the Yukon rather than Quebec. Similar results appear possible by analyzing chemical isotopes in feathers; birds from different parts of North America pick up differing ratios of isotopes, depending on the type of rocks that underlie their territories.

Advances in microelectronics have also ushered in a new age of migration study. Through the 1970s and '80s, smaller and smaller radio transmitters, some lighter than a penny, made it possible to follow birds as tiny as thrushes, with the biologists racing behind in cars or airplanes, trying desperately not to lose the beeping signal; two friends of mine once spent eleven grueling days following a hawk from Pennsylvania to North Carolina, living behind the wheel, always afraid that an ill-timed bathroom break or lunch stop would sever the link and ruin the project. For a week and a half, they lived on junk food and soda, their hair going greasy and the inside of the car piling up with hamburger wrappers. Now, sophisticated transmitters communicate automatically with satellites, which store the coordinates, so that all the researcher has to do is log on by computer and download the data.

"You can sit at your gawdamn desk and tap on your gawdamn keyboard without ever once going outside," grumbles a lean, grizzled ornithologist of my acquaintance, weathered from years in the sun and wind. "The ultimate in fat-man biology, that's what it is." However dissolute its effect on the character of ornithologists, the new technology has produced some astounding results, from unmasking the wintering grounds of the spectacled eider to tracking individual peregrine falcons from western Greenland to Bolivia, and Swainson's hawks from northern California to Argentina. Now researchers are talking about transmitters so precise and rugged that they will pinpoint a bird's location to within a fraction of a meter, give exact readings on its flight altitude, body temperature, and heart rate, and work for many years, rather than a few months.

Fat-man biology is here to stay. But a simple metal band can still shed light on the mysteries of migration, so I'll keep climbing my mountain and setting my nets, seining birds from the cool winds of autumn.

Migration is the overwhelming rule among temperate birds, and as we learn more about the ecology of tropical species, we may find that it is the rule among birds, period. Of the roughly 650 species that breed north of the Mexican border, only about 100 are completely nonmigratory, and the great majority of those live along the Gulf Coast, in the extreme Southwest, or in parts of coastal California—environments where seasonal swings in temperature and food supply are less pronounced and thus the reasons for migration less pressing. Another dozen or so species, like the white-tailed ptarmigan, Steller's and gray jays, Clark's nutcracker, and Strickland's woodpecker, are altitudinal migrants in at least part of their ranges, dropping to lower elevations in winter.

Another 100 or so species winter at sea, leaving 407 types of land-based migrants. An analysis by biologists Douglas Stotz, John Fitzpatrick, Ted Parker, and Debra Moskovits showed that in 233 of those species, most individuals stay north of the Mexican border. The other 174 pass the nonbreeding season in the tropics—but not just anywhere, the researchers found. The richest region, attracting more than 200 species of northern migrants (including some that also winter in the United States), were the coasts of Mexico and Central America; mountainous inland areas held between 100 and 200 species, as did the islands of the West Indies, and Venezuela and Colombia in northern South America. Surprisingly, the vast interior of South America— the Amazon basin, the Pantanal marshes, the pampas of Argentina—attract fewer than fifty species of northern migrants, perhaps because of the distance, perhaps because, with so many species of resident birds, the competition is simply too great.

This pattern was even more pronounced when only songbirds were examined. For this group, more properly known as passerines, the most important wintering areas are clearly the lowlands of southern Mexico and Central America, followed by the West Indies and northern South America. Fewer than ten species, among them the bobolink and eastern kingbird, regularly migrate beyond the Amazon. And within this group, competition between migrants and residents may be especially strong. While a Caribbean island may have only five resident species of small, insect-eating songbirds, an equivalent area of Amazon forest may have 100, nearly a sixth of all its bird diversity.

The importance of Middle America as a wintering ground places special pressures on migrant songbirds. North America forms a giant funnel, with Central America as its narrow neck; this sets up a tremendous disparity between the amount of land available for breeding and that available for wintering. The magnolia warbler, a dapper creature wearing a yellow breast streaked with black, is fairly typical. It nests in conifer forests from the Northwest Territories to Newfoundland, and down the Appalachians as far as Tennessee, but in winter it migrates to the lowland forests of Mexico and Central America, a small fraction of its northern range.

By some estimates, the ratio of breeding range to wintering range is for many songbirds as great as six to one—in other words, for every six square miles of nesting habitat there is just one square mile of Central American forest. Little wonder, then, that conservationists have been so worried about deforestation in Central America, a region with one of the worst environmental track records in the tropics. Some countries, like El Salvador, have

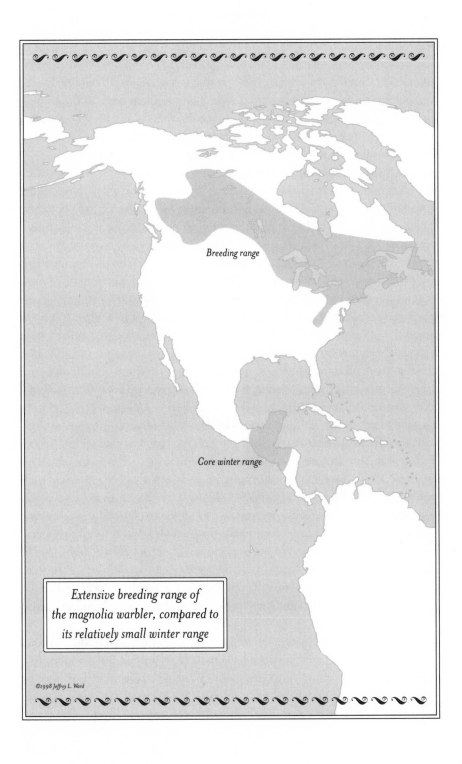

Breeding range

Core winter range

Extensive breeding range of
the magnolia warbler, compared to
its relatively small winter range

©1998 Jeffrey L. Ward

been almost completely stripped of their forests, while in others the pace of destruction is accelerating, despite land protection efforts. (Stotz and his colleagues do point out that, while northern migrants have relatively small wintering ranges, they are still better off than many resident birds in the tropics. Nearly half of all tropical bird species are restricted to a single ecological region—and thus at greater risk from habitat loss—compared to just 12 percent of migrants, which show a bit more ecological flexibility.)

From inside my mountaintop blind, I watched two Canada warblers, which would probably pass the winter in Colombia or Venezuela. They were feeding in a clump of red maple saplings that shimmered in the sunlight; a clutch of spider eggs had hatched in it and the minute spiderlings were "ballooning," extruding long strands of silk that the breeze would catch, blowing them off to new places—those, that is, that weren't eaten first by the hungry warblers.

The morning was turning warm, but the birds showed no dimming of their migratory urge. That's because migration isn't sparked by temperature—most migrants leave long before truly cold weather arrives—or by hunger. After all, a bird is at its fattest just *before* beginning its migration, a fact recognized as far back as the eighteenth century.

After more than a century of research, scientists think they understand some, but not all, of the pieces that interlock to produce migration. Before a bird migrates, there must be two elements present—a genetic predisposition and one or more environmental triggers. The most important trigger appears to be the changing length of daylight through the year, known as photoperiod.

Except at the equator, the daily ratio of sunlight to darkness changes at a predictable rate through the seasons, balanced at exactly twelve hours of each at the spring and autumn equinoxes. Almost every living thing, plants and animals alike, uses the local photoperiod as an external timekeeper, in sync with an internal, biological clock in the pineal gland. Interestingly, the photoreceptors in birds are not the eyes, as one would expect (and as is the case in mammals), but are lodged deep inside the brain, in an area known as the ventromedial hypothalamus. Unlikely as it sounds, the receptor cells react to the extraordinarily low levels of light that penetrate the thin skull and surrounding brain tissue.

Photoperiod controls much of a bird's life, from the production of sexual hormones and the development of its gonads to its annual feather molt. In a classic experiment in the 1920s, biologist William K. Rowan showed that by controlling the amount of light his caged juncos were exposed to, he could control when and how often they came into breeding condition. Changes in the photoperiod also bring on a condition known as hyperphagia, in which migratory species become gluttons, eating as much as they possibly can and laying down thick deposits of fat to serve as fuel for their impending flight.

One result of a changing photoperiod is *Zugunruhe*, a term coined in the 1950s by the German ornithologist Gustav Kramer. It describes the nocturnal restlessness that European birdkeepers noticed centuries ago in their caged nightingales and other captive songbirds. In spring and fall, the birds began fluttering in their cages just before sunset, continuing until a few hours after midnight—the same period, it turns out, as the peak of nocturnal migration each night. Many researchers since then, on both sides of the Atlantic, have shown that changing the photoperiod to match spring or fall light levels brings on *Zugunruhe*, and that the birds orient themselves toward the direction in which they should be traveling—north in spring, south in fall.

The apparatus scientists use for these studies, known as an Emlen funnel, after its inventor, is elegant in its simplicity—a large, funnel-shaped cone of blotter paper, topped with wire screen, and with an ink pad for a floor; more recent versions substitute typewriter correction paper for the blotter and ink. The bird, seeing the night sky through the screen, orients itself according to the stars, and repeatedly jumps toward the correct direction, trying to take off, leaving an inked or scratched record of each attempt on the paper with its feet.

The link between day length and the onset of migration has been shown repeatedly in experiments with captive songbirds, both in Europe and in North America. Keep a caged passerine in a room with a light/dark ratio that matches the outside world and *Zugunruhe* kicks in right on schedule. Keep the same bird in a room with a constant photoperiod that mimics either summer or winter and it shows no pre-migratory restlessness, even if other variables, like air temperature, are altered.

Changes in the photoperiod pull many strings in a bird's body, and for years, biologists sought a single "migration hormone" that controlled the whole shebang. At one time, it was thought that prolactin and adrenocortical hormones were key elements in synchronizing the physical changes that lead to migration, but research now suggests that a suite of biochemicals—testosterone, thyroid hormones, and glucocorticoids among them—play various roles in coordinating restlessness, hyperphagia, and other aspects of migration.

That isn't to say that weather has no effect on bird migration; it clearly does. Had I climbed the Kittatinny Ridge two days earlier, when there was little wind and hazy, summerlike temperatures, I would have seen few migrating hawks, if any. Since the 1930s, birders in the Appalachians have known that the heaviest hawk flights occur after the passage of a cold front, with lowering temperatures, rising barometric pressure, and a brisk, northerly wind that provides plenty of updrafts as it strikes the ridge system and deflects upward. Among songbirds in fall, cool temperatures intensify *Zugunruhe* and warm temperatures diminish it, and the birds usually wait for nights with steady northerly winds to give them a push south. In spring, the reverse is true; cool temperatures retard their restlessness, while balmy south winds provide an impetus to move.

Somehow, perhaps by sensing changes in atmospheric pressure, birds can perceive the approach of major weather systems, the "instinctive perception" about which Aristotle wrote. An especially strong storm may send them fleeing en masse, as was the case with the exodus of waterfowl that shut down those Midwestern airports. The day before the storm hit, observers in North Dakota noticed that mallards and snow geese were pouring south under sunny skies. In spring, the northward migration of both American robins and Canada geese is closely tied to something known as the 2-degree isotherm—the weekly progression of above-freezing temperatures (2 degrees Centigrade) that marches steadily across the continent, arriving in southern Indiana by about March 1, in central Michigan by April 1, and in James Bay by April 30. Geese need the thaw for open water, robins for soft soil and worms; both time their migration to match its progress.

But environmental triggers like photoperiod and weather will not, of themselves, prompt a bird to migrate. There is a genetic component as well, although ornithologists are discovering that the urge to migrate is a lot more plastic than they once thought. This discovery, in turn, is shedding new light on the evolution of migration routes.

Proof that genetics, not just environment, are responsible for the itch to migrate came from studies in the 1940s of dark-eyed juncos in northern California, where migratory and nonmigratory races of the same subspecies occur. Captive juncos were exposed to an artificial photoperiod; when the amount of light was gradually increased to mimic spring, juncos of the migratory race grew restless, while the nonmigratory birds ignored the cue.

Similar experiments have been conducted with blackcaps, Old World warblers more closely related to kinglets and gnatcatchers than to the colorful wood warblers of the Americas. Sparrow-sized and gray, with a thin, pointed beak and a charcoal cap (reddish-brown in females), the blackcap thrives in hardwood forests and parks across Europe. Researcher Peter Berthold went

one step further than the simple experiments done with juncos by his prede-
cessors. Taking blackcaps from a partially migratory population in southern
France, among which roughly three-quarters of the birds migrate to Africa
while the rest remain for the winter, Berthold selectively bred the songbirds
based on their migratory urges. To his surprise, he found that after just three
generations, he could produce a line of either completely migratory or completely sedentary blackcaps. It was, Berthold noted, the fastest genetically
based behavior change ever recorded in wild vertebrates.

"We do not need to look at the migration pattern of a species as though it
were a piece of sedimentary rock formed through eons of slow change,"
Smithsonian researcher John H. Rappole wrote in 1992. "Migration is a
dynamic process, surely subject to rapid evolutionary change." At almost the
same time Rappole made that statement, dramatic evidence for rapid evolution of a new migration path came from Europe, once again involving work
with blackcaps by Berthold and his colleagues.

For as long as people have studied blackcaps, the species migrated south
to the Mediterranean and Africa—those from Norway and Western Europe
taking a route to the Iberian Peninsula and across Gibraltar. In the 1950s,
however, British birdwatchers started finding more and more blackcaps in
England during the winter. At first they were simply oddities, presumably lag-
gards from the local breeding population that hadn't migrated—an unexpected spice to a weekend of birding but nothing more. But within a few
decades, it was obvious that something strange was happening; the number of
wintering blackcaps in England continued to swell to an estimated 10,000.
More surprising, banding studies showed they were not local birds at all but
were coming from Germany and Austria, 800 miles to the southeast. The
trend continues, and today one German blackcap in ten winters in Britain,
not Africa.

Was the change the result of local environmental cues, or was it something
more profound, an alteration of the blackcaps' genetic code? To find out,
Berthold and his associates went to Britain and captured forty wintering
adults, then bred them in captivity back in Germany, along with a separate
group of blackcaps from the normal migratory population. Both sets of offspring were exposed to the same environmental conditions, given the same
view of the night sky—yet when they were placed in Emlen funnels, the English birds oriented on a compass heading of 273 degrees—toward London—
while the chicks of German-caught blackcaps tried to fly on a heading of 227
degrees, toward Spain, their traditional way station on the route to Africa.

Why the change? A few migrant blackcaps always strayed to Britain, the victims of a bad mutation in their genetic code controlling orientation; in centuries past, natural selection weeded these out. But in this century, the researchers noted, winters in Britain have grown warmer, and vast numbers of berry bushes—a source of winter food for the blackcaps—have been planted to protect coastal dunes. Because there are fewer of them, there may be less competition among blackcaps in Britain than in the Mediterranean, and Britain, like America, is a nation awash with back-yard bird feeders, providing even more winter food. Finally, the trip to England is shorter (and presumably less hazardous), and English-wintering blackcaps return to Germany several weeks ahead of the Mediterranean counterparts to claim the prized nesting territories—all factors that appear to give British birds a clear survival advantage.

Beyond its obvious lesson on how migration routes evolve, the blackcap study may have much greater significance. The world is changing rapidly, and not necessarily for the better, as far as birds are concerned. Mounting evidence suggests that global climate change is an impending reality, promising a radical alteration of the habitats on which birds depend, while many of the crucial links in the migratory net, like forests, wetlands, and food, are being fragmented by human population pressures. The blackcap study hints that birds may be more flexible than we had realized, that they may be able to adapt to changing conditions better than conservationists had dared hope.

Then again, this is but one species, responding not to the destruction of its traditional wintering grounds but to the availability of a previously unsuitable area. Nor does it address the fundamental issue of habitat preservation, one of the most pressing in migratory bird protection. "If the change is simply destructive, if a bird is given no alternatives, it doesn't matter how much genetic variability you have or how rapidly you adapt," observed ornithologist Frank Moore of the University of Southern Mississippi, commenting on Berthold's study in *The New York Times*.

Interestingly, a similar situation may be developing in the rufous hummingbird of the American Northwest. A male rufous hummer looks like a newly minted penny, dazzling copper when the sun hits its back and red throat, and even the green female has red-gold streaks in her tail and on her sides. The most northerly of all hummingbirds, the rufous nests from the Alaskan panhandle to Idaho and Northern California and each autumn migrates south along mountain ridges to its wintering grounds in Mexico.

As far back as 1909, a handful of rufous hummingbirds were found in the Southeastern United States in fall, the apparent victims of faulty instincts. The distance from the Pacific Northwest to Alabama or Georgia is roughly the same as to Mexico. All it might take is a hitch in the bird's orientation system to send it in the wrong direction. Instead of heading due south, it slews off to the left and winds up in Dixie instead of Michoacán or Chihuahua.

As the sport of birding became more popular in the 1970s and '80s, records of wintering rufous hummers increased sharply east of the Mississippi; other Western species, including black-chinned and buff-bellied hummingbirds, were also sighted along the Gulf Coast in growing numbers. The reason, it was assumed, was more (and more knowledgeable) observers, and the increasing popularity of hummingbird feeders. But there may be more to the situation than that. Hummingbird researchers Geoffrey Hill of Auburn University and Bob and Martha Sargent of the Hummer/Bird Study Group in Alabama believe the surge may indicate a new, rapidly evolving migration route, analogous to that of the blackcap. They point to the number of rufous hummingbirds involved (nearly 900 reports in the five years ending in 1994) and the fact that some banded hummers have returned year after year. Nor have reports of other vagrant Western species, like Anna's, calliope, and black-chinned hummingbirds, risen at nearly the same rate.

The Southeast has been transformed in the last century or two; the virgin forests have been replaced by neat subdivisions, each with gardens that bloom through the mild Southern winters, many now stocked with hummingbird feeders filled with sugar water. Once a death trap for hummingbirds, the suburban Southeast is now a land of milk and honey—and those hummers that wind up there easily survive to pass on their "faulty" genes to the next generation.

Not everyone buys this argument, however. Some ornithologists, noting that the rufous hummingbirds leave the Western United States in early September but don't arrive in the Southeast until November or later, believe that they pass the missing eight weeks in Mexico. Perhaps, they speculate, winter jet-stream winds crossing Mexico push hummingbirds out of their traditional wintering grounds, especially younger birds without established territories. Others suggest that the Gulf Coast birds are simply wandering up from Mexico, prodded by scarce food supplies, or that rufous hummers are indeed evolving a new migration route, but one that is shaped like a U, curving south from the West Coast through Mexico and then up into the Southeast.

Interestingly, the easy living in the Southeast hasn't attracted the region's one breeding species of hummingbird. Each fall, the ruby-throated hummingbirds that nest in the East decamp from the United States entirely, migrating across the Gulf of Mexico to Central America; only a handful, all very young birds apparently from late broods, linger in Louisiana, perhaps because they are not strong enough to make a cross-Gulf flight. Why don't rubythroats take advantage of the same smorgasbord as the rufous hummers?

It may be that rufous hummingbirds, which nest so far north, are better adapted to colder climates than the rubythroats; they can easily survive temperatures in the single digits, if food is available. What's more, rubythroats would have to substantially trim their journey of several thousand miles, including a 600-mile flight over open water, and alter their physiology in terms of fat deposits and molt schedules—a complex series of changes. For a rufous hummer, on the other hand, all that would be necessary to shift the wintering grounds is a simple and common navigational error, since the travel distance is the same either way.

While the European blackcap made news because its migration change was so dramatic and sudden, many other birds have altered their migratory behavior over the years, usually in response to human activities. Back-yard bird feeders keep purple finches, pine siskins, and other seed-eaters farther north than was once the case, and with them the sharp-shinned and Cooper's hawks that prey upon them. Canada geese from the eastern Arctic traditionally wintered along the Outer Banks of North Carolina, where market gunning and (later) guiding waterfowl hunters were a way of life for generations of local residents. But starting in the 1960s, the geese increasingly wintered farther north, in the Chesapeake Bay region, where they had an abundance of soybeans, corn, and other field crops close to sheltered tidal marshes and bays. Many of the fields were specifically planted by refuges and gun clubs to encourage the geese to stick around.

This phenomenon is known as "shortstopping," and the classic example occurred over a twenty-five-year period in the Midwest. Canada geese from the central Arctic always wintered along the lower Mississippi, but in 1927 the state of Illinois converted thousands of acres of rich bottomland into a waterfowl refuge, complete with crops; within a few years, half the geese on the central flyway were stopping there, instead of continuing south. Then, in 1941, the federal government opened the enormous Horicon National Wildlife Refuge even farther north, in Wisconsin, employing the same mix of ponds, lakes, and crops to shortstop the fickle geese that so recently had

favored Illinois. The original migration to Louisiana and Arkansas, meanwhile, had dried up so completely that goose seasons were closed in both
states, to the vociferous complaints of hunters.

In the East, shortstopping also knocked the stuffing out of the hunting
industry on the Outer Banks and Lake Matamuskeet, but created a new one
along the Chesapeake. By the 1980s, half a million Canadas were wintering
on the bay, and the autumn economy of the Eastern Shore revolved around
geese. In November and December it was hard to get a hotel room, for the liberal bag limits drew hunters from across the United States, all willing to pay a
hundred dollars per day, per man, to hunt on leased farms with local guides.

I love goose hunting—the cold dawns of stinging sleet and north winds,
the art of setting a decoy spread just right to fool these wary birds, the way a
good caller can make a simple wooden tube sound like a flock of raucous
birds. Mostly I hunt on my own, near home, but a few times I joined friends
on expeditions to the Eastern Shore of Maryland for a couple days of waterfowling.

At four-thirty in the morning, the diners would be crowded with men (and
a few women) in camouflage clothing, stoking up for a long, freezing day huddled in some soggy pit blind, dug into a windy knoll in the middle of a vast
cornfield. Cold rain or snow showers were ideal weather for goose hunting, so
forget pancakes; you need serious food for those conditions—a heaped platter
of scrambled eggs, home fries, and a large bowl of homemade blue-crab chow-
der on the side, the kind of artery-clogging stuff that warms you half the day.
You could tell the guides from their clients at a glance; the guides were the
ones wearing worn, insulated coveralls in brown camo patterns, stained with
mud and patched at both knees. The guides nursed coffee and sat together in
a haze of cigarette smoke, trading information on how they'd been faring.
They jingled whenever they moved: the goose-call lanyards around their
necks were strung with tarnished leg bands, taken from the waterfowl their
clients had shot.

The degree to which the Eastern Shore economy focused on geese each
winter can hardly be overstated. From farmers leasing hunting rights on their
land to guides for thousands of dollars, to the rural families who charged two
bucks per goose to pluck the day's bag, hunting money infused the area, as
hunting infused the local consciousness.

I recall one December evening when, after a long, frigid day crouched in a
rain-soaked blind, my friends and I dragged ourselves back to the old inn
where we were staying. Down in the dining room the staff was nearly finished

serving the weekly seafood buffet, and there was no time to change if we were going to eat. The thought of smoked bluefish and fresh crabcakes had kept us going all day, so we trudged in as we were. I found myself shuffling past the warming trays beside an elegantly dressed older couple, he in suit and tie, she wearing pearls. Anniversary dinner? I wondered. The woman looked me up and down, taking in my mud-smeared clothing, plastered hair, and the tufts of white goose down clinging to my pants like tropical fungi. I smiled wanly, wishing I could disappear.

"D'you get any today?" she asked brightly. "I was just saying to George here, I bet this big cold front will bring in a lot of flight birds from up north. They'll be hitting 'em hard tomorrow in the cornfields." I stammered an agreement. Only on the Eastern Shore, I thought in surprise, do grandmothers speak fluent Goose.

The irony is that, just as geese shortstopped in the Chesapeake, leaving the Outer Banks high and dry, so now the birds are shortstopping still farther north. Even worse, the Atlantic flyway population of Canada geese, which migrate to the bay, started declining in the late 1980s, largely because of poor weather on the Arctic nesting grounds. To protect the breeding stock, daily hunting limits on the Eastern Shore were cut back steadily—from four to three, and finally to one, and guides went begging for business.

The coup de grâce came in 1995, when the federal government stunned the region by closing the fall goose season entirely for several years, citing the precarious state of the Arctic goose flock—an irony, given how explosively nonmigratory Canada goose flocks were growing. Some guides shifted to hunting ducks, or to snow geese, whose numbers were also at historic highs. Adding insult to injury, now the Chesapeake's snow geese are shortstopping, too; last winter, nearly a quarter million camped out in southern Pennsylvania.

I banded all the rest of that day on the Kittatinny, catching a dozen and a half more hawks, finally rolling up my mist nets as the sun dropped below the cloudless horizon. It was dark by the time I hiked down the mountain to the car, the pack chafing against my shoulders, and when I got home, the moon was rising again. Tired as I was, I set up the scope once more on the front steps, peering at the noticeably slimmer face of the waning disk. I left the clipboard inside; this was merely for fun, not science.

The wind had died with the sunset, and there were fewer birds aloft, fewer of the sweet, tinkling calls dropping like snow. To me, the night seemed featureless and blank, flecked with pale stars that I could assemble into constellations only by an effort of will. Yet to the birds this same sky, the very air around them, hummed with information that was denied to me—clues that allow them to navigate across thousands of miles. We've made great progress toward learning where the travelers go, and when, but we are just broaching the next, more exciting horizon: How do they find their way?

3

The Way South

Let's say you want to visit a friend. How do you do it?

That depends. If the friend lives just across town, you'll follow a series of well-remembered landmarks, probably without any conscious thought at all. But instead, let's assume this old acquaintance has retired and bought a backwoods cabin in scenic Wildcat Hollow, in a state you've never visited. You pull out the road atlas for a multistate overview, running your finger across the map; looks like the most direct route is I-95 and I-85 south across Virginia into North Carolina, then west on I-40 to Asheville. Now you unfold the AAA road map, which has more detail. You see that from Asheville you must head west on Route 19 through the mountains. But, of course, there's no sign of tiny Wildcat Hollow on the big road map. Fortunately, your friend sent you a hand-drawn map, which tells you what landmarks to watch for and shows how you must turn off Route 19 to reach the little dirt road that snakes its way up through the valley. The cabin, conveniently, is marked by a red X.

That's how most of us get from place to place—by using maps, road signs, landmarks, and other navigational aids, in a hierarchy from the most general (a map of the entire country) to the most specific (a scribbled drawing of Wildcat Hollow or your memory of particular landmarks). But what of a broad-tailed hummingbird, migrating from northern Idaho to the central Mexican plateau? Or an Arctic tern, which will draw a 22,000-mile figure eight on the Atlantic Ocean? Here we enter poorly charted waters—which is an ironic turn of the phrase, for even the smallest migratory hummingbird, with a brain scarcely larger than a dried pea, can plot a flawless course spanning immense distances.

How birds find their way has long been the most baffling aspect of migration, ascribed—for lack of a better explanation—to enigmatic "instincts,"
mysterious and inexplicable. The truth is far more wonderful than that,
although after half a century of research we have glimpsed only a piece of the
puzzle. We do know that they can track the sun, the moon, and the stars,
compensating for their apparent movement to use them as compasses. But
birds can also apparently perceive a host of sensations that are beyond our
unaided senses—weak magnetic fields, faint odors, polarized light, barometric
pressure, even extraordinarily low-frequency sound waves that echo halfway
around the world. Combined with the genetically programmed urge to head
in a certain direction at a certain time of the year, these clues allow birds to
cross continents, oceans, hemispheres.

So I suppose Lake Erie isn't such a big deal, especially not with a tailwind
blowing. It was an early autumn day that felt closer to winter, with a north
wind that turned the lake gray and white with froth. The wind flogged the
cottonwoods and ashes that grew in the swamp at Magee Marsh, just west of
the Marblehead Peninsula in northern Ohio, at the western end of Lake Erie.
I suspect the boardwalk that threads its way through the swamp and woodland would have been nearly deserted on an ordinary weekend, but like an
occupying army, more than a thousand people had descended on the area for
the biennial Midwest Birding Symposium, a four-day conference. Since the
indoor sessions wouldn't start until late morning, everyone—being birders—
had been out at the crack of dawn, anxious to see what the powerful cold
front had blown into town.

The organizers of the symposium had chosen the Marblehead because it is
a migrant funnel, a narrowing of the usually wide path of travel. Like water,
birds often take the path of least resistance, threading the narrow isthmuses
between the Great Lakes, or leaving Canada near Point Pelee, crossing over a
series of islands in 75-mile-wide Lake Erie, and alighting along Marblehead
Peninsula. Unless they hit head winds, even the smallest songbird can make
the trip easily, landing in the hardwood forests and swamps on the Ohio
shore, where we waited.

There was a real range of birding abilities and experience evident along the
boardwalk, from the hotshots of the hobby, mostly young men with a zealot's
gleam in their eyes, to rank beginners unsure of the identity of starlings and
robins. Yet there was little sense of competition, only a spirit of sharing that
I've found is typical of birders almost anywhere in the world. On this cold,

blustery morning, the birds were relatively few and hard to find, staying deep in the thickets to escape the wind, and when someone found a species out of the ordinary, they corralled passersby to point it out.

Back in the 1960s, at a roadside rest area near Patagonia, Arizona, someone found rose-throated becards and thick-billed kingbirds, both very rare species. Word spread, birders from all over the country descended on the area, and in the process they discovered the first black-capped gnatcatchers ever seen north of Mexico. That enticed even more birders to visit the spot, and they found even more rarities, which attracted more visitors, *ad infinitum, ad avis,* until Patagonia was synonymous with good birding. This became known to birders, in all seriousness, as the "Patagonia Picnic Table Effect."

Here we had the same phenomenon, writ small. An older couple spied a little brown creature moving at the center of a tangle of fallen logs—a winter wren, a diminutive, almost mouselike bird. Other folks stopped to watch; for some it was a "life bird," a species they'd never seen, always cause for excitement. Then someone in the gathering crowd noticed a gray-cheeked thrush perched quietly in a jungle of dogwood. More people paused, one of whom saw movement in the deep shadows and whispered, in an urgent voice, "Mourning warbler!" Binoculars swung away from the lesser beings, and word that this sulking, seldom-seen bird had been spotted rippled out along the boardwalk, until forty people were jockeying for a view of the warbler— which had, quite predictably, vanished from sight. But at that point, someone in the crowd saw a worm-eating warbler on the other side of the trail and the binoculars swung in unison once again . . .

I kept watching the thrush. I like warblers well enough, but thrushes are particular favorites of mine. For one thing, they are lovely, well-proportioned birds, with large black eyes that give them the appearance (if not the reality) of intelligence and personality. Their colors are muted browns and buffs that, like the fallen leaves of autumn, run the gamut from the rust of the wood thrush and veery to the sepia monochrome of this gray-cheeked thrush, the spots on its dusky breast smudged like rain-smeared paint.

Gray-cheeked thrushes are the most northerly of the genus *Catharus,* whose name comes from the Greek word *katharos,* meaning pure or clear. That describes the song of the graycheek, a descending spiral of crystalline music that surges up at the end, and that I will always associate with the shadowed forests of Newfoundland and Alaska. In autumn they move south across the United States, western birds trekking overland through Mexico, others crossing the

Gulf to the Yucatán, and still others hopscotching across the islands of the Caribbean, all eventually converging on northern South America.

For a gray-cheeked thrush hatched in the spruce woods of Newfoundland to migrate to the humid forests of eastern Colombia, it must accomplish two distinct, but complementary, actions: orientation and navigation. Orientation provides the bird with a directional bearing, which may have to change several times during the course of the journey, while navigation allows it to locate itself in the landscape and correct for outside forces that might alter its course. The first is the compass, while the second is a sort of map, and one without the other is useless.

The thrush is born with a genetic predisposition to react to the changing length of day, the photoperiod, and to orient itself in a certain direction based on the time of the year. This direction will be different, depending on where the thrush was born. A graycheek from Newfoundland will orient to the south-southwest, along the Atlantic Coast, eventually switching to an almost due-south orientation when it reaches the Gulf States. But those headings would send a thrush hatched in the Alaska Peninsula out into the Pacific Ocean; graycheeks from that region orient first to the southeast, then the south.

How birds migrate long distances, and how they find their way back to a particular place, like a homing pigeon zeroing in on its loft, has intrigued people for centuries. Homing ability is almost universal in nature, found in organisms as diverse as salamanders, wolf spiders, snails, spiny lobsters, salmon, snakes, mammals, and birds. Charles Darwin, who studied pigeons as part of his research on natural and human selection, wondered if crated birds, hauled far from their lofts and released, somehow memorized the twists and turns in the road as they felt them, replaying and reversing this memory to find their way home. Even Darwin admitted that was an unlikely explanation, but inertial or retracement navigation theory, as it was known, wasn't disproved until this century. As early as the 1850s, one or two European scientists had suggested that animals might use the earth's magnetic field as a framework for navigation—but that idea was bludgeoned by the leading minds of the day. "I had no need to declare my disbelief in Dr. von Middendorff's magnetic hypothesis," sniffed Alfred Newton, an 1870s zoology professor at Oxford, "for I never met with any man that held it."

The first real breakthrough in orientation research didn't come until the 1950s, when Gustav Kramer in Germany, studying *Zugunruhe*, noticed something striking in his flocks of captive starlings, which in Europe are migratory.

On heavily overcast days the starlings fluttered against their cage walls randomly, but on sunny days they clustered in particular directions—northeast in spring and southwest in autumn, the headings they would travel if they were free to migrate.

Through a long series of now-classic experiments, Kramer and his associates proved that the birds were using a sophisticated form of solar compass—comparing the position of the sun with their internal, twenty-four-hour biological clock, and compensating for the sun's apparent movement across the sky. If the starlings were kept indoors under artificial light, fooling their bodies into thinking it was earlier or later in the day than was really the case, their ability to orient themselves when shown the sun was affected. For example, if the starlings' internal clock was set six hours behind the actual time, they misinterpreted the sun's position when taken outdoors, rotating their orientation 90 degrees from where it should have been. Likewise, by using mirrors to alter the apparent location of the sun, the orientation of the caged starlings could be changed at will. Unlike photoperiod, which controlled timing, this use of the sun controlled direction.

For a time, biologists thought they'd found the central mechanism of bird orientation. But further experiments, many with homing pigeons, showed that the sun compass was only part of the story. For one thing, steering by a solar compass proved to be a learned ability rather than an innate one and wasn't essential for orientation; pigeons raised out of sight of the sun and exercised only on heavily overcast days lacked the ability to use the sun, but were still able to travel easily; by contrast, pigeons that learned to track the sun had a harder time navigating on cloudy days. The sun was obviously one piece of the puzzle, but it is not the only cue that migrating birds use.

At the same time as their solar experiments, Kramer's team found that night-flying songbirds might also use the stars to orient themselves. Over the years, a number of researchers followed up on this finding—Franz and Eleanore Sauer in Germany in the late 1950s, and the father-and-son team of John and Stephen Emlen in the 1960s at the University of Wisconsin. The Sauers showed that European garden warblers oriented to the night sky just as Kramer's starlings did to the sun, and that by switching the direction of the sky in a planetarium—aiming the North Star to the south, for instance—they could change the orientation of their study subjects.

The Emlens, for whom the Emlen funnel was named, built on this work in an elegant series of planetarium experiments, using captive indigo buntings.

The Emlens confirmed the Sauers' findings, then went on to show that, while the buntings did not orient on a single star, like Polaris, neither did they read the map of the entire night sky.

First, the researchers eliminated Polaris, the North Star around which the rest of the constellations appear to rotate and which some scientists believed was the most critical clue for nocturnal orientation. Surprisingly, the loss did not affect the buntings' ability to orient at all. Then they removed some of the stars around Polaris; still no change. Finally, when they had blanked out all the stars within 35 degrees of Polaris, the buntings lost their ability to orient, showing that the birds used the rotation of the stars within that zone, rather than particular clusters, as their benchmark.

But was there something inherently important about the star patterns in the northern sky? Were the buntings, in essence, glancing up and saying, "Well, there's Cassiopeia over there, so I better hang a left"? No; the Emlens rearranged the planetarium so that the sky wheeled around Betelgeuse, the bright star in Orion's shoulder. Young buntings raised under this substitute sky ignored Polaris and its neighbors and oriented on the spinning stars within 35 degrees of Betelgeuse. As with the sun compass, it seems birds are born with an ability to learn the night sky, but not with a specific star map imprinted on their genes.

Early human mariners used the stars, too, but they also relied on "lodestones"—pieces of a magnetic iron ore called magnetite that allowed them to create crude compasses. More than a century after "Dr. von Middendorff's magnetic hypothesis" was scorned, German researchers showed that birds can, indeed, navigate using the earth's weak electromagnetic field, and when their cages were placed within devices known as Helmholtz coils, which reversed the polarity of the field, the birds would flip their orientation. In the 1970s, Cornell University ornithologist Charles Walcott and his coworkers even fitted homing pigeons with tiny Helmholtz coils that sat on the birds' heads like little dunce caps, allowing the scientists to show that they had the same effect on the navigational sense of free-flying birds.

The earth's core is molten iron, and its slow, roiling movements generate an electromagnetic field that encompasses the entire planet, arching out into space at one pole and diving back in at the other. Although generally quite weak, the intensity of the magnetic field differs, depending on latitude and underlying geology—strongest in northern Canada, northern Asia, and the fringes of Antarctica; weakest near the equator. Deposits of iron and other minerals can create local "magnetic anomalies," like the ones in Massachusetts that have flustered generations of experimental pigeons. Adding to the complexity of the magnetic landscape, the high-speed jet-stream winds carry charged ions from the sun through the atmosphere, especially after explosive solar storms, further altering local magnetic fields.

Interestingly, birds seem to distinguish not between the north and south magnetic pole, the way a man-made compass does, but between the nearest pole and the equator; this would be especially helpful during one of the earth's regular magnetic back flips, when the polarity of its field reverses, as has happened thirty times in the past 5 million years. In addition to the intensity of the magnetic field, birds appear to be sensitive to what geophysicists call the "dip angle," the pitch at which the magnetic lines intersect the ground, which become steadily steeper the closer a bird goes to the pole and would provide a handy way of assessing latitude.

Even though researchers showed that birds could sense the planet's magnetic field, exactly how they did it remained a mystery until the 1970s, when it was found that some magnetically sensitive bacteria possess tiny magnetite crystals in their cells. Since then, magnetite has been found in the nasal cavities of several species of wild birds, including bobolinks, white-crowned sparrows, and pintails, as well as honeybees and other organisms with a geomagnetic sense. (Magnetite is found as well in human brains, but no one has proven that humans have an innate directional sense. And some of us have even less than others.)

Yet there is provocative evidence that magnetic reception is tied, somehow, to vision, an idea first proposed (over much skepticism and ridicule) in 1977 by Oxford physicist Michael J.M. Leask. In 1992, research was published by biologists in Indiana showing that among red-spotted newts, common North American salamanders, there was a correlation between the wavelengths of light and the amphibians' ability to orient magnetically. The very next year, German and Australian researchers studying a tropical bird called the silver-eye found that it could orient itself magnetically in white, green, or blue light, but that it lost its magnetic sense when bathed in red light, a discovery they later repeated in European robins. Like so much else involving animal navigation, exactly how light, vision, and electromagnetic fields interact is a mystery.

So which is more important to a migrating bird—a celestial or a magnetic compass? This was a bitterly contested question for years, with expert opinion swinging from one to the other like a pendulum. More recently, the consensus has been that birds use the various navigational clues available to them in a hierarchy, with visual guideposts like the sun and stars a clear favorite. American biologists Kenneth and Mary Able announced in 1993 that the savannah sparrows they had been studying essentially calibrated their magnetic sense by checking it against visual cues, especially the band of polarized light that moves across the sky at a 90-degree angle to the sun. Humans can see this dark band only by looking through a pair of polarized sunglasses, but the sparrows can somehow see it easily. For the many songbirds that migrate at night, the highly polarized band that stretches across a clear sky at sunset, just as they take off for their nocturnal flight, may be the most important road sign of all.

"It appears that the magnetic navigational sense in migrating species is used as a backup when celestial clues are lacking," Ken Able told reporters when the study was published. But just three years later, a team led by Peter Weindler in Frankfurt, Germany, showed that the interplay between visual and magnetic cues is much more subtle and complex than that. Weindler studied garden warblers that migrate from Europe to Africa, a trip requiring them to fly first southwest via the Iberian Peninsula, then southeast through Africa.

By raising some warblers in cages that provided celestial and magnetic cues and others in cages where the magnetic field was blocked, the researchers showed that, to navigate successfully, young warblers must experience both celestial landmarks and magnetic guidance before setting out to migrate. Under some conditions, the garden warblers (still confined in their cages) could orient just fine using only celestial markers, aiming themselves southwest for what would have been the first leg of their trip. But in the absence of a magnetic field, none later made the crucial course change to the southeast, suggesting that some navigational decisions are based on magnetic landmarks alone.

While most of the research into bird navigation has centered on celestial and magnetic orientation, they are not necessarily the only techniques available. Birds, for instance, can detect infrasound, extremely low-frequency waves of the sort generated by wind, ocean surf, volcanic eruptions, earthquake tremors, and thunderstorms, and which can travel for hundreds or thousands of miles. In theory, at least, a bird migrating down the Great Plains could keep the sound of the Atlantic in one ear and the Pacific in the other, and thus stay on course.

Some seabirds have a finely honed sense of smell, which is discriminating enough to guide storm-petrels back to their own nest burrows, in complete darkness and in the midst of colonies that may number in the millions; it may also play a role in longer distance movements. While a sense of smell seems poorly developed in most other birds, there is conflicting evidence regarding homing pigeons—those tested in Italy suffered major navigational problems when deprived of their sense of smell, while those from Germany were unfazed. Charles Walcott's birds at Cornell fell somewhere in between these extremes, suggesting that birds from different regions may learn to use different suites of navigational clues. It seems clear that some birds can use these various techniques as mutual fallbacks, switching from stars at night to the sun by day, to magnetic fields when the sky is cloudy, or to faint odors, polarized light, and low-frequency sound as the situation demands.

Compasses based on the stars or magnetic fields may give birds a directional bearing, but they must also be able to maintain that bearing in migration—an undertaking that includes many roadblocks and distractions. For instance, a bird flying due south with a crosswind can maintain a steady bearing and yet be pushed far off course by the sideways force of the breeze. Coping with wind drift or other kinds of displacement seems to come with experience, perhaps by learning landmarks or by cross-checking with other orientation cues. Young birds on their first migration are often pushed off course by wind, but by the next spring some of them begin to show an aptitude for true navigation, compensating for wind drift and barriers like lakes and mountains.

The difference between how adults and immatures navigate was made clear in an experiment with Dutch starlings in the 1950s. Normally, starlings that nest in Holland migrate south to northern France and Germany. Nearly 11,000 were captured in Holland, banded, and taken about 400 miles south to Switzerland, where they were released. Young starlings, making their first migration, simply oriented themselves and took off in their innate direction, southwest, winding up in southern France, Spain, and Portugal. The juvenile starlings flew on the correct bearing and for the correct distance, but they didn't compensate for the fact that they had been moved. While a few of the adults also made that mistake, most of them somehow corrected for the displacement and flew northwest, back to their normal wintering grounds. What's more, many of the displaced youngsters kept returning to their "incorrect" wintering area in subsequent years—an indication that birds become faithful to winter locations by learning, as well as by instinct.

The ability to home in on a known location, even when taken far from the birds' normal range, is another puzzling aspect of bird navigation, one that has been studied through displacement experiments like the Dutch starling project. In one famous case from the 1950s, scientists caught a dozen and a half Laysan albatrosses on Midway Island near Hawaii, banded them, then shipped them off on military airplanes headed as far away as Japan and Washington State. All but four of the birds made it back, and they wasted no time about it—the bird deposited in Washington averaged 317 miles per day on the flight home.

For several decades after World War II, ornithologists were forever shanghaiing birds, hauling them off great distances, then seeing how fast they could find their way home—if at all. Here's a very small sampling of the experiments:

—A Manx shearwater, a small, stocky seabird resembling a gull, was taken from its nest burrow on Skokholm, off Wales, and moved to Boston. It completed the trip home in just twelve days—one day faster than the airmail letter sent from the United States confirming the bird's release.

—Nearly 900 hooded crows were captured in the Baltics during their spring migration north into Russia and Finland and shipped by train to Denmark and Germany, up to 600 miles west. Like the Dutch starlings that didn't compensate for the displacement, most of the crows migrated north parallel to their original course, winding up not in Russia but in Sweden.

—Seventeen gannets were caught on their breeding grounds at Bonaventure Island in Quebec and moved 200 miles southwest to inland Maine, an alien habitat for these deep-sea hunters of fish. "I tried to anticipate what a gannet might do when it suddenly found itself over the potato fields of northern Maine, and it seemed possible that the drainage pattern might guide the bird downstream along the first river it encountered," wrote one of the biologists who conducted the 1947 experiment. Instead, the gannets wandered aimlessly until they hit the coast, which was presumably familiar to them, then arrowed back to their nests in Quebec.

—White-crowned sparrows caught during the winter in San Jose, California, may have had the roughest times of all. Hundreds were shipped to Louisiana and released; from there, they eventually flew north to their breeding grounds in Alaska, then migrated back to San Jose the next fall—a bad move. Fifteen of them were trapped a second time and shipped all the way to Maryland, from where they labored back to Alaska once more the following summer. This time, when six of the original cadre reappeared in San Jose for a third winter, the researchers finally left them alone.

So let's put it all together. That gray-cheeked thrush that I was watching along the shores of Lake Erie must get from northern Newfoundland to Colombia. It has inherited a genetic program that kicks it into migratory restlessness when the photoperiod diminishes to a certain point and that urges it to fly south-southwest, a direction it can determine by any of a half-dozen clues. Along the way (if this is its first migration) it will learn to correct for wind drift and other obstacles. But if it continues to fly on that southwesterly bearing for the whole trip, it will wind up in Acapulco, far off course. Partway through the journey, the thrush must change course to the southeast, if it is to reach its ancestral wintering grounds. In other words, it must navigate—it must have a map. Finding out how birds accomplish this aspect of migration has proven much more difficult than learning how they orient.

"A migratory bird follows its inborn instinct, or can we perhaps call it an automatic-pilot system, which gives information on compass heading and degree of migratory activity as the internal seasonal clock ticks on," Swedish ornithologist Thomas Alerstam observed. "If an innate orientation programme contained a sequence of different compass headings which covered the entire duration of the migration period, it could direct the birds to follow a more or less complicated and winding migration path."

That sequence of compass headings is what biologists call a time-distance or time-and-direction program, and it is how they think birds like the gray-cheeked thrush accomplish long, zigzagging migrations, especially on their first trip, with no adults to guide them. Also known as vector navigation, a time-distance program is simply a series of genetic instructions that direct the bird to fly in a certain direction for a certain length of time, then change to another heading for another, preset period of flight. (Tests on caged birds show it is the passing of time, rather than the distance flown, that triggers each course change.) In a way, it's like a Boy Scout following the legs of an orienteering course: Follow bearing 270 degrees for 173 feet, then take bearing 35 degrees for 535 feet, then 118 degrees for 90 feet, and so on. The Scout doesn't know where the course will lead, but if he follows the directions correctly, he will come to the proper spot at the end.

The first time a thrush migrates, it has nothing but the time-distance program to go by; that may be why so many young birds end up far off course, to the delight of birders. When it reaches its wintering grounds, the thrush may wander widely until it finds a suitable place to stay. But on subsequent flights, it can refine its navigation, relying more and more on landmarks and navigational techniques it has learned, taking the most direct route in transit, and finding its breeding and wintering locations each year with stunning accuracy. Here again, a hierarchy of orientation clues may come into play, with geographic landmarks, and perhaps olfactory and local magnetic clues, playing an increasing role the closer the bird gets to its destination—the difference between following interstate highway signs and looking for a familiar corner restaurant near home.

"There is increasing evidence that the navigational portion of bird migration comprises a system integrating a number of redundant mechanisms," Berthold writes. "Most likely, the primary framework is a time-and-direction program, the magnetic compass acts as a primary compass (possibly with limited effectiveness under low visibility) [and] the star compass develops as a largely independent mechanism but later functions, like other mechanisms, as . . . part of one complex system."

But while it is assumed that birds share a fairly universal talent for orienting themselves, not every bird is born with a compulsion to follow a detailed genetic map. For them, migration is more a matter of experience than instinct.

In 1955, Canadian ornithologist H. Albert Hochbaum published a book about the migration of ducks and geese, based on his years of study at the Delta Waterfowl Research Station in Manitoba. The title Hochbaum chose, *The Travels and Traditions of Waterfowl,* may at first seem strange. Tradition, after all, is the handing down of culture or custom from one generation to the next, while migration, as we've seen, is precoded in a bird's DNA before it even leaves the egg. That is certainly tradition, but only of a very limited sort.

And yet Hochbaum was precisely right. For a few groups of birds, most prominently waterfowl, migration is largely a learned behavior, passed down from parent to offspring, or modified by social bonds within the flock—a tradition in every sense of the word. Not only does this exception to the general rule shed light on the larger mysteries of orientation and navigation, it poses some unique challenges—and opportunities—for protecting migratory birds.

Ducks, geese, and swans certainly share many of the same orientation and navigational abilities as other birds—use of solar and celestial compasses, an awareness of magnetic fields, and so on. But it seems that the kind of time-distance program common in other birds is either missing, or—and this is more likely—it is superseded by a learning process in waterfowl. The timing of migration seems to be innate, as is a very general compass bearing; a duck raised in isolation knows *how* to migrate, but not specifically *where*. Among these birds, and unlike almost all others, the details are bound up in tradition, passed on by older generations.

Ducks and geese migrate in flocks, and unlike most other birds, in which the adults and immatures migrate separately, the family groups stay together for the southbound flight. It is then that youngsters are introduced to specific routes, stopover sites, and wintering grounds, memorizing landmarks along the way. Among puddle ducks like teal and mallards, the family dissolves during the first fall or winter, the young simply following the local flocks south and the newly mature ducks pairing up on the wintering grounds. While females have a strong fidelity to their birthplace, returning in spring to the same area where they were hatched, males do not, and they tend to follow their mates—usually a different hen, and a different destination, each year. Among geese and swans, which are slower to mature than ducks, the family remains together through another full migration cycle, not splitting up until the second winter, thus further reinforcing those traditions. Males pick their mates on the wintering grounds, the same as ducks, but once they mate, these birds tend to be monogamous for life.

The trumpeter swan is the largest waterfowl in the world, weighing as much as thirty-eight pounds, with wings that span almost eight feet. Trumpeters were named for their ringing *koo-hoo!* calls, so often compared to the notes of a French horn, and made particularly resonant by an extra loop in the bird's trachea. That signature call once sounded from the Bering Sea east as far as Lake Erie during the summer breeding season, and from the Sacramento Valley of California to the lower Mississippi and the Chesapeake Bay in winter. Unlike the slightly smaller and vastly more abundant tundra swan, which nests in the remote Arctic and subarctic, its temperate range meant the trumpeter swan was vulnerable to human pressure year-round, and their huge size, flavorful meat, and incredibly soft feathers made them immediate targets of commercial hunting.

Trumpeters vanished from the Atlantic seaboard shortly after European colonization, and by the latter half of the nineteenth century their breeding colonies in the Great Lakes, upper Midwest, and prairies were going fast. In the 1820s and '30s, the Hudson Bay Company in Canada was exporting as many as 8,000 swan skins a year, most of them trumpeters', which were used for quill pens, down quilts, ladies' hats, and powder puffs. Even as the swans became rarer, the kill remained disastrously high; between 1853 and 1877 alone, the company sold more than 17,000 swan skins. Although the Migratory Bird Treaty Act of 1918 gave the swan federal protection in the United States and Canada, it was almost too late. By 1932, only 69 trumpeters were known to survive in the wild, most of them in the remote Centennial Valley of southwest Montana, close to the Idaho and Wyoming borders.

To protect the swans, the area was declared Red Rock Lakes National Wildlife Refuge, and over the years the population slowly grew, aided by handouts of grain and natural hot springs that kept parts of the refuge ice-free even in the frigid Rocky Mountain winters. Other small flocks existed in Yellowstone National Park and in western Canada, but the species was considered in peril until the 1950s, when a large and previously unknown trumpeter swan population was discovered in Alaska. The trumpeter swan was taken off the federal Endangered Species list in the early 1970s, and today their numbers are estimated at more than 16,000, of which 13,000 nest in central Alaska and migrate to the Alaskan panhandle, coastal British Columbia, and Washington State.

I saw my first wild trumpeter swans at Red Rock Lakes back in the early 1980s, on a monthlong camping trip through the northern Rockies. At daybreak I would sit by the edge of one of the shallow alkaline lakes and watch these almost supernaturally regal birds gliding through the reflected sunrise, splitting the painfully bright orange with rippling bands of black and sparkling light. Those Rocky Mountain trumpeters, which now number more than 2,000, have done well in the past sixty years, but their success has come at a price. Unlike their Alaskan cousins, which retain their traditional migration route, the 500 or so Red Rock/Yellowstone birds have become essentially nonmigratory, tied to their breeding grounds by generations of handouts. By the late 1980s, they and most of the 1,600 Canadian swans wintered within a thirty-mile radius of Red Rock Lakes, getting along on a government subsidy of wheat. The old migration system, which had served trumpeter swans for tens of thousands of generations, had collapsed.

There are two reasons why this was an unfortunate development. One is the loss of the cultural tradition that perpetuated the migration—as much a part of the swans as their snowy feathers and echoing calls, separating them indefinably from city park mallards or half-domestic pigeons. The other problem was more concrete. Concentrating so many swans in a small area through artificial feeding made it much more likely that a disease epidemic or a natural disaster could kill many of them at once. This hazard was no secret, but it took such a near-calamity before managers started addressing the situation. Bitter temperatures during the winter of 1989 sealed off even rivers and spring-warmed waters, and more than a hundred swans died of starvation. The next winter, the U.S. Fish and Wildlife Service, which operates the refuge, embarked on an ambitious program to restore the trumpeter swan's long-lost migratory habit.

The supplemental food was stopped, and ponds that were used for feeding were drained. Over the next several winters, hundreds of adult swans were

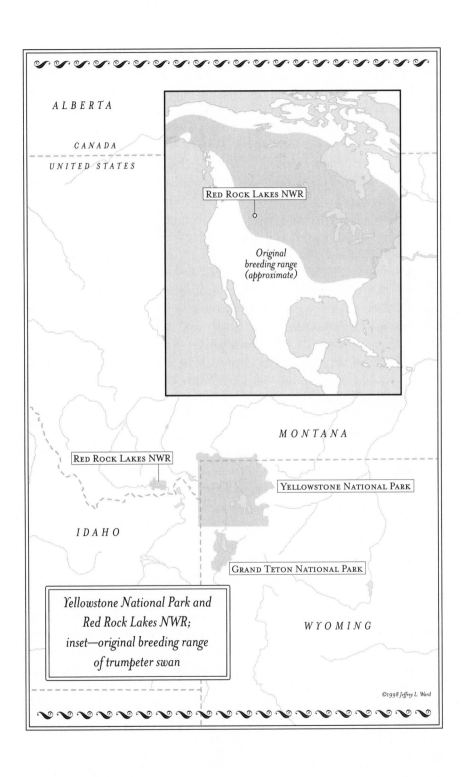

ALBERTA

CANADA

UNITED STATES

RED ROCK LAKES NWR

Original
breeding range
(approximate)

MONTANA

RED ROCK LAKES NWR

YELLOWSTONE NATIONAL PARK

IDAHO

GRAND TETON NATIONAL PARK

Yellowstone National Park and
Red Rock Lakes NWR;
inset—original breeding range
of trumpeter swan

WYOMING

©1998 Jeffrey L. Ward

netted and trucked to Utah, Wyoming, Oregon, and Idaho; younger swans were rounded up earlier in the fall and shipped out, too. Think of it as a forced-busing plan for birds. The idea was to introduce the resident swans to new wintering grounds; their own homing abilities would lead them back to Red Rock in the spring, but with luck, they might migrate to their new winter homes the next year voluntarily. Migrants coming down from Canada would be forced to keep moving, dispersing over a much wider swath of their historic range.

It seemed to work. While many of the evicted swans (which were tagged, dyed, or otherwise marked) disappeared, and presumably died, some did seem to learn new migration routes, including immature swans shipped to Oregon, which then flew on their own to California for the winter. Instead of migrating back to Red Rock Lakes, the swans returned to Oregon, raising hopes that a new, and above all migratory, breeding population might be started.

Biologists are also trying to use the abundant tundra swans as migratory tutors of a sort. Tundra swans gather by the tens of thousands in Utah each fall, en route from the Arctic to California. Trumpeter swans from the Rocky Mountain population are to be released at Bear River Migratory Bird Refuge along the Great Salt Lake, in the hope that they will join the tundra swans migrating to the Sacramento Valley in California, once a major wintering ground for the trumpeters.

While the Rocky Mountain swan population has been growing, the species has returned to many other places in its former range, aided by humans. Decades ago, swans from Red Rock Lakes were moved to Nevada, Oregon, and South Dakota—not in an attempt to re-create migratory routes, but in an effort to spread the nonmigrants around some of their old haunts. Now more than half a dozen other states and provinces—Minnesota, Iowa, Wisconsin, Michigan, Ohio, Missouri, and Ontario—have reintroduced trumpeters, using either captive-bred birds or eggs taken from the large Alaskan and Red Rocks populations. Overall, the programs have been a success, if by success one means simply getting warm bodies back into the wild. But from a migratory standpoint, the birds leave a lot to be desired; they are at best indifferent migrants, not moving until the winter ice forces them south. They have no traditions to fall back on, no culture of migration from the original trumpeters that carried their bugling calls across the East, but were killed off a century or more ago. We know (at least in general terms) where those swans used to migrate; we know what places along the way would be good stopover sites. A pity there is no such a thing as Migration 101, a way for humans to teach swans where to fly.

Perhaps there is, although I will confess that when I first learned of Bill Lishman, I rolled my eyes heavenward: Spare us, O Lord, from more kooks. Flipping channels one night, I happened on a television news report about the Canadian artist's attempt to lead a flock of geese south from Ontario to the mid-Atlantic states behind an ultralight aircraft. The reporter was extolling the project in conservation terms, as a way to save the "threatened" geese. Threatened? Good heavens, I remember thinking, Canada geese have become the starlings of the late twentieth century, hundreds of thousands of them on every golf course and farm pond in the East, and now this guy wants to lead more of them down here?

Fortunately, some people have more vision than I. By imprinting geese on the plane, teaching them from birth to follow it as though it were their parent, you can (in theory) lead them anywhere you like. Lishman and his colleagues weren't the first to train wild birds to follow them around—filmmakers Jen and Des Bartlett did that with snow geese back in the early 1970s, shooting some stunning footage from moving vehicles—but Lishman was more ambitious and got a lot more attention, even inspiring the movie *Fly Away Home*.

Hollywood aside, the technique, if perfected, is one of the most promising for reestablishing lost migration routes among waterbirds like swans. Lishman and his associates are currently experimenting with trumpeter swans, trying to teach them to fly a hundred miles from Virginia to the Eastern Shore of Maryland, as a preliminary step to establishing a flock in upstate New York that would migrate to the Chesapeake. A number of state agencies are trying or considering the same thing, as a way to reestablish migratory routes in the Midwest. Meanwhile, in 1997 USFWS biologists led a flock of imprinted whooping cranes—which also must learn their migration routes—on a 700-mile trip from Idaho to New Mexico, where they later integrated with common sandhill cranes and migrated with them back to Idaho. (Lishman's group is also working with cranes, experimenting first with sandhills as a surrogate for whoopers, hoping to eventually build a migratory flock between Manitoba and the southeastern United States.)

While encouraging, the method has serious drawbacks. Swans have proven much less pliable than Canada geese, less willing to blindly follow the plane, while the very large, leggy cranes have caused crashes and near-crashes by flying too close to the ultralight, at one point prompting the USFWS to reconsider the safety of the project; the cranes also had trouble keeping up with the ultralight, since they are built for slow, soaring flight rather than sustained flapping. Other possibilities being tried by various groups include training the birds to follow a truck, or creating large, motorized model cranes that would be flown from the ground by radio, like model airplanes. Most important, no one yet knows if birds imprinted on a plane—or a truck or a model—will function socially once they reach maturity. An attempt to start a migratory whooping crane flock in Idaho by fostering chicks with sandhill cranes failed because the whoopers didn't pair up and mate with each other, having imprinted on their sandhill parents.

It is fine to say that trumpeter swans "traditionally" wintered along the East Coast or in California—but how did that tradition start in the first place? Scientists have long struggled with how bird migrations evolve, and what forces spark and shape them.

Unlike bones, migration routes don't fossilize—although they sometimes do retain echoes of their past, like the wheatears discussed earlier, which still retrace their colonizing routes from Canada back to Europe and from Alaska back through Asia. Studies like Peter Berthold's, on the way blackcaps changed their wintering grounds from Africa to England, show how rapidly birds can evolve a new, genetically based migratory route. But there is still no convincing, overarching hypothesis that wraps it all up neatly, encompassing everything from hummingbirds to albatrosses, no Unified Theory of Migration Evolution. Instead, we have a firm starting point, a pretty good notion of the process that would lead from a sedentary population to a migratory one, and a lot of explanations—some plausible, some less so—for why.

First, the firm starting point. "The availability of food is the driving force in the evolution of migratory patterns," American ornithologist Paul Kerlinger writes. "A bird that can find more food will live longer and produce more offspring than one that finds less food. If by moving from place to place a bird can find more food, migration will evolve." It isn't cold, it isn't snow, it isn't summer heat or monsoon rains that drive birds to travel, but the pursuit of a full belly.

Nor should a bird move unless conditions change at home. Not being psychic, birds have no way of knowing that things are better somewhere else, and species certainly do not "decide" to become migratory, in any conscious sense. Ornithologists agree that migration probably evolves in a three-step process, starting with a nonmigratory population. Some environmental or ecological pressure—perhaps a change in local climate that creates more pronounced seasons, thus affecting the food supply—forces part of the population to move for part of the year into adjacent regions where conditions are better, then shift back to their ancestral range in winter, creating a partially migratory species. If the pressure continues or intensifies, so this line of reasoning goes, the original sedentary population may be eliminated completely, eventually leaving only the migrants.

This three-stage scenario makes logical sense, is backed up by a lot of circumstantial evidence based on partially migratory species, and is widely accepted by the experts. That's the "how." The fur starts to fly, however, over the "why"—over what triggers the evolution of migration. The earliest attempts to find a cause focused on large-scale environmental calamities like climate change. (Suggestions that continental drift were responsible died quickly when it was realized that the land masses reached roughly their current positions long before most modern birds evolved. But before we conclude that bird migration is a recent development, consider the flightless, toothed seabird Hesperornis, whose fossils were found in the Arctic, where even during the Cretaceous the climate was vile in the winter. Likewise, the fossils of its ternlike contemporary Ichthyornis have been found in deposits formed hundreds of miles from land, in the great inland ocean that once covered the Great Plains—certainly the equivalent of modern long-distance migration, as evolutionary biologist Alan Feduccia has pointed out.)

For a long time, Pleistocene glaciers of the last ice age were the favorite explanation for migratory evolution. Two major schools of thought developed, nicknamed the "northern home" theory and the "southern home" theory. According to the former, migrants were originally native to the northern latitudes and were forced south as the glaciers advanced; after the ice melted, the birds followed an instinctive racial memory back to their original range, but for some reason continued to migrate to the tropics in winter. The southern home theory is the reverse: Migrants started out as tropical species hemmed in by glacial cold toward the poles. When the climate moderated, they were able to forge north, exploiting the newly available land but returning to their southern range each winter. (One more recent idea, put forth in 1985 by George Cox of San Diego State University, splits the difference. Cox suggested that the Mexican Plateau served as a "staging area" for evolving migrants, which pushed to the north in summer and south, into the tropics, in winter to avoid drought as the region dried out during the Pleistocene.)

Of the two, the northern home theory was obviously weaker; the idea that birds could retain a memory of the north, lasting tens of thousands of generations in the tropics, was hard to swallow. And if they originated in the north, why not simply become nonmigratory there again once the climate warmed, instead of trekking all the way back to the tropics each winter? After all, some birds have adapted to the cold and live there year-round. The southern home theory makes more intuitive sense, since many migrant birds—especially in the Western Hemisphere—have large clans of close relatives in the tropics, suggesting that is where their families originated.

But while ornithologists generally accept that many Nearctic migrants originated in the tropics, they no longer credit the last glaciers with triggering most of their migrations. Scientists now realize that the Pleistocene ice sheets were much too recent to account for migration paths, few of which (the wheatears being an exception) show any correlation with the patterns of glacial retreat anyway. In fact, as we'll see in a moment, some of the more remarkable migration routes today may have evolved during the peak of the glaciation, not in its aftermath. Pleistocene glaciers, incidentally, have also been dethroned as a force in the evolution of many North American bird species. For more than a century, it was common wisdom among ornithologists that the advancing ice isolated eastern and western populations of many species, leading to their evolution as separate but closely related forms. Now DNA analysis shows that most of these "species pairs," like Bullock's and Baltimore orioles, split as much as 5 million years ago, long before the glaciers arrived.

Unlike these early climate frameworks, most of the current theories on how migration evolves center on natural selection—the idea that migration confers advantages, however small, on the reproductive success of the birds undertaking it, outweighing the costs it exacts. While some of these hypotheses still depend on climate change as the motive force, such as a shift to more pronounced seasons, others look to ecological factors like increasing competition for breeding sites or food. One related set of theories suggests that migration to a temperate breeding range reduces nest predation (which is extremely high in the tropics) while taking advantage of a summertime abundance of insects to sate all those hungry, chirping mouths. Winter food shortages then require a return to the tropics, which may be a smart move, despite the trials of migration. As Smithsonian ornithologist Russell Greenberg has pointed out, resident northern birds like chickadees and song sparrows usually raise more chicks in a single season by laying larger clutches of eggs or bringing off multiple broods, but neotropical migrants come out ahead because they tend to live longer, not having to survive the rigors of winter.

One problem with trying to decipher how neotropical migration evolved is that, until recently, science knew relatively little about the tropical end of the equation. That is finally changing, and one of the realizations is that many supposedly sedentary species of birds migrate within the tropics themselves. Biologists Douglas Levey and Gary Stiles looked at seasonal movements among tropical birds in Costa Rica and discovered some surprising parallels with migrants from North America. Instead of eating insects, intratropical migrants tend to eat fruit and nectar, which are patchy, seasonably scarce resources. The same holds true for a large number of northern migrants, nearly half of which—like thrushes, tanagers, and vireos—eat a lot of fruit while in the tropics. Levey and Stiles also found that northern migrants tend to winter in either seasonally dry forests, in the canopy of mature forests, or along the edges of woodland—the same habitat choices that tropical fruit- and nectar-eaters usually make. While the theory fits the evidence for some migrants, the authors admit, it misses the mark on others, especially warblers and some flycatchers, which are largely insectivorous.

"Our major point is that particular diets and habitats preadapted some [birds] for long-distance migration," Levey and Stiles write. "It is not coincidental that many migrants share taxonomic and ecological affinities with tropical residents that depend on plant reproductive resources or live in seasonal habitats—it is from this group that they were drawn." Fruit-eaters in particular would have benefited from expanding their breeding ranges north into the temperate zone, with its summertime flush of mosquitoes, black flies, and other bugs, because chicks need protein, something lacking in fruit. "In short, we suggest that overwintering migrants do not switch 'to' fruits . . . but rather switch 'back to' fruits—a view that agrees with current theory on the Neotropical basis of Nearctic migrants."

But if the end of the ice ages didn't provoke the evolution of migration, surely this period of geological history had an effect on the phenomenon. Indeed, Timothy C. Williams of Swarthmore College, a leading researcher into transoceanic bird migration, and Thompson Webb III of Brown University have proposed that some of the most remarkable migrations evolved not at the close of the Pleistocene but at its height, as a way of crossing the continental glaciers.

By analyzing fossil pollen (nearly indestructible stuff preserved in annual layers on muddy lake bottoms), Williams and Webb were able to reconstruct the pattern of plant communities across North America 18,000 years ago, at the height of the last glaciation, and 9,000 years ago, when the glaciers were in rapid retreat. They also used computer simulations, created by the National Center for Atmospheric Research, to model winds during these two periods. The team wanted to know if the current orientation system used by many birds—a southeast compass heading of about 155 degrees—would have worked then, under different ecological and meteorological conditions. This heading is the same one followed by birds that leave the Northeast coast and arc across the western Atlantic to South America, including songbirds like the blackpoll warbler, and many shorebirds.

The fossil pollen shows that, while much of the northern half of the continent was under ice 18,000 years ago, there was still a great deal of breeding habitat for songbirds in Beringia and parts of western Canada, while shorebirds could have nested across the High Arctic coast, which was also ice-free. The ice cap would have produced a powerful high-pressure center sitting over central Canada, pumping strong northwest winds across the Northeast—precisely the conditions that would aid migrants on a 155-degree heading. Birds breeding north of the ice would have to make two nonstop jumps, Williams and Webb believe. The first, lasting 90 hours for the slowest songbirds and 50 hours for the speediest shorebirds, would take them over the glacier to the Northeast coast, which included the continental shelf, then above sea level and ice-free. From there, the birds would set off again over the ocean on a flight lasting between 60 and 110 hours, as they do today.

"Of particular interest are flights from Alaska," the researchers wrote. This region "constituted the great majority of breeding area [18,000 years ago] and . . . fixed heading flights at about 155 degrees, coupled with favorable winds [then], could have provided a rapid and efficient route to wintering areas in South America, Central America and the Caribbean." They also note that, because of lower sea levels during the ice age, there was more land along what is now an oceanic route: "For the great majority of migrants to South America, these tracks would have taken birds over the (probably) much expanded land areas of the Caribbean and Central America, a logical initial route between North and South America. Migrants at this time might have used fixed headings not to make long flights over open ocean or inhospitable arid lands, but to direct movements over well-vegetated continental areas or to guide relatively short flights over water between land areas."

Given that for almost all of the past 900,000 years North America has been under the thrall of glaciers, "present conditions might be considered a rare period in avian evolution, posing a relatively brief stress on a system largely evolved under different conditions," Williams and Webb concluded.

Whatever natural selection forces first prompt birds to become migratory, the wind ultimately shapes many—perhaps most—of their paths, especially those that cross large expanses of water. Reliable south winds in springtime and powerful north winds in fall help songbirds make their annual crossing of the Gulf of Mexico. And the powerful low-pressure systems that haunt the Aleutians in autumn, hammering them with strong storms every few days, aid a whole suite of birds heading out of Alaska. In his work with Alaskan shorebirds, Bob Gill told me, he's seen enormous flocks of bar-tailed godwits wait for the worst of a storm to clear, then move out en masse on their 6,800-mile trip to New Zealand.

"It's after the rain has passed and the clouds are starting to tear apart that the godwits get real antsy. They'll fly up, spiral way up, then spiral back down, then back up again. Then they take off, all together, hundreds of them, and get up to three or four thousand feet before I lose them."

By catching the backside of a low-pressure system, with its counterclockwise flow, the godwits enjoy tailwinds up to 80 miles per hour, which carry them more than a thousand miles. Without that assist from the wind, Gill and others calculate, the godwits would run out of energy reserves long before they reached land. Other birds, also departing from Alaska, use the same weather systems in slightly different ways, Gill has found. The small "cackling" race of Canada geese, brant, and several shorebirds must cross the Gulf of Alaska and the northeast Pacific on their way to the mainland. Since they are heading southeast, instead of south-southwest like the godwits, these birds time their departure a little later, after the front has passed, when the winds have shifted more from the southwest, or else they pick storms that follow a slightly more northerly track.

Many researchers have also noted the strong link between prevailing winds and the epic migrations of seabirds, like the Wilson's storm-petrel. A weak, fluttery flier not much larger than a swallow, this petite bird nonetheless departs each April from its breeding islands off Antarctica and moves up the coast of South America on the southeast trade winds, then clockwise across the North Atlantic during our summer, this time riding the sea breezes known as the westerlies. Then the storm-petrels are pushed south off Gibraltar and Africa, recrossing the Atlantic to South America on the northeast trades, and back to their islands by November, the start of the Antarctic summer and another breeding season. Over this whole, clockwise trip, which may span 18,000 miles, the petrels rarely have to buck a head wind.

And so it is for many other seabirds, which crisscross the globe the way we navigate our back yards. Their journeys are molded by the global, interlocked system of wind currents, which carry them along, and sea currents whose upwellings and gyres concentrate their food—a support network that seamlessly links all the world's oceans and permits the greatest migrations of all.

4

Riding the Sea Wind

It was dark in the wheelhouse of the *Desiderata*, save for the dim red glow of a large compass, its S aimed straight ahead. The other four people aboard were dozing as best they could, wedged in narrow berths below or simply stretched out in sleeping bags on the lower deck, having been pitched there repeatedly by the heavy seas and recognizing the futility of argument. It was 4 A.M. and, feet braced against the swells, I was standing a solo watch at the wheel of the 51-foot ketch, running before the wind 60 miles off the coast of Nova Scotia.

With the sails trimmed, a hundred fathoms of water beneath us, and no shoals to hit for hours in any direction, my only job was to keep the yacht on course and to watch for fishing boats. This late-August trip was my idea; we were heading to offshore waters to look for migrating seabirds, and the others had come along to enjoy a pleasant four-day excursion—and also because, unlike them, I know absolutely nothing about sailing and couldn't hope to do it alone. The crashing seas were making it anything but a holiday, however. The boat bucked through a series of especially strong rollers, and I clung to the wheel for balance. From the galley below came the banging of an unsecured cabin door, the clatter of metal cooking implements stuttering in confusion, the clink of bottles rattling nervously in their racks. Then a heavy thud and a muffled curse: one of the sleepers had been dumped out of bed again. Moreover, ever since we'd left shore in this rented ketch, we'd been plagued by mechanical problems—clogged fuel lines and water-tainted diesel that killed the engine before we even got out of the harbor; a faulty starter that meant we had to jump the motor each time it died by short-circuiting it

with a screwdriver; worrisome leaks and a bilge pump that worked only spo-
radically; and trouble with the electronics systems that periodically left us
with no radar, depthfinder, or radio.

Nevertheless, it was a perfect night for a solitary vigil; when I looked up
through the glass hatch above my head, I saw the silhouette of the mainsail
and wooden mast swinging back and forth, cutting a path through the heart
of the Milky Way. To the west, Jupiter had just set, while to the east, Venus
crowned Orion as he climbed out of the sea. Straight ahead there was little to
ease the darkness except the blush of the running lights on the bow—red on
the left, green on the right, spilling their faint shimmer onto the waves.

Yet every so often, like the flicker of a ghost in the corner of my eye, I
thought I saw movement out there—movement that was not the roll of a
breaking wave or the flash of phosphorescence. I stared straight ahead, ignoring the compass and the wind and the swells, watching. There was nothing
but the sheen of the water. My eyes teared from salt spray and strain, and I
wanted to blink, but I did not.

Then a small shape suddenly flashed through the diffuse beam in a split
second's time—a pale body and pointed, dark wings with white chevrons.
Several more scattered up from the water, disappearing into the immensity of
darkness. Like a restless dream, *Desiderata* was flushing birds from their sleep
on the rolling waves.

Migration is not solely a land-bound phenomenon; in fact, the hardiest of
all migrants are those that roam the open sea. Each spring gray whales travel
in slow procession 7,000 miles up the Pacific coast from Baja California to the
Beaufort Sea, while in the Atlantic, humpback whales do the same between
the Dominican Republic and the icy Davis Strait. Half-ton bluefin tuna, slashing through the water at speeds exceeding 50 miles per hour, move from
Brazil to the Gulf of Mexico and thence north to the rich summer feeding
grounds off Newfoundland; some cross the Atlantic from the Bahamas to
Norway in just two months. Eels descend Eastern streams and swim to the
Sargasso Sea to spawn and die; sockeye salmon emerge from the inky depths
off Russia's Kamchatka Peninsula to ascend the rivers of Bristol Bay in Alaska.
And before dams choked them off, chinook salmon once climbed from the
Pacific a thousand miles up the Snake River to breed.

The open ocean would seem a foreign place for birds, yet the greatest of all
bird migrations also take place far from the sight of land, far from the eyes
and—until recently—the curiosity of humans. Landbird migration in the
Western Hemisphere is a largely linear affair, north and south across the

joined continents, but beyond the coastlines, the offshore waters of North and South America draw seabirds from almost every corner of the globe—from Tasmania and Antarctica, New Zealand and the Pribilofs, the Mediterranean and Azores, and from tiny islands marooned in the empty Pacific and Atlantic. Their migrations scribe circles, loops, and figure eights, drawn on a planetary scale.

Westerners have noted the migration of land birds in this hemisphere since the fifteenth century, when Columbus—approaching the West Indies in October 1492—encountered great flocks of passerines and shorebirds and took them as a sign that the Orient was somewhere close. After nearly five centuries, many questions about their movements have been answered. Yet all but the most fundamental aspects of seabird migration remain sheathed in mystery. Even the breeding or wintering grounds of some species have yet to be discovered, and we know next to nothing about where many of them go when they depart their nesting territories, and how they make their living on the open sea.

These so-called pelagic birds—the albatrosses, petrels, shearwaters, storm-petrels, skuas, jaegers, and a few shorebirds, among others—spend almost their entire lives over water. Unlike pelicans, cormorants, and most gulls, which haunt coastal waters but return toward land each night to roost, the pelagics may wander thousands of miles from shore, spending years at a stretch at sea. When a young black-footed albatross, hatched on Midway Island in the Hawaiian chain, opens its seven-and-a-half-foot wings and lumbers into flight for the first time, it will not touch ground again for three years, during which it wanders the Pacific from Japan to the Aleutians, and south as far as Mexico.

I now realized that the birds beneath *Desiderata*'s bow were red-necked phalaropes, diminutive shorebirds with needlelike bills, not much larger than sparrows, which breed on the soggy tundra of the Arctic coastal plain. In summer, they are painted with russet and blue-gray, the females more brilliantly than the males. (Phalaropes are one of the few bird genera in which females are more colorful than males, a reflection of reversed sexual roles in which males incubate the eggs and tend the chicks.)

By late summer, however, the phalaropes of both sexes have molted into their winter plumage, frosty gray and white, with a robber's mask of charcoal around each eye. They desert the Arctic and spend the winter far out at sea, as at home in this unforgiving environment as any albatross or storm-petrel. Those that breed in Alaska and western Canada travel to the cold, nutrient-rich waters off Chile, but no one knows where these eastern birds, disturbed by our passage, go each year. In years past, millions have been seen gathering in immense flocks on the Bay of Fundy, using it as a staging ground before departing in September, but the wintering area—perhaps in the South Atlantic, perhaps east toward Africa—has never been found. Some researchers believe the phalaropes leave Nova Scotia on the prevailing northwest winds, swing out over the Atlantic toward Bermuda, then curve back across the Caribbean on the trade winds, crossing Central America and joining the western birds off South America. Trouble is, no one has seen large numbers of phalaropes crossing the Caribbean or Central America, and the mystery remains. "This level of ignorance about a North American breeding bird is virtually unique," wrote one ornithologist recently.

The cockpit of the yacht was wide, but if I gripped the wheel with my right hand, braced a knee on the port locker, and craned my neck far enough through the doorway, I could see over the hull and down to the waterline, where we trailed an endless stream of fire. Observed closely, what first looked like a sheet of light resolved into countless pinpoints, like luminescent dust, each flashing with the same chilly, blue-white light as the stars overhead.

Humans have been fascinated by such "phosphorescent seas" since before the days of Aristotle, who considered the phenomenon akin to lightning. Sir Isaac Newton believed marine luminescence was a sign that all bodies emit light when disturbed, either "by Heat, or by Friction, or Percussion, or Putrefaction [sic], or by any Vital Motion." Benjamin Franklin, who (like Aristotle) first believed the glow was connected to lightning, had by 1753 decided otherwise: "It is indeed possible, that an extremely small animalcule, too small to be visible even by the best glasses, may yet give a visible light."

The "small animalcules" are dinoflagellates, not quite animals, not quite plants, but protists, members of the third great kingdom of life, along with protozoans, photosynthetic algae, and some fungi. A large number of dinoflagellate species produce light, glowing dimly when darkness falls but flashing brilliantly when disturbed by a wave or a passing boat; it may be a defense mechanism, because any animal trying to eat them will be lit up, and thus be made more vulnerable to its own predators. Because of their light, it is easy to see the dinoflagellates, but they are only the most visible form of plankton— that astonishingly complex community of minute, floating organisms that are the basis of marine food chains the world over. But while some plankton can be found in almost any patch of ocean, they are not distributed evenly, which is why we were making this midnight run off Nova Scotia. To find pelagic seabirds, we had first to locate the great, seasonal blooms of plankton. Our depthfinder (which was, at the moment, working) had been showing more than 110 fathoms most of the night—better than 700 feet of water—as we crossed a depression on the continental shelf known as the LeHave Basin. But as the first measure of daylight began to obliterate the eastern stars, erasing Orion's torso, the digital figures changed dramatically. The bottom rose hundreds of feet toward the surface as we reached the edge of the LeHave Bank, an area the size of Rhode Island where the depth is just 250 feet or so. Soon the red and green running lights of harpoon boats looking for swordfish twinkled in the murky distance, and we flushed not only the occasional small flock of phalaropes but larger birds as well, and more frequently.

Submerged banks like the LeHave shadow the edge of dry land from the Canadian Maritimes through southern New England. During the last ice age, when glaciers lowered the world's sea levels, these banks were dry land, the tundra home of mammoths and mastodons, whose fossilized teeth are still dredged up by trawlers. The banks were the New World's first claim to fame, the legendary fishing grounds like Georges Bank off Cape Cod and the Grand Banks off Newfoundland. Around the banks, waters of different temperature and salinity mix—the frigid Labrador Current dropping down from the Arctic, the relatively fresh Nova Scotia Current pouring from the Gulf of St. Lawrence, and the warm, very salty Gulf Stream curving north just beyond the continental shelf, spinning off eddies of blue, subtropical water toward shore. Where the currents collide with each other, and with the sunken topography, they create upwellings that drag nutrients from the ocean bottom to the surface, where sun-loving phytoplankton can feed upon them.

The result is a riotous food web knitted on featureless waves. The plankton attract small fish like capelin and herring, which draw bigger predators to feed upon them, like the much-depleted northern cod. What's more, the plankton feed large vertebrates directly, including the largest of all, the great baleen whales like humpbacks, fins, right whales, and blues. Boat captains who take paying customers to see the leviathans learned long ago to watch the horizon, not just for vaporous spouts but for wheeling clouds of seabirds, feeding on the same slurry of protein as the whales. Plankton sets the table, and all the ocean comes to eat.

Daylight grew, giving shape to the deep swells that pummeled us all night, their tops torn to spray by a freewheeling northwest wind. As my eyes roamed the horizon, I saw black silhouettes parting from the waves, rising on the breeze—solid bodies like cigars, with long, thin wings that stretched to even

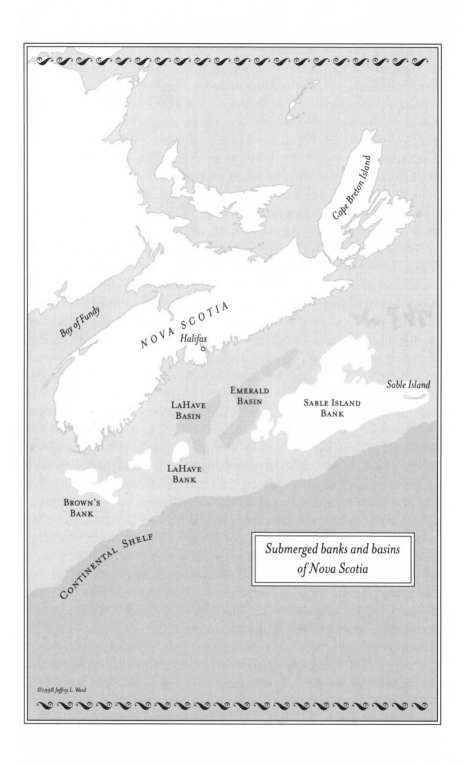

Cape Breton Island

Bay of Fundy

NOVA SCOTIA

Halifax

Sable Island

EMERALD
BASIN

LaHAVE
BASIN

SABLE ISLAND
BANK

LaHAVE
BANK

BROWN'S
BANK

CONTINENTAL SHELF

Submerged banks and basins
of Nova Scotia

tapers, held stiffly down. A bird swung across the bow and along the westward side; it pirouetted, one wingtip barely clearing the heaving waves, the other pointed skyward, and recrossed our path. Its body was about the size of a crow's, with a short tail and a fairly long, hook-tipped beak; the white undersides glowed yellowish in the newly risen sun, the back was a warm brown, and a white collar edged its black cap. What I noticed wasn't color or pattern, though, but the amazing speed and grace with which it flew. There was none of the languid, loose-limbed rowing of a gull; this bird had a gymnast's form, with crisp, snappy wing beats, smooth glides, and fluid turns.

It was a greater shearwater, one of the most common birds in the western Atlantic in late summer. In many respects, the shearwater, named for its wave-skimming flight, is typical of pelagic species, combining incredible numbers, a minute breeding range, and wanderings that scribble half the planet. Aside from a handful of individuals in the Falkland Islands off Argentina, greater shearwaters breed in just three locations in the South Atlantic—more than 5 million pairs on Nightingale and Inaccessible, islands in the Tristan da Cunha group, and up to 3 million more on nearby Gough Island. ("Nearby" is, of course, a relative term. Gough is more than 200 miles away from Tristan da Cunha, but given that these lonely islands are almost exactly halfway between South America and Africa, that's practically next door.)

None of these islands is large; Nightingale is barely half a mile long. Yet as restricted as its breeding territory may be, the greater shearwater's oceanic range is vast, encompassing virtually the entire Atlantic basin, from Drake Passage off Tierra del Fuego to the Cape of Good Hope off southern Africa, and up to the Labrador and North Seas. After leaving their breeding islands, the shearwaters drift along the eastern coast of South America, feeding on planktonic crustaceans. When these sink to deeper waters in May, the birds push quickly north, apparently crossing the tropics in a rush, since they appear off the Carolinas just a few weeks later. By June and July they have reached the Gulf of Maine and Nova Scotian waters, just in time to exploit the spawning schools of capelin; many continue as far north as the Davis Strait between Canada and Greenland. Moving with the prevailing westerly winds, the shearwaters swing clockwise toward Europe by autumn, rocketing back across the Atlantic from northwestern Africa to the nesting grounds for the start of another breeding season in October, the austral spring. Conservatively, a greater shearwater following this route will cover more than 13,000 miles in about nine months.

More and more shearwaters were visible, many crisscrossing our wake—looking, no doubt, for food, as they have learned to do with the fishing boats that scatter by-catch and offal. Under wild conditions greater shearwaters feed on a variety of marine animals, with small fish and squid topping the list, and I watched those nearest our boat plunge again and again into the water. One slid alongside and paced us, curious, and I could see clearly its long, thin bill, tipped with a somewhat bulbous hook. What was remarkable wasn't the tip, though, but the nostrils. Shearwaters, petrels, albatrosses, and storm-petrels belong to a group of seabirds known as "tubenoses," and for good reason—their nostrils are encased in tubular, external sheaths that sit atop the beak, an arrangement unique among birds.

The nostrils are a visible indication of an invisible talent—for tubenoses, unlike the great majority of birds, have a well-developed sense of smell. It has been known for some time that storm-petrels use their noses to find their nests at night. Now experiments in the Antarctic suggest that some petrels can detect tiny concentrations of dimethyl sulfide, a chemical released when zooplankton feed on phytoplankton—in fact, researchers estimate they may be able to home in on such olfactory signals from as far away as four kilometers. Some scientists even believe that albatrosses may recognize each other at least partly by scent, much as dogs do. Tubenoses all share a peculiar, musty odor that stems from a foul, greasy stomach oil that they produce during digestion and feed to their chicks. Certainly they smell to the human nose; one birder I know, who's been studying pelagics for more than twenty years, claims to be able to locate big storm-petrel flocks in the fog by sniffing them out.

Spending months at sea raises a significant difficulty for pelagic birds: finding something to drink. Seawater is saltier than the bird's own blood, so much so that drinking a quart of ocean water requires a quart *and a half* of freshwater to flush out the toxic salts—a quick route to death by dehydration. But tubenoses, gulls, and many other seagoing birds have specialized glands, located above each eye, which filter excess salt from the bloodstream, expelling it down the nostrils. They have perpetually runny noses, from the endless drip of concentrated salt that trickles down their beaks.

While birders rarely venture out to sea, fishermen do so every day. Seabirds are their constant companions, often the only sign of life in a lonely, dangerous place. The day before we sailed, we stopped to buy chum—fish livers and herring roe, to be chopped up and ladled over the side to attract birds—from a local fisherman. Randy, in his early thirties, with a mop of thick, dark hair and a drooping mustache, had the guts in two 10-gallon plastic buckets; the heavy, oily smell of fish seeped out despite the tightly sealed lids.

Nova Scotian politeness is legendary, but the notion of someone chartering a boat and sailing all the way to the fishing banks to *watch birds*, of all things, at last overcame any strictures against nosiness. "What sort of birds are you looking for?" Randy asked doubtfully. But the names—shearwaters, storm-petrels, fulmars—obviously meant nothing to him, and the skeptical furrows on his forehead grew deeper. Finally I dug a battered field guide out of my gear, thumbed it open, and handed it to him.

"Oh! You mean the noddies!" Randy said, brightening immediately and tapping a finger on the painting of a greater shearwater. He turned the page, finding a rendering of several black-brown, slender-winged birds. "And the Careys! Yeah, we see them all around out there, too."

Sailors and fishermen have had their own names for birds for centuries, and many species, including shearwaters and several tropical terns, are called "noddies." The name comes from the Middle English word *nodde* and means a simpleton; the birds' unwary nature around humans, a result of millennia of breeding on predator-free islands, led mariners to consider them stupid.

The "Careys" Randy referred to were the storm-petrels, swallow-sized seabirds that look as though God dipped them entirely in dark chocolate, save for a white thumbprint left on their rumps. They flutter and swoop like bats, reaching out with their long, jackstraw legs and webbed feet to patter about the surface of the water, as though walking on it. This behavior struck early sailors as a biblical echo, and they called them "Mother Carey's chickens"—a corruption of *Mater cara*, the Virgin Mary. (Indeed, the very word "petrel" is a diminutive meaning "Little Peter"—a reference to St. Peter, who walked on the water with Christ's help.) The arrival of Mother Carey's chickens was taken by sailors as a sign that gale winds were coming, the root of the modern name storm-petrel.

Seabirds are the basis of much maritime superstition. Storm-petrels were known as "soul birds" in some cultures, thought to be the incarnation of evil captains banished for their cruelty to an eternity on the waves. Their landing on a ship was an omen of profound bad luck, and touching one brought death to a sailor or his family. And as Coleridge's *The Rime of the Ancient Mariner* made famous, it was believed that albatrosses were terribly unlucky beings to have around, that at a whim they could bring gales or leave ships becalmed:

And I had done a hellish thing.
And it would work 'em woe:
For all averred, I had killed the bird
That made the breeze to blow.

Albatrosses are rarely seen in the North Atlantic, although yellow-nosed albatrosses, which breed with the greater shearwaters on Tristan da Cunha, are sometimes seen off the U.S. and Canadian coasts. Slightly more common are sightings of black-browed albatrosses, stunning birds with slender wings spanning more than seven feet, brown-backed and ivory below. These birds nest off Cape Horn, South Georgia, and other islands in subantarctic waters, and generally roam the oceans below 40 degrees south latitude, where the winds grow in ferocity the farther south one goes—the "roaring forties," "howling fifties," and "screeching sixties," as sailors dubbed the region.

After a four-hour shift, I was finally relieved of my duty at the wheel, and as the smells of breakfast rose from the galley (making me wonder how anyone could cook in such a lurching environment), I was free to devote all my attention to the birds. Here and there I spotted what appeared, at a casual glance, to be gulls, but were not—gray birds the color of backlit fog, with the stiff-armed flight and odd bill shape of a tubenose, and a barrel-chested, bull-necked profile that made them look like wrestlers with wings.

These were northern fulmars, which nest on sea cliffs right around the Northern Hemisphere and fan out after the breeding season across the cold oceans. Few birds have enjoyed such unbridled success as fulmars, which have exploded in both range and numbers over the past two centuries, despite a low reproductive rate and pressure from Icelanders and others, who relished them for their oil and meat. Originally restricted to the high latitudes, like the cliffs encircling Baffin Bay in Canada and Greenland, and Eurasian islands in the Arctic Ocean, the fulmars began pushing south in the early eighteenth century. They colonized most of Iceland by the 1750s and marched on Europe from there—the Faeroe Islands by 1820, Scotland by 1878 (with the rest of Britain shortly thereafter), Norway in 1924, and French Brittany in 1956. A less dramatic, but still notable, southward push also occurred in the western Atlantic, with fulmars setting up shop in Labrador and Newfoundland. According to the best estimates, there are between 4 and 16 million *pairs* of northern fulmars in the world, and that number is still climbing.

Soon the *Desiderata* had attracted an entourage, fulmars and greater shearwaters toiling along behind us, joined by a few young black-backed gulls, all of them hoping for an easy meal. We slopped some chum into the water, the slick of fish oil spreading a gleam on the waves, and immediately the flock thickened. Whether the birds came because they could smell the oil or because they recognized from afar the plunging, jerky movements of others busily feeding, I could not say, but the cloud of birds grew by the moment.

I spotted the first sooty shearwaters of the day—slim charcoal-colored birds that flapped more and glided less than the larger greater shearwaters. Sooties are most common in the Pacific, where they breed by the millions on the coasts of Chile, New Zealand, and Tasmania. A relative few also nest in the Falklands, and these travel a circular route around the Atlantic, much like the greater shearwater's—up the coasts of South and North America in April, arriving in the Maritimes in early summer, east with the prevailing winds in August and September toward Europe, then cutting across the tropics in essentially nonstop flight back to the breeding grounds.

We started encountering small flocks of storm-petrels, too, moving in groups of two or three dozen, dark against the Prussian blue of the water, with their rump patches glowing white. They were hard to watch; Houdinis of the sea, they appeared and disappeared in the troughs of the waves. Some looked brownish, and flew in a way that combined the stiff-winged flap of a moth with the erratic movement of a pinball; these were Leach's storm-petrels, and when they finally came close to the boat I could see their forked tails, the best field mark. Others looked blacker at a distance and flew a fairly direct course, with a constant mix of shallow, fluttering wing beats and short glides; these were the nearly identical Wilson's storm-petrels, which sported squared-off tails. (This question of tail shape may sound like a subtle difference, but in the world of pelagic seabirds, where dozens of species share the same bland mix of gray, black, white, and brown, it passes for a glaringly obvious characteristic.)

So similar to the eye, these two species of Mother Carey's chickens nest at opposite ends of the world. The Leach's breed along the northern rim of the Atlantic and Pacific, including monstrous colonies in Newfoundland. Baccalieu Island alone, four miles long and half a mile wide, supports at least 3.5 million pairs, with an additional 1.2 million pairs in eight other colonies around the province. Wilson's storm-petrel, on the other hand, nests along the coast of Antarctica and the subantarctic islands, in even greater numbers; while reliable figures are hard to find, it is often said to be the most numerous wild bird on earth.

In the nonbreeding season, the storm-petrels swap oceans. Leach's go south toward Antarctica, and Wilson's come north toward the Arctic. It is an annual ebb the length of the Atlantic, a tide of Careys spread on the waves. As the Wilson's crowd onto their nesting grounds in November and December, the Leach's advance to fill the now-vacant South Atlantic Ocean. When the Leach's return north to breed in May and June, the Wilson's follow on their heels, taking over the offshore waters between Canada and Britain, beyond the Leach's daily travel range. During the in-between seasons, like August, when neither species is tending eggs or chicks, both mix on the sea.

So engrossed was I in the storm-petrels and shearwaters whizzing close to the surface that I almost forgot to look up. When I did, I saw scraps of white crossing the blue sky—fork-tailed and slender, a line of Arctic terns beating south before the wind. These are the most famous of all pelagic migrants, perhaps the most storied migrants of any kind.

The Arctic tern's breeding range encompasses the highest latitudes of the Northern Hemisphere, including the most northerly dry land of all, Cape Morris Jesup in Greenland. In autumn all the terns funnel down two main flyways. One traces the west coast of North and South America, but the main flyway seems to be the Atlantic route. A tern hatched, say, on Ellesmere Island in the Canadian Arctic will head east, crossing the North Atlantic to Britain, where it is joined by terns from northern Asia, Europe, Iceland, and Greenland. The flocks swing south, most following the coast of Africa, hooking east past the Cape of Good Hope and on toward Antarctica; others split off at the bulge of Africa and recross the Atlantic, coming to Antarctica from the coast of Argentina. Either way, they have the prevailing westerly winds at their backs as they cross the screeching sixties.

By November or December, the terns reach the limit of the south polar ice pack, which is shrinking in the constant sunlight of the austral summer. They congregate in a main wintering area south of the Indian Ocean, hunting krill on the edge of the ice floes until early March. Many, however, keep moving east, circumnavigating Antarctica over the course of several months.

With austral winter approaching, the adult and immature terns part company. Some of the adults retrace their route up the coast of Africa and across the North Atlantic, but others cut northwest from the Cape of Good Hope, moving diagonally across the ocean back to eastern Canada. The first-year terns, on the other hand, follow the easterly winds in the polar zone to the tip of South America, and migrate north along that continent's western shore, spending the northern summer in the cold Humboldt Current off Peru and Chile.

When all is said and done, an Arctic tern may cover more than 22,000 miles in a single year. Watching the flecks of white disappear into the crisp sky, I shook my head in disbelief—the only thing an earthbound human could do.

A week later I had, like Mother Carey's chickens, swapped oceans, trading the spruce-girt coast of Nova Scotia for foggy central California. A hundred miles south of San Francisco lies Monterey Bay, a perfect half-circle embraced by hills of pine and live oak, where warm afternoon breezes carried the heady, cinnamon-spice aroma of chaparral and sagebrush.

Rimmed by nearly forty miles of coastline, Monterey Bay is an aquatic wonderland. Underlain with submarine canyons that plunge nearly two miles below the surface and bring deepwater habitats close to shore, the bay features an unparalleled mix of marine life. Forests of giant kelp edge the shallows, home to sea lions and harbor seals; female sea otters float languidly on their backs, clutching fuzzy infants to their chests. Brown pelicans, loons, and cormorants speckle the sky and water. Farther out, rich plankton blooms support anchovies and squid, tuna and albacore, as well as humpback and blue whales, the latter using the bay as a weaning area for their calves.

Not surprisingly, Monterey Bay also attracts a phenomenal diversity and abundance of migrant seabirds. Some come a short distance, like the thousands of ashy storm-petrels that nest on the Farallon and Channel Islands just to the north and south; others, like northern fulmars and fork-tailed storm-petrels, drop down from Alaska or the Pacific Northwest. Most, however, make stunning transoceanic journeys, coming from breeding colonies in Tasmania, New Zealand, Japan, Korea, Chile, Hawaii, Australia, and Antarctica, pulled like filings to a magnet by the largesse of Monterey's waters.

Along the Atlantic coast, the wide continental shelf keeps the deep water—and the truly pelagic seabirds—up to a hundred miles from land. But because Monterey Bay's submarine canyons come so close to the coast, it is possible to see dozens of species of seabirds just a short boat ride from shore. For more than twenty years, a remarkable woman named Debra Love Shearwater (no, it is not her given name; this is California, after all) has been guiding pelagic birding tours on Monterey Bay, showing tens of thousands of people the globe-wreathing migration that focuses here.

At 4:30 A.M. on an early August morning, Debi and I drove down from her home in the dry hills forty minutes east of the ocean, across the flat fields of artichokes that back the high dunes, and through the deserted town. Autumn on Monterey Bay is generally a placid season, a gentle prelude of flat-calm waters and sunny skies before gray winter storms lacerate the sea. But even in the predawn gloom as we unloaded our gear, we could feel the strong northwest wind that made the flags above Fisherman's Wharf snap and crackle. The boat—no leaky sailing yacht but a 75-footer with powerful diesel engines, built for the bay's popular sport-fishing industry—was rocking at the wharf, where two dozen sleepy-eyed birders waited with their packs, raingear, and binoculars. Fishermen shuffled onboard a neighboring boat, thick rods in hand, and stared with undisguised bewilderment at the rest of us.

Debi hopped up on a bench and launched into a safety-and-orientation spiel she's given more than a thousand times. A dental hygienist by training, Debi has been guiding natural-history tours on the bay since the 1970s, first as a sideline and later, when repetitive motion disorder ended her days as a hygienist, as a full-time occupation. After she finished her introduction, we all found protected nooks in the lee of the cabin, and as the boat surged out beyond the Coast Guard breakwater—its boulders buried beneath heaps of querulous sea lions—we felt the wind-driven rollers coming in from the open Pacific. Cold, damp wind rippled over me, and I squirmed deeper in my layers of thermal long johns, flannel, wool, and Gore-Tex, musing on the irony of wearing not one but two pairs of gloves on an August day in California.

For the next hour or so I dozed in the chilly darkness. We were heading for the albacore fishing grounds, a 45-mile run to an area that birders visit only a few times each year. Today's trip would take us over submarine canyons nearly 6,000 feet deep, and the birdlife would be much different than on the previous day, which we'd spent in relatively shallow inshore waters.

The moment the darkness receded, we began to see sooty shearwaters, and there was hardly a time during the next twelve hours when this species was out of sight. Estimates place the seasonal population along the California coast at more than 2 million, many of which swarm across Monterey Bay in summer and autumn; their molted feathers sometimes blanket the beaches in thick gray windrows. Unlike the sooties I'd seen in Nova Scotia, which nest off Argentina, the Pacific sooty shearwaters come largely from breeding colonies near Australia and New Zealand. Like so many seabirds, they trace a clockwise migration, borne by prevailing winds—north to Japan and the

Kuril Islands, east to Alaska, and then south along North, Central, and South America, finally cutting west across the Pacific again. In the process, they travel from subantarctic to subarctic waters and back each year.

During the voyage of the *Beagle* in 1835, Charles Darwin witnessed a great movement of sooty shearwaters off the coast of Chile: "I do not think I ever saw so many birds of any other sort together, as I once saw of these . . . Hundreds of thousands flew in an irregular line for several hours in one direction. When part of the flock settled on the water the surface was blackened, and a cackling noise proceeds from them as of human beings talking in the distance." In 1919 along the coast of Peru, Robert Cushman Murphy, whose pioneering studies of seabirds remain a classic of ornithology, had a similar encounter: "There were certainly upwards of 10,000 in a single dense flock that passed us, the long-winged, graceful shearwaters flying high and low in the howling wind over the choppiest areas of water. Do they seek such whipped-up places because the wind and waves bring their food to the surface? Whatever the reason, they were marvelously at home in the terrific gusts."

Out on the open bay, exposed to the full force of the north wind, the boat hammered through ten-foot waves, slamming down into each trough with a twisting lurch that had about half the passengers prostrate with nausea. We were followed from shore by a ragtag bunch of gulls, mostly young western gulls still wearing the dirty brown plumage of adolescents. No one was even slightly interested in them, western gulls being common as kelp along the California coast, but their presence would serve to attract more exciting species. So throughout the morning, one of Debi's two volunteer guides, Jim or Jennifer, stood at the stern of the boat tossing frozen anchovies high into the air, or dipping into trash bags stuffed with stale popcorn and flinging the white puffs into the trailing flock.

The feeding calls and wheeling birds worked; Jennifer shouted and pointed to a brown shape racing toward us on impossibly long wings, like an eagle stretched to the breaking point. It was a black-footed albatross, and the sight of it riveted me—ribbon-thin wings spanning nearly seven and a half feet, dark brown above and below, and a long, heavy-beaked face gone gray, like the muzzle of an aging hound. Indeed, black-footed albatrosses do gray as they age, and this one could easily have been twenty or thirty years old; I could only guess at the uncountable miles that had passed beneath its limber wings.

Blackfoots nest squarely in the middle of the Pacific, using remote beaches in the Hawaiian archipelago like Laysan Island, Kure Atoll, and Midway Island; a few also nest near Japan. At one time, they and the pale, closely related Laysan albatross also bred on islands in a wide band across the western Pacific—Wake Island, the Northern Marianas, the Marshall Islands, and others. What the Japanese feather hunters of the nineteenth century didn't exterminate, the occupying forces on both sides of World War II almost did, both incidentally and deliberately. On Midway, home to a huge military airport, 54,000 albatrosses were killed between 1955 and 1964 by the U.S. Navy to minimize collisions with planes. I recall, as a kid in the 1960s, seeing documentaries about the "gooney birds" of Midway, condescendingly funny films full of slow-motion footage of the birds' cumbersome attempts to land—belly flops, head-over-tail tumbles, and other pratfalls—or their lumbering, laboring takeoff runs, wings akimbo. Hilarious stuff, and when the narrator solemnly explained the necessity of killing the gooneys to make the islands safe for planes, which made the world safe from Communists, it was hard to feel sorry for a bird so patently ill suited to real life. (The albatrosses should enjoy a gentler future now that the air station has closed and the atoll has been transferred to the National Wildlife Refuge system.)

Drop a human from a plane and an albatross might find that amusing, too, for we would be as far out of our element as a land-bound gooney. On the rolling surface of Monterey Bay I saw the truth, graceful and buoyant on the wind. The albatross quartered behind us, swooping low to the surface, one wingtip flicking a bare inch above the heaving waves, yet never touching the water. Then it rose high into the wind, stalling and rolling to the other side, slipping back across our wake, never lifting a wing to flap. We were being treated to a display of what ornithologists call dynamic soaring, a flight technique mastered by the albatrosses, enthralling to watch but difficult to explain in words.

We'll start with the fact that the wind almost always blows on the open sea, and it blows less quickly down near the waves, which obstruct and slow its passage. Gliding twenty or thirty feet above the surface, an albatross dives forward at a shallow angle with the wind at its back, building momentum. As it bottoms out in the lower, slower-moving air near the water, the bird pivots and pulls up, turning into the wind. Its momentum, aided by the sea breeze under its wings, lifts it to its original level. Then the albatross turns downwind again into another shallow dive. In this way, much like a sailboat tacking against the wind, the great bird zigzags across the ocean, hardly stirring a muscle.

The trouble with dynamic soaring, of course, is that it is wholly dependent on the wind. Albatrosses cannot flap their outsized wings for any length of time, can scarcely get themselves airborne on calm days, in fact. The largest species, like the wandering albatross with its eleven-and-a-half-foot wingspan, are kept effectively penned in the far Southern Hemisphere by the doldrums, the wide band of relatively windless air that blankets the tropics. (Shearwaters and storm-petrels, which engage in more energetic, flapping flight than albatrosses, are not stymied by the calm air, and black-footed albatrosses nest in the North Pacific, above the doldrums.) Those southern albatrosses that do penetrate the doldrums may find themselves permanently trapped on the wrong side; at least, that is the explanation usually given for birds like the Atlantic black-browed albatross that returned to the Faeroe Islands between Scotland and Iceland each summer for thirty-four years in the late nineteenth century, and another that appeared annually in the Shetlands from 1972 to 1987.

Two more albatrosses joined the first, knotting the air around us with their movements. At first, every binocular was turned on them, but within a few minutes some of the other birders lost interest, scanning the waves for new curiosities, and when a small bunch of black storm-petrels turned up, the albatrosses were forgotten entirely. Only Jennifer and I still watched them threading the wind. "Instant desensitization," Jennifer muttered, glancing at the other birders. "One second it's the most beautiful bird they've ever seen, the next second they're bored with it."

For most of the morning the action was spotty. We would cruise for long stretches, conscious only of the boat's lurching motion, the pewter horizon, and the breath of salt spray, and with only our entourage of ratty gulls for entertainment. Then, without warning, we would find ourselves surrounded by seabirds—meandering lines of Sabine's gulls, their semaphore-flag wings coded in black and white; hundreds of shearwaters like smoky ghosts; a few albatrosses coursing the periphery, royalty among commoners. People would point and shout, "Long-tailed jaeger at one o'clock!" "Arctic terns crossing the bow!" "Big flock of storm-petrels at three o'clock—I think they're fork-tails!" You weren't sure where to look first, worried that while you were drinking in the sight of one rare bird, two others were slipping past you unseen.

And then, just that suddenly, the sea and sky would fold the melee into itself, wheeling it off quickly in our wake and leaving us to watch empty air once more. This wasn't caprice, of course, but a vivid illustration of the patchy manner in which resources are distributed in the ocean. Where nutrients well up from below, there the phytoplankton bloom. Where phytoplankton grow, zooplankton come to eat them. Where there are zooplankton, there are a host of predators like anchovies and squid—and so on, higher and higher through the food chain.

Up ahead, a small flock of shearwaters buzzed and flared, feeding on a school of something below the surface. The boat slowed, and Debi's voice crackled with excitement over the loudspeaker from the wheelhouse. "South polar skua at eleven o'clock!" she shouted, and there was a general rush to the port rail that made the boat list slightly. Sitting on the water fifty yards away, calm amid the swirl of shearwaters, was a squat gray-brown bird, the size of a large gull, but with a heavier, more predatory look. Its beak had the long profile of a gull's, but with a wicked hook at the tip. The boat closed in, and the skua lifted off and sailed around the bow, showing wide, hawklike wings marked with slashes of white near the bends. As it circled the boat twice, a chorus of contented "aahs" followed behind it like dust.

For many people on board, this rather nondescript creature was a big reason they had paid for twelve hours of bone-chilling cold and acute sea-sickness. As their name implies, south polar skuas breed on the Antarctic coastline—one of the few birds, besides penguins, to nest in that harsh environment. The austral winter sends them scattering across the northern Pacific and Atlantic Oceans, usually far from land. For serious birders interested in adding species to their North American life lists, the south polar skua is a difficult prize, and such long-haul pelagic trips are the only reasonably sure way to see one.

People wax poetic about songbirds or raptors, but skuas engender different emotions, not always positive. Aggressive hunters and efficient scavengers, they are tough to love, and impossible to forget. (I will always recall the great skua in Scotland that nearly scalped me for approaching its nest.) Robert Cushman Murphy voiced a typical reaction. Skuas have left, he said, "a more vivid impression in my memory than any other bird I have met. The skuas look and act like miniature eagles. They fear nothing, never seek to avoid being conspicuous, and, by every token of behavior, they are lords of the far south. In effect, they are gulls which have turned into hawks."

Murphy was speaking specifically of the brown skua, but the description is apt for the closely related south polar skua, similar in plumage, habit, and demeanor. The heavy street-brawler body and hooked beak allow the skua to dominate the Antarctic scene. Scavenging the carcasses of beached whales, snatching fish or squid from the oceans, even filching placentas from crab-eater seals in mid-birth, or stealing milk from the oozing mammaries of nurs-ing elephant seals, they let nothing go to waste. In a pinch, seaweed and moss will do, if meat is unavailable. But first and foremost, skuas are hunters, patrolling the great penguin colonies. "Eggs are stolen whenever left exposed, and can be picked up almost without a pause in the skua's flight," Murphy wrote of the south polar skua. "In victimizing young penguins, they attempt to cut one off when it is somewhat separated from its fellows, contriving to urge or drag it little by little away from the center of the population, then strike it on the head and gradually worry it to death . . . There is, indeed, no defense against the appetite of the skuas except a fitness and willingness to fight back."

While the Arctic tern has the greatest regular migration of any bird species, a south polar skua actually holds the record for the longest known flight by an individual bird. Banded as a chick in 1975, while still in its nest on an island off the Antarctic Peninsula, the bird was shot six months later along the western coast of Greenland, a straight-line distance of some 9,300 miles. Interestingly, another skua banded in the same area, at about the same time, was recovered later off Baja California, thousands of miles up the Pacific.

More and more birds shadowed us, and up ahead we could see the bushy spouts of humpback whales. Then, without warning, we found ourselves surrounded by cetaceans—six humpbacks rising and diving to either side, scalloped tail flukes lifted high in the air, hundreds of Pacific white-sided dolphins skipping between waves, and literally thousands of northern rightwhale dolphins, lissome and black as night, leaping out of the water as far as we could see to the west and north.

I scrambled to the bow of the boat to look down. Seven rightwhale dolphins and four whitesides were riding our bow wake, playing in the pillow of water created by the boat's ten-knot speed. The whitesides were powerfully built creatures, with short beaks and fluid swaths of pallid gray on their sides. Others, racing effortlessly beside the boat, leaped repeatedly from the sea, rooster tails of water streaming behind them; at times they smacked their tails on the surface in mid-leap, making a resounding crack.

By contrast, the rightwhale dolphins hardly resembled the typical image of a dolphin at all. Eight or nine feet long, they were slender, almost petite, with narrow tail flukes and pectoral flippers so dainty they seemed misproportioned. They lacked dorsal fins, and those that leaped from the water looked, in profile, more like sea lions than dolphins. They were surpassingly elegant, like penguins, but without the comedy. The numbers of both species around us were amazing; by Debi's reckoning there were more than 1,200 whitesides and 3,100 rightwhale dolphins—the latter by far the largest group she'd ever seen in Monterey Bay.

The scene was one of utter chaos, whales and dolphins mixed with California sea lions, a few northern fur seals, gulls, pelicans, albatrosses, storm-petrels, phalaropes, skuas, and jaegers. We spotted pink-footed shearwaters, which breed off the coast of Chile, and one beautiful flesh-footed shearwater, a rarity from Australia and New Zealand, the hue of black-walnut wood. A Buller's shearwater, also from New Zealand, left me breathless—spotless white below, a dark M written boldly across the gray of its clean, knife-blade wings, the finest sight on the ocean.

We were always on the watch for a bird with a difference—a seabird that didn't quite match any of the expected silhouettes, color patterns, or flight maneuvers. New species are recorded frequently on pelagic trips, making such outings both exciting and scientifically important. Just a week before, one of Debi's tours had spotted a dark-rumped petrel, an exceptionally rare bird of Hawaii and the Galápagos seen in North American waters only three times before.

They also encountered an odd dark petrel that didn't match anything normally sighted off California. At first, the bird was identified as a Murphy's petrel, a little-known South Pacific bird that shows up rarely along the American coast. Yet others weren't convinced; perhaps it was a Solander's petrel, they said, a threatened Australian species similar in plumage and equally rare in this hemisphere. Photos of the bird were sent to experts across the country, and a few weeks after my visit they reached the stunning conclusion that it was a great-winged petrel—a New Zealand species that neatly matched the mystery bird's appearance, save for the uncomfortable fact that it's never been recorded beyond the extreme southern Pacific. But who knows? Maybe that's because no one ever looked for it.

Anyone can pick up a pair of binoculars and wander outside to watch land-birds, but to study pelagic seabirds requires a sturdy boat, lots of time, and a strong stomach. The handful of commercial tour operators like Debi Shearwater, and the even smaller cadre of ornithologists specializing in pelagic birds, have only scratched the surface. Imagine how limited our knowledge of land-bird migration would be—the nuances and diversity and complexities we would miss entirely—if we could visit only a few times a year, within only a day or two's hike inland from the sea, and if we could stay only a few hours. No wonder pelagic birding is considered one of the last frontiers of ornithology.

For the number of birds in view around our boat, the noise level was rather low—some squawking and croaking, but nothing like the cacophony that accompanies nesting. Many seabirds come ashore to the breeding grounds only at night, in screaming, raucous multitudes—the loud calls perhaps serving as a sort of proximity alarm to avoid collisions with other birds, or perhaps for mate recognition. In any event, the din is, by all accounts, ear-splitting, as suggested by this description of a sooty shearwater colony on two islands near Cape Horn in South America: "The nights we spent at Deceit and Wallaston were made hideous with the noise of petrels [shearwaters] passing overhead. I attribute the most ghastly sounds to this species . . . noises like choking cats—'cha-whee-whoo'—grating and choking with noise like gurgling intake of breath and lancing, working often to a climax. The sound is not altogether unpleasant at a distance, but uttered by those birds passing at close quarters it is appalling."

With a world of empty water over which to roam, the pelagic seabirds would appear safe from mankind's damaging ways. Not so. Humans rarely pass up an easy meal, and seabirds—with their incredible abundance, concentrated breeding habits, and lack of wariness—have been convenient targets down through the centuries. Almost 2 million fulmars were killed in Iceland in the first half of the century, and cod fishermen in the North Atlantic once used baited lines to catch greater shearwaters. "During the 'seventies of the last century it was not unusual to see from 200 to 500 of these birds hanging from the rigging of a Grand Banker," wrote Murphy in the 1930s. While that practice has died out, more than 50,000 shearwater chicks are taken each year from their nest burrows on Nightingale Island for food and fishing bait, a rate of harvest that conservationists believe could lead to a population collapse.

Given their ponderous reproductive rate, most seabirds are painfully slow to recover from overhunting. The short-tailed albatross was once among the most abundant tubenoses in the Pacific, breeding on eleven islands near Japan and migrating to the western coast of North America. Feather hunters all but exterminated the species in the late nineteenth and early twentieth centuries, reducing the shorttails to a tiny remnant on the volcanic slopes of Torishima Island. In 1939 and 1941 the volcano erupted, obliterating the species—or so it seemed. A handful of birds wandering at sea were spared, and returned in 1950. Yet with almost fifty years of protection, the total short-tailed albatross population still numbers only a few hundred, and spotting one off the American coast is a major event.

Legal protection now covers many seabird colonies, but some are still heavily exploited for food, including sooty shearwaters. Known as "mutton-birds," their chicks and eggs are collected by Maoris on several islands off New Zealand, with roughly a quarter million taken each year for sale as food or to be rendered into oil and soap. Short-tailed shearwaters (not to be con-fused with short-tailed albatrosses) are also known as muttonbirds, and some 200,000 chicks are taken each year from the vast rookeries on islands in Bass Strait, between Tasmania and Australia. The harvest provides seasonal income for the small community of Bass Islanders, including mixed-blood descen-dants of Tasmanian Aborigines, who otherwise face crushing poverty.

The muttonbird season in Bass Strait is short, lasting five weeks in late March and April. The chicks, pulled from their nest burrows on the grassy islands, are killed with a snap of the neck; later, in the plucking sheds, their stomach oil is drained for use as a medicinal additive, and they are cleaned for market. Coming from a total population of 23 million, the short-tailed shear-water harvest is, wildlife managers believe, easily sustainable, and the com-mercial cull may actually have preserved the nesting islands, which might otherwise have been used for cattle grazing, ruining the grassy cover and col-lapsing the nest burrows.

Actually, humanity's indirect actions have had far greater impact on pelagic seabirds. They swallow the innumerable bits of plastic junk we loose upon the oceans, they drown in gill nets spread for fish and swallow the baited hooks set by long-liner boats, they accumulate in their tissues a cocktail of toxic chemicals. Commercial fisheries, including increasing harvests of krill and other plankton, suck up the very creatures upon which they feed. And the introduction (intentional or otherwise) of rats, dogs, cats, pigs, goats, sheep, and cattle to predator-free nesting islands may have been the worst calamity of all. Cats ate the Guadalupe storm-petrel out of existence by 1912; mongooses are blamed for the precipitous decline of the black-capped petrel of the Caribbean, and the presumed extinction of an all-dark subspecies, the Jamaican petrel. Almost every nesting island around the world is now besieged by one carpetbagging mammal or another.

The lovely Buller's shearwater, which captivated me in Monterey, had a dreadfully near-brush with extinction. It nested only on Poor Knights Island off New Zealand, where the Maoris collected it for food. When the Maoris were massacred by Europeans, their domestic pigs went feral, ballooning in numbers and rooting out the defenseless shearwater nestlings. Fortunately, in 1936 the last wild pigs were killed, and about one hundred pairs of shearwaters, which had escaped the carnage by nesting on offshore rock stacks, returned to breed. By 1981, 200,000 pairs were again breeding on Poor Knights, now a bird reserve, and the total population was put at 2.5 million.

Here and there around the world, conservationists are battling to take the islands back for the birds. It is not a pretty war, being fought as it is with traps and enormous quantities of poisoned bait, nor is it cheap, but people have managed to sweep some islands clean of rats, cats, and other predators. If we can keep the oceans themselves healthy and functioning, the future may be brighter for these global pilgrims.

By late afternoon we were motoring back to land. A few miles from shore, the low clouds and mist that had blanketed us all day parted, and warmth returned with the sunlight to my chilled fingers and toes. The rollers subsided, and the amber hills of the Monterey Peninsula beckoned. Then the throb of the engine rose a few notches, and the boat turned back toward the open sea. Those who had been green and prostrate with seasickness all day, and who had begun to stir hopefully as land neared, looked positively stricken.

Out ahead a couple of miles, a whirling gray smudge fogged the horizon, a great boil of birds feeding on an anchovy school. As we approached, we could see they were sooty shearwaters—more than 10,000 of them, diving and splashing after fish. We reached the trailing edge of the flock, where birds were lifting off to leapfrog to the front, thus keeping up with the moving school below; the musty smell of shearwater hit our noses, a cloying odor of old fish and locked rooms.

Soon we were within the goliath flock, which was taking flight—a clockwise wheel of birds with us at the very hub, a gyre of black, whispering wings mixed with the staccato cries of gulls and the angular, plunging forms of diving pelicans. Running along the surface, the multitudes of shearwater feet made gentle splashing sounds like wet applause, and as more and more took wing, the noise rose to a thunderous waterfall of an ovation—a tempest of birds and sound around which, for a moment, it seemed that the world itself revolved.

5

Rivers of Hawks

Open up a relief map of North America, the kind that shows the ripples of mountains and the flat wash of plains, and look south into Mexico, where the continent narrows like a horn. Most of the country is highland—the long line of the Sierra Madre Oriental running down the eastern flank, the Sierra Madre Occidental down the west, framing a jumble of desert. As the two ranges taper together, a third, a line of snowcapped volcanic mountains running across the country like a jeweled belt, joins them almost from coast to coast. The mountains do not quite reach the Gulf of Mexico, although when you stand on the bone-white beach at Paso Doña Juana and look across the great dunes you can see them, hazy in the distance. Here, in the state of Veracruz, the coastal plain that was a hundred miles wide at the Texas border shrinks to a narrow band maybe twenty or thirty miles across, nearly pinched off by the mountains.

It is along this constricted plain that migrating hawks come each year. Flowing down from the eastern two-thirds of North America, they gather in greater and greater numbers—Swainson's hawks from the prairies of Alberta and Nebraska, broad-winged hawks from Ontario and Maine, peregrine falcons from Greenland and Baffin Island—more than a dozen species of agile hunters.

They weave around the Great Lakes or cross the plains, thread their way down the Appalachians or trace the Atlantic shore. They form a broad, thick front that overspreads the Texas flatlands, hooks around the Gulf and down the edge of Mexico, where in Veracruz they find themselves caught in a bottleneck, a squeeze play between the mountains and the sea. They come by the millions each autumn, the greatest migration of hawks in the world—but only recently has science even noticed them.

The sun had set, and a gentle breeze took the edge off the blistering heat of a late September day, cooling Cardel, a small town of aging buildings and several thousand residents a few miles from the ocean. A hotly contested basketball game was under way in the central *zócalo*, the court surrounded by bright lights and flowering trees filled to overflowing with flocks of complaining great-tailed grackles. Bird-sellers peddled cages with pathetic white-fronted parrots to motorists stopped for a red light at a nearby intersection.

I walked past the ball game and around the corner, into dimly lit streets lined with closed shops and storefronts, past a church where the ghostly white shape of a barn owl drifted out of the stone bell tower and into the darkness. I was late for a meeting with my friend Ernesto Ruelas Inzunza, who was trying to organize the first survey of eastern Mexico's incredible fall hawk migration.

I'd met Ernesto two years earlier, when he was serving an internship at Hawk Mountain Sanctuary in Pennsylvania. A lithe man in his mid-twenties with a quick smile, he grew up near the city of Xalapa in the cool mountains west of Cardel. We worked together on a banding project that autumn, and I was spellbound by his stories of the migration that pours through Veracruz state each year.

Hints about the extent of hawk migration through Mexico date at least to the late 1800s, just passing mention of large flocks in the journals of early naturalists. In the 1960s, American and European biologists working on other projects in Veracruz reported random sightings and informal counts suggesting that heavy flights pass over the region. But in the 1970s attention switched to Panama, when biologists reported seasonal totals that approached 900,000 birds, making it second only to Eilat, Israel, as the biggest hawk bottleneck in the world. To anyone who spared the question much thought, it was obvious the hawks migrating through Panama had to be crossing Mexico somewhere.

Ernesto knew where they were; he'd seen the huge flocks passing north in the spring up the mountains near Xalapa and had heard that great numbers followed the hot, humid coastal plain south in the fall. A hawk enthusiast from an early age, Ernesto recognized that Veracruz offered conservationists a unique opportunity—a way to monitor, at just one site, a large percentage of North America's migratory raptors.

He also saw a chance to begin to change public attitudes. Rural Mexicans are most likely to view raptors as competitors, pests, or gunnery targets, while rampant habitat destruction and pesticide use are ever greater threats to the birds. Ernesto secured funding to both count the migration and launch a public education program, with support from the National Fish and Wildlife Foundation, Hawk Mountain, HawkWatch International (a Utah-based raptor group), and the Mexican nonprofit organization Pronatura Veracruz. The result was a project that has become known as *Rio de Rapaces*—River of Raptors.

I found Ernesto at the team's headquarters, a stark, unfurnished house at the edge of Cardel. Most of his crew was assembled: fifteen young Mexicans, mostly university students trained in biology or education. Along with them were Jeanne Tinsman, a biologist from Hawk Mountain, and John Haskell, a college student from Montana representing HawkWatch. While teams of counters monitored the flight from four sites between the mountains and the coast, and others attempted to capture hawks for banding, a crew of educators would begin the crucial task of trying to change negative attitudes toward raptors by working with local farmers, townspeople, and schoolchildren.

The project was timed to begin in mid-September, just as the first big flights were expected to arrive. The hawks that migrate through Veracruz are among the most highly migratory birds in the Western Hemisphere, and their journey south lasts for weeks, even months, linking widely separated—and hugely different—worlds.

Consider, for example, the broad-winged hawk, a chunky, crow-sized bird of the eastern hardwood forests, where it drifts beneath the canopy of oaks and maples on wide wings or sits quietly along the edges of small meadows and streams, waiting for frogs, snakes, mice, or other small vertebrates. It has an extensive range, nesting from New Brunswick to the edge of the Great Plains, and as far south as the Gulf Coast; broadwings also nest in a narrow band of aspen and birch forest across southern Canada as far as Alberta. In autumn, though, the broadwings decamp into Central and South America for the nonbreeding season. They leave behind the cool, dappled light of the deciduous woodlands for the humid shade of the rain forest. Some make it as far as Brazil and Peru, and others stop as far north as southern Mexico. (A handful, almost all inexperienced immatures, wind up in south Florida, having apparently made a wrong turn and showing an unwillingness to cross the Caribbean or the Gulf.) But the heart of the broadwing's winter range stretches from southern Costa Rica down through Venezuela and Colombia. Depending on where it starts and where it ends, a broad-winged hawk may travel 4,800 miles one way.

A Swainson's hawk goes considerably farther. This may be the loveliest of western raptors, a graceful predator with long, tapered wings like candle flames, a creamy breast, and wing linings contrasting with sooty flight feathers: a bird made of light and shadow, at home in the pale blue bowl of the prairie sky. Like the broadwing, Swainson's hawks are buteos, soaring hawks with fan-shaped tails, engineered to hitchhike on the gentle lift of rising air. Inveterate travelers, Swainson's hawks may migrate from British Columbia or Alberta all the way to Argentina, a straight-line distance of some 7,000 miles. Even those from the southern extreme of the species' nesting range, in the American Southwest, face a journey of more than 5,000 miles to the Argentine wintering grounds. They leap across the rain-drenched waist of the hemisphere, stopping only to sleep along the way in the thick forests of Mexico, Central America, and the western Amazon, until they come again to grasslands and wide horizons on the Argentine pampas.

At least a dozen more species of raptors migrate through Veracruz in the autumn. Besides broadwings and Swainson's hawks, the most abundant are turkey vultures and Mississippi kites. The kites are particularly beautiful—small, delicate creatures, pearl gray below and smoky above, with squared-off tails and long wings, and eyes the color of blood. When traveling, all these raptors depend on thermal updrafts, the invisible bubbles of rising air generated when the sun heats the surface of the ground.

The broadwings and Swainson's hawks, kites, and vultures all move across the landscape by sliding from thermal to thermal, forming enormous kettles that swirl and seethe with wheeling birds. Yet that name suggests nothing of the scale that is involved. When a kettle of hawks rises on a thermal, there may be thousands of them roiling the sky, etching great, chaotic globes that drift against the clouds, until the birds bleed off the top in a narrow stream, gliding for miles on locked wings to the next thermal.

The migrants pass through eastern Veracruz because of geography and aerodynamics. In Veracruz in autumn, the hottest sun, and thus the strongest thermals, are found down on the coastal plain, where conditions for Ernesto's counters can be brutal. One day I helped Jeanne Tinsman and Carlos Zavala Blas, a university forestry student, at the Rio Escondido count site, one of four strung across the thirty-mile coastal plain like a picket line. We perched on a bluff that sat truck-high along the busy road leading to Xalapa, jammed all day with exhaust-belching traffic; we often had to yell over the roar of down-shifting tractor-trailers simply to be heard. Below us, miles of sugarcane fields, cattle pastures, and fruit orchards baked in the 100-degree sun.

The only handy shade was a spindly acacia tree sapling, and as the day passed I inched in a wide semicircle, trying to stay in its meager shadow; I drowned in sweat, and gray dust clung to me like a shroud. Even worse, the hawks came through in a dribble that died away to nothing by lunch, so that by midafternoon I simply gave it up for a bad effort and lay back with a bandanna over my eyes, trying to ignore the ants crawling through my hair.

Since I had grown up in the Appalachians, where hawk-watching is something done on scenic mountaintops, Veracruz took some getting used to. At the village of Chichicaxtle, the counters set up shop at the community soccer field, sitting on the team bench; at Chachalacas on the coast, where the greatest numbers of falcons and accipiters were expected, they worked on a hotel balcony. The main count station in Cardel was the least bucolic of all. There, the biologists sat on the roof of a vacant five-story building, one of the tallest structures in town, crouching in the razor-thin shade of the stairwell or sitting beneath a tattered beach umbrella on old school chairs that someone had salvaged.

At street level, Cardel had an unpleasant grittiness—litter and garbage stacked beside buildings, open sewers along some of the back alleys—but from above we saw little but the tops of the trees that lined every street; the weathered cement buildings that poked up here and there reminded me of the ruins of an ancient city overrun by jungle. Our perch overlooked the flower-drenched *zócalo* a block away, where crowds of children played in the evening. From below rose the constant blare of car horns, taped music, and the chatter of the neighboring marketplace, a cacophony that made this sound less like a research station than a bus stop—and, in fact, the battered old bus did stop just a block away.

By the third week of September, the first significant movements of Mississippi kites and broad-winged hawks were finally under way, with daily counts of ten or twelve thousand at Cardel, consistently the best location. One day late in the month I tagged along with Jeanne and María Liliana Vanda. We were joined by a bird-mad Frenchman named Christian, who'd heard rumors of the migration and—despite a lack of Spanish, much English, or a clear notion of exactly where he was going—had come to Mexico and wandered around Veracruz until he wound up in Cardel. Ernesto had found him the day before in the middle of the square, peering through a spotting scope to the great amusement of passersby, and invited him to join us.

The day started slowly, with a smattering of hawks in the hot blue sky—the Mississippi kites, with their long, falconish wings and lighter-than-air flight, the dumpy broadwings, an occasional peregrine or kestrel. One peculiar hawk with charcoal plumage and paddle-like wings stumped me for several moments, until I realized it was a hook-billed kite. This tropical species, which sports a grossly oversized bill for crushing the shells of land snails, was always considered nonmigratory. The counters at Veracruz, however, have found that the hookbills do migrate through each fall in small but significant numbers, just one of many surprises the project was producing.

Soon there was little time for seeking out rarities. By midmorning, hawks were pouring through by the thousands, engorged kettles that rose and fell on the hot northeast breeze against a backdrop of distant thunderheads. It took our full concentration to count the passing birds—broadwings for me, the multitudes of kites for Liliana, and turkey vultures for Jeanne. The vultures are another species considered largely nonmigratory across North America, yet each autumn more than a million pass through Mexico and Central America, making them the most abundant species on the Veracruz count— and suggesting just how little we know about even the most common birds.

For more than twenty years, I've been watching and counting hawks in the East, but I had trouble keeping up with the punishing pace of the migration. It simply isn't possible to count the swirling birds when they are kettling, any more than it would be to count leaves caught in a vortex of wind; besides, if you stare too long and hard at the boiling movement you'll get either a headache or motion sickness, or both. The trick is to wait for the hawks to start streaming out of the top of the thermal, which they do in a fairly narrow, orderly band. Then you balance your binoculars with one hand and the heel of the other, pick a point in space, and, as each hawk passes that imaginary marker, use your thumb to click one of those little hand-held counters, which you hold in the crook of your fingers.

This works just fine back home in the Appalachians, where a busy day may produce four or five thousand hawks, but we were seeing kettle after kettle after kettle, with more than that in each one. By eleven o'clock my thumb muscles were complaining, and I had to keep switching the counter from hand to hand to avoid cramps. Every time a kettle passed we'd quickly glance at our clickers, transfer the figure to a count sheet, and reset to zero. We were also recording environmental data—temperature (hot), cloud cover (none), wind speed and direction (zip)—as well as information about how high the birds were flying and their heading. Soon, though, there was almost no time for anything except more counting, more clicking.

At the peak of the day's activity, a small group of local schoolteachers taking part in the project's education program arrived for a visit. They gaped at the spectacle overhead, and they gawked, more discreetly, at us—a Frenchman, a *gringo*, a *gringa*, and one Mexican *señorita*, all maniacally sweeping the sky with binoculars and clicking like crickets on our counters, too busy for anything beyond the barest courtesies. Tall, lanky Christian was a particular attention-grabber, with scarecrow legs poking out of blue nylon gym shorts, a torn green T-shirt, a Stetson hat, and sandals over Argyle socks.

Watching the migration was a bit like seeing some great, slithering snake, for the main stream of the flight shifted sinuously from east to west as the day wore on, moving like a sidewinder tossing its looping body across the land—a serpent of hawks that for hours had no head and no visible tail. Within this migratory river the birds eddied and swirled, forming sheets and curtains, billowing up into kettles, gliding out in streams, so that at any given moment there were many thousands of hawks in sight, all moving with unified purpose to the south. Staring up for so many hours through the tunnel vision of my binoculars, I developed the persistent (and disquieting) illusion that the birds were fixed and motionless and that I was racing steadily beneath them.

By late afternoon, the falling sun no longer fueled the churning thermals, and the flocks dwindled; out in the countryside, the hawks were looking for their evening roosts, groves of trees where they could safely pass the night. Watching one small flock of a few hundred broadwings, I saw them fighting to stay aloft, banking and soaring in the diminishing lift, batting at the air with their wings. Then, as if on a signal, they folded their wings and fell like black raindrops, vanishing into the scrubby woods beyond town. The sky, pregnant with life and movement all day, was suddenly barren.

This one day, I found myself thinking, I have watched more hawks than I've seen in—how long? Twenty years of sitting on Pennsylvania mountaintops? My entire lifetime? I couldn't even begin to guess. But when Liliana added up the neat columns of numbers on her count sheet, we found that an astounding 88,000 raptors had passed us.

That night we celebrated at a small restaurant in Cardel. The walls were painted a lurid, unappetizing shade of yellow, but the kitchen produced platters of sublime *camarones al mojo de ajo*—long-legged, whole jumbo shrimp drenched in garlic butter. We shucked them with greasy fingers, piling their shells like the skulls of conquered enemies on a battlefield, and mopped up the sauce with soft flour tortillas. The entire crew was jubilant, confident that the best was yet to come.

After supper, even though it was pitch dark, we again went looking for hawks. One of Ernesto's goals for the autumn was to find a nighttime roost, net some of the birds, and perhaps collect data to find out whether or not the hawks that pour through Veracruz eat on the way. It seems odd to even ask this question. Birds are metabolic dynamos, requiring food at regular intervals and in significant amounts; few old sayings are as completely false as to claim that a finicky person "eats like a bird." If I ate, by proportion, the same amount as a chickadee, I would have to consume about fifty pounds of *camarones* every day, and even raptors, with their somewhat slower metabolism, must eat about 10 percent of their weight to stay alive.

Yet there is a long-standing suspicion that the two most social species of migrant raptors, the broad-winged and Swainson's hawks, do not eat as they pass through Mexico and Central America. The evidence is admittedly sparse, and mostly negative. Despite the huge numbers of hawks that travel through the region, few are seen actively hunting, and beneath the roosts where thousands may spend the night, there is an almost complete absence of droppings and castings—the wads of fur, bone, and other indigestible material a hawk regurgitates several hours after eating.

On the surface, fasting would make sense, as Ernesto had pointed out a day or two earlier. "Think about it," he said, as we bumped down a potholed road, sugarcane fields stretching in every direction. "Where are you going to find enough grasshoppers and frogs for a flock of 400,000 broadwings?" Even before the native tropical forest in Veracruz was cleared for farming, the land would have had a hard time supporting such an overwhelming annual influx of birds. By fasting, the hawks could avoid intense competition and speed their journey. This could be particularly beneficial for Swainson's hawks, those creatures of open grassland that must transverse huge expanses of lowland forest on their journey.

Not everyone buys these arguments. A hawk, critics say, would be unable to lay on enough fat—up to 50 percent of its lean body mass, by one estimate—to survive a weeks-long fast. But the debate is largely theoretical. Proponents of the fasting theory have produced elegant computer models in their defense, and opponents have done equally complex equations to show that fasting hawks would have to drink copious amounts of water to survive, something they do not appear to do.

What's missing is evidence from the field—actual observations. So after we pushed ourselves away from the restaurant table, stuffed and groaning, we set off in the pickup truck, piles of mist nets, aluminum poles, and banding supplies rattling in the back. There was a raucous shindig going on at the church around the corner, a major religious festival featuring music and fireworks, which we could hear fizzing and booming in the distance as we left town. Soon we were in darkness, creeping down dirt roads that skirted the cane fields and pastures, shining our flashlights up into mango groves and stream-side thickets, looking for raptors.

Despite hours of searching, we didn't find any. This was particularly frustrating since these were the same areas out of which clouds of hawks rose like insects around streetlamps each morning. We decided to try the nearby river canyon, where remnants of the native forest remain on the steep, rugged slopes. We were also nervous, because marijuana plantations are secreted throughout this area, and their heavily armed protectors were unlikely to believe that we really were just birdwatchers and not American-funded narcotics agents. I kept mentally practicing, *"No soy del DEA! No soy del DEA!"*— even though I doubted it would do me much good.

The road was rutted and pocked with an equal number of deep holes and sharp, protruding rocks; several times the underpinnings of the gray truck smacked sickeningly against an especially high spot. Through it all, we stopped every forty or fifty yards and swept the trees with our flashlights; crickets chirped and an owl whinnied off in the distance, but still we found no hawks. Only when we managed to get the truck stuck in a ditch for nearly an hour did we call it quits.

But then, just when everything seemed to be falling into place, the flight dried up, as though the migration had exhausted itself in that one spectacular eruption. The following day the count station at Cardel recorded about 30,000 broadwings, and the day after that, just 1,500. On September 29 the counters sat on the roof all day and saw a pitiful 120 birds.

At the team meeting that night, in the old house that served as the crew's spartan headquarters, everyone had a dejected, fatalistic attitude. One of Ernesto's assistants, Octavio ("Tavo") Cruz-Carretero, normally full of sly jokes, sat slumped against a wall, staring straight ahead; Liliana and Jeanne fidgeted with their notes from the day, saying little. Ernesto wore a pinched, worried look almost all the time, and his skin had a jaundiced tinge from too much stress.

A great deal, not least Ernesto's reputation, was riding on this project, for which he had lobbied support from some of the most important conservation groups in the United States. The logistics had been much more difficult than anyone had imagined; money transfers had been delayed by weeks, and acquiring equipment was a chronic problem. The team was splitting into bickering factions, and Ernesto had to contend with all these headaches while still holding down two part-time jobs for private conservation groups. He'd only recently returned from a quick trip to Chiapas, three hundred miles by bus to the south, where he was conducting bird surveys in the highland cloud forests. Dark smudges ringed his eyes from an ongoing lack of sleep, and his doctor had just told him he was, at age twenty-six, developing an ulcer.

It was possible bad weather had temporarily shut down the flight—winds had been blowing hard from the north for two cloudy days—but almost everyone reluctantly agreed that the broadwing migration had probably peaked with the 88,000 birds we counted three days before.

By American standards, of course, it had still been a stunning season. A really good day at a hawkwatch in Massachusetts or Pennsylvania is five or six thousand birds, and Hawk Mountain regulars still talk with awe about the "Miracle Day" in 1978 when 21,000 were counted. Most count sites record about 20,000 birds a year, and until recently the one-day record for any Eastern location was set in 1986 at Quaker Ridge in coastal Connecticut, when 30,535 broadwings passed by. Veracruz had handily crushed all those old marks.

Yet these young Mexicans and Americans were hoping for much more, and their faces reflected that as they reviewed the staffing schedule for the next day. No one had ever tried to quantify the Veracruz flights, but everyone on the project felt sure it would prove to be the greatest raptor migration in North America, maybe (if they were being honest with themselves) the greatest in the world, eclipsing the massive flights at Panama and Eilat, Israel, where nearly a million hawks a year pass by.

They may still have believed it—after all, just because you don't count a bird doesn't mean it isn't there. Even with four count sites, there were still enormous areas of countryside that weren't being covered, through which the hawks might have slipped. But with the season more than half over and broadwings falling far short of expectations, the team members knew they would have to wait for another year to prove their convictions.

I volunteered to help the banding crew the next day, and in the morning I rode north with Tavo to the town of Palma Sola, hoping to buy white-winged doves; these we would harness in leather jackets and use to lure sharp-shinned hawks and Cooper's hawks into our mist nets. Tavo knew of several families who caught whitewings each year, but the dove migration was late in coming and no one had any live birds to sell.

It was an overcast and breezy day again, but patches of blue were starting to appear in the solid banks of clouds. As we left Palma Sola, we both noticed large kettles of hawks forming, and Tavo pushed down hard on the pickup's accelerator. The sky was swarming with hawks when we careened into town, and as we parked by the building where the counters watched, people gestured to us frantically from the roof.

"The flight just started an hour or so ago, and we already have nearly 75,000 birds," Jeanne said when we got up there. "All of a sudden there were hawks *everywhere*. We can hardly keep up." As she spoke, John Haskell and Carlos Armenta Contreras, one of the Mexican crew, were methodically quartering the sky with their binoculars, too busy counting and clicking to say anything. Christian, still wearing his Stetson and his Argyles, kept watch through his worn spotting scope.

It was an astounding sight. Huge kettles of broadwings rose against the gray sky like tree trunks, connected by wide, thick sheets of hawks; the flocks were so big that even those that were a mile away and a thousand feet high looked like dark thumbprints in the air. When a kettle formed right overhead, the feeling of vertigo was overwhelming—tens of thousands of hawks swirling in a great globe, riding the invisible air bubble up and sheeting off the top in a broad ribbon, like someone peeling a vast, ethereal apple.

Hard as it was to believe, this was only the prelude. After lunch—not that anyone had much time to eat—the number of hawks rose to insane levels. The kettles grew to almost frightening proportions, twenty or thirty thousand hawks in each one. Wherever we looked, there were layers of broadwings, some so low that we could count the black tail bands on the adults, others so high that we could barely distinguish them against the hazy sky. It was almost easier to believe that the atmosphere was generating them from sunlight and wind than to think that all these birds were *real*, coming from the cool, damp woodlands of New England or the Midwest, assembling in their hundreds of thousands to march across the world.

We paired off into teams of counters and recorders and divided the sky among us. I took the northwest, sinking down against the brick stair shaft and bracing my binoculars on my knees. Much of the time there was no question of counting individual broadwings, so we started clicking them off by tens, then by hundreds, mentally chopping the endlessly flowing streams of hawks into manageable chunks. Sometimes even that wasn't enough, and we were forced to count by blocks of about a thousand, all the while noting the other migrants that were swept up in the great torrent—ospreys, accipiters, peregrine falcons, turkey vultures, Swainson's hawks, and others.

"Cooper's hawk," John Haskell said, not pausing as he clicked broadwings. "Two ospreys. Another Coop. Sharpie. Somebody get this bunch of TVs up here." Overhead, a disorganized current of turkey vultures was sliding under the main flow of buteos, and Jeanne clicked them off hurriedly, then went back to logging the other migrants as John barked them out. "Two more sharpies. Osprey. Peregrine. *Shit*—where did that stream of broadwings come from?"

A few minutes later there was a brief respite, and Haskell glanced at his clicker. "We had 13,500 broadwings in that kettle," he said, his voice—like all our voices—ragged with mingled awe, stress, and exhilaration. "You know, when all this started, I was just thinking I needed to clean the dust off my binoculars. Then I realized I was looking at hawks, not dust, and thought, Oh my God."

After an hour of counting, my arms ached and my eyes felt gritty; after two hours, there was a blister on my right thumb from punching the button on the metal clicker. Yet even harder was the mental strain of trying to concentrate on dry, lifeless numbers in the face of such surpassing beauty. Nothing in a lifetime of birdwatching had prepared me for this spectacle; I wanted to stand, head back and jaw slack, and simply drink in the sight of a sky electric with birds. But I could not; there was work to be done. Each time a kettle exhausted itself I read the number off the clicker dial and quickly multiplied by whatever estimation block I was using, reset the counter to zero, yelled out the figure to Jeanne, who kept patient track of the numbers and the weather data, then raised my binoculars to yet another kettle unfolding overhead.

The numbers were astounding. From two to three o'clock we counted 126,516 broad-winged hawks, nearly 46,000 of them in one ten-minute period. The next hour there were even more: 166,790 broadwings, along with nearly 11,000 turkey vultures. Whenever anyone spared the time to look to the east, a steady stream of kestrels was moving low along the horizon, passing at a rate of five per minute, their wings flashing in the sun like a thin reflection off the distant Gulf.

Our numbers, as precise as they may sound, were not perfect; surveys of wild animals almost never are. Experiments have shown that when experienced counters are tallying a passing stream of hawks by multiples of ten, they overestimate very slightly, by about 4 percent. When counting by hundreds, however, they generally *underestimate* by about 20 percent, and when forced to work with blocks of thousands, as we were, the final figure is usually about 40 percent too low. We knew that, and we knew that we were counting only the lowest of three discrete layers of hawks, some so high they were barely visible in our powerful binoculars. All we could do was try to estimate what we saw as painstakingly as we could, striving for an unattainable degree of accuracy. Our estimates would set the "at least" level—that at least this many hawks passed us, and more.

And it wasn't just hawks. Lines of white pelicans and ibis snaked overhead, weaving and undulating; clouds of anhingas formed their own precision ballet, wheeling in perfectly synchronized harmony, each long-necked, long-tailed bird like a flying cross. Hundreds of scissor-tailed flycatchers zipped past us, pearl-gray, with a rouge of pink under each wing, dragging their comically long, forked tails behind them. Wood storks, looking huge and ungainly and prehistoric, came in ragged chevrons of fifty or sixty at a time—long, stick-figure legs trailing behind, their flinty, naked heads drooping on skinny necks that seemed barely able to support them. Something about wood storks excited Christian even more than the hawks. Each time he saw another approaching flock his marginal English would desert him, and all he could do was point and shout, "*Woodwoodwoodwoodwood!*"

And so it went for hours, until the sun dropped toward a thin band of rusty clouds on the western horizon. The flight ebbed suddenly, and by five o'clock the only hawks in sight were a peregrine falcon perched for the night on a radio tower a few blocks away and a merlin that was hunting among the rooftops, scattering flocks of grackles and parakeets from the shade trees in explosive fits.

Word trickled in from the other count sites—20,000 birds counted at Chachalacas along the coast, 55,000 at Chichicaxtle, 65,000 at Rio Escondido. And from our hot, dusty rooftop in Cardel, we had counted more than 435,000 raptors. As our sense of numbed disbelief gave way to comprehension, we realized we had witnessed—by far—the heaviest hawk migration ever recorded, anywhere in the world. Yet as impressive as the numbers sounded, we knew we had only scratched the surface of what had actually passed us, had counted only a portion of the magnificent floodtide of birds.

Why bother to count the hawks of Veracruz? Partly simple curiosity, and the human compulsion to catalogue the world around us. But there is a more practical, and increasingly critical, reason. Almost all the world's broad-winged hawks, a large percentage of its Swainson's hawks, and hefty portions of a dozen other species crowd through this narrow portal each year, affording conservationists their only opportunity to monitor the populations of these widespread, secretive birds.

Swainson's hawks are already in sharp decline in California and some parts of Canada, and the fragmentation of both tropical and temperate forests may imperil the woodland-nesting broadwings. By combing the sky, affixing numbers to the multitudes, researchers in Veracruz could provide an annual snap-shot of a species' health and serve as an early warning system if its numbers begin to drop.

Ornithologists have long speculated why a few species of hawks, fiercely antisocial through most of the year, gather in the kind of immense migratory flocks we witnessed. They have noted that the flock migrants tend to be those, like Swainson's hawks and broadwings, that must travel the farthest between winter and breeding grounds; also biologists have noticed that all depend almost exclusively on thermals rather than deflection currents (created when wind strikes the side of a mountain or high beach dune) for their lift.

The hawks must derive a clear benefit from flocking—but what? Predator avoidance, an obvious advantage for smaller birds, is unnecessary for raptors. Nor is it a case of older, more experienced birds leading youngsters south; in fact, among most birds of prey the immatures seem to migrate first, before the adults. And as the controversy over fasting indicates, the birds aren't feeding in groups, so shared information about food sources—another common advantage of flocks—doesn't appear to apply.

The fact that all the flock migrants depend on thermals may provide an answer. A thermal has the power to lift a hawk rapidly for hundreds of feet, but it is invisible, and a lone bird may waste a lot of time and energy simply trying to find one. And once inside, the hawk may easily stray out of the core of the thermal column, where the lift is greatest.

A flock of several thousand hawks in a thermal, on the other hand, func-tions like dirt caught up in a dust devil, making the invisible visible. They can watch one another for clues about where to fly and gain information that shaves time and saves precious calories. This benefit may be further magnified when there are many separate kettles in the air, for hawks in one thermal may be able to watch those beyond for clues about where the next thermal is to be found.

The toughest job for a hawk migrating through eastern Mexico might not be finding a thermal but finding a place to spend the night. Originally, the Veracruz lowlands were covered with dry tropical forest, but in the centuries since the Spanish conquistador Cortés landed in 1519, less and less of this natural landscape has survived.

More than 75 percent of Veracruz's land area has been converted to agriculture, and most of the remaining native vegetation is found in the mountains. On the coastal plain, where the hawks migrate each fall, only scraps of forest linger, the rest being given over to pasture for goats and cattle and vast expanses of sugarcane. We visited one small sugar operation—we could smell the heady, sweet aroma from far off. While a band of sweaty men fed bundles of cane into a press to extract the sap, others worked over steaming, bed-sized vats, stirring the viscous orange syrup with long-handled wooden paddles. Finally, the goo was ladled into wooden molds, where it hardened into cones of pure brown sugar.

Although a far cry from its original condition, the Veracruz countryside is nonetheless remarkably beautiful. I spent a week helping Ernesto's crew build the banding station on a ranch near Chichicaxtle, and the view brought me to a standstill repeatedly—rolling, vividly green pastures stitched with wooded fencerows and small groves of trees, the jagged, forested volcanic peaks blue in the distance, small groups of women or solitary men on old bicycles moving slowly along the dirt road.

We built the canvas-and-wood blind beneath a sprawling, ancient acacia tree, flat-topped and wide as a barn, and the hot afternoon wind made the long grass ripple like a ground fog. We shared the pasture with a herd of red-brown goats and a single, monstrous Brahman bull, a gray monarch always attended by a court of white cattle egrets. This behemoth had a shoulder hump that rose as high as my head and curved back like a breaking wave, and we always gave him a wide, respectful berth. Our caution was misplaced, however; the bull was docile, and there were other, less obvious dangers about.

One day, four men with rangy, tired-looking dogs came through the pasture, plastic mesh bags slung over their shoulders and powerful, homemade slingshots tucked in their belts. Ernesto struck up a conversation, but all I could decipher with my awful Spanish was that they were hunting. I recognized the word *armadillo* easily enough, and then one of the men swept his arm across the pasture and said, *cascabel,* holding his hands about four feet apart. I felt a chill up my spine, for the *cascabel* is the neotropical rattlesnake, a massive pit viper of short temper and lethal toxicity. It turned out that the long pasture grass—through which we had been blithely tramping for a week—is loaded with them. Thereafter I ignored the bull and watched my step closely.

While farming provides a livelihood for the people, roosting hawks are forced to use what little tree cover is left, including plantations or inaccessible canyon walls where mature forest remains. "Most of the land here in central Veracruz has been converted to grassland or sugarcane, and you can see how scarce and patchy the remaining trees are," Ernesto said as we drove between count sites. "This forces the hawks into marginal roost sites like steep canyons, even banana and mango plantations, and makes them more vulnerable to predation, to humans, to fires. After a heavy flight, each tree looks like a busy hotel."

While broad-winged hawks need tall, mature trees as roost sites, Mississippi kites seem to prefer the lower, scrubbier roosts that mango plantations afford; such sites also usually have plentiful cicadas, which the insectivorous kites eat. This, however, raises another troubling matter, the question of chemical contamination. While there is doubt that broad-winged and Swainson's hawks feed during migration, most of the other migrants hunt and eat daily, including kites, and bird-eaters like sharp-shinned and Cooper's hawks, kestrels, merlins, and peregrines. That probably places them at risk of picking up toxins, because dangerous pesticides like DDT that have long been banned in the United States are still widely used in Latin America, poisoning the food chain.

Mexico affords legal protection to raptors, but it is a paper law only, poorly enforced and widely ignored in the field. In October, when the big migration of white-winged doves surges south in Veracruz, dove hunters exact a heavy toll on hawks and other birds as well. Ernesto once found the remains of sixteen protected species at a popular gunning site near Palma Sola, including ibis, magnificent frigatebirds, laughing falcons, American kestrels, Cooper's and sharp-shinned hawks, swallows, scissor-tailed flycatchers, and several kinds of gulls. Such killing goes on with impunity.

There is also a thriving illegal trade in birds of prey, although nominal enforcement has at least pushed it slightly underground. Walking the streets of Xalapa, with the snowcapped cone of the volcano Orizaba towering in the distance, we saw bird sellers on the sidewalks with the other vendors, stacks of cages full of wild-caught songbirds like cardinals, painted buntings, black robins, and slate-colored solitaires. Raptors are available, but not out in the open—nor always for local purchase.

Down amid the great dunes at Playa Paso Doña Juana, I spent an afternoon with Tavo trying to track down the man known as "the Hawker," one of several shady characters who trap falcons—mostly peregrines—for the black market. The technique they use is as simple as it is effective. A pigeon is harnessed in a leather jacket festooned with nooses made of fishing line and tied to a weighted cord. When the hidden trapper sees a peregrine flying down the beach, he tosses out the pigeon, which flutters madly; the falcon swoops in, grabs the bird, and finds its feet hopelessly tangled.

Posing alternately as a friend of the trapper and as someone who had arranged a purchase, Tavo asked around the small village of Doña Juana, learning that we'd missed the man by a day. I confess I was somewhat relieved; at least one of the hawk trappers is alleged to have ties to Mexico's brutal drug trade, and I had been practicing my line about the DEA again under my breath. Despite Tavo's jaunty bravado, I wasn't sure exactly what we were going to do if we confronted the man. Waggle our fingers at him? Threaten to report him? Run like hell?

Driving south along a sand road between the dunes and the sea later that afternoon, we watched kestrels zipping by, and on a dead snag jutting up from the man-high patches of prickly pear we spotted an aplomado falcon. This spectacular, nonmigratory species once was found as far north as Texas and Arizona, but today it is uncommon even in its core range in Latin America. Aplomados are elegant birds—lead gray above, with a black "cummerbund" and cinnamon belly. The falcon sat calmly facing the sea breeze as I photographed it. Those pictures are now bittersweet for me, for a week later, while two women from Ernesto's team watched helplessly, the Hawker returned and trapped this bird, spiriting it away from its world of beach cactus and ocean horizons.

Veracruz, with its litany of environmental problems, points up one of the most difficult and important challenges in protecting migratory birds—the need to save migration corridors. If, as seems likely, virtually every individual of several species of birds of prey migrates through eastern Mexico, this narrow ribbon of land may be the single most important parcel of real estate in raptor conservation. Certainly it is every bit as crucial as their breeding and wintering grounds, yet until recently little attention was paid to identifying and protecting such migratory bottlenecks.

While most of the attention on the *Rio de Rapaces* project is understand-ably focused on raptors, the survey inevitably highlights the region's impor-tance for migratory birds of all sorts. Some days we counted four or five thousand wood storks, which are an endangered species in their U.S. range. While it is unclear what percentage of the storks we saw were coming from the States, and how many were from northern Mexican breeding colonies, Veracruz may prove a valuable monitoring site for them as well.

White pelicans were a constant presence, wanderers from the northern prairies of the United States and Canada en route to the Gulf of Mexico. There was something quietly regal about them, soaring in lockstep on down-bowed wings, necks tucked and bright orange bills shining in the sun. Like the wood storks, they reminded me of ancient reptiles—flights of pteran-odons gliding out of the Cretaceous. The pelicans flew in layered chevrons and long lines that rippled and snaked, as though playing giant games of crack-the-whip against the sky.

Northern migrants surged through in waves, absent one day, everywhere the next. One day, wherever we looked we saw swarms of chimney swifts heading for the Amazon, filling the air with their high-pitched twittering, fly-ing cigars with buzzing wings. Scissor-tailed flycatchers, icons of the southern plains, arrived in late September, and I was thrilled to see the first one—an old friend the color of a worn dime, its forked tail, twice the length of its body, dangling from behind. The next day I saw hundreds, and two days after that Jeanne counted nearly 30,000 of them in just a few hours.

The migrants weren't even all birds. Each day, tremendous clouds of green darner dragonflies, probably numbering in the millions, would stream by us while we were counting hawks, the dry rattle of their wings sounding like sleet on dead leaves. No one is sure where they spend the winter or where the flights originate, although huge swarms of darners are occasionally seen in the northeastern United States in August and early September heading south. Whatever their origins, these would seem to be true migrants, for in April multitudes of dragonflies again pass back north through Veracruz with the hawks.

Ernesto grew up in the middle of this international flyway. In his early teens, an interest in natural history channeled itself toward birds—specifi-cally raptors, after a hunter gave him two kestrels he'd shot and injured. The birds died after a few days, but the spark grew quickly.

For a few years, Ernesto and two or three friends like Tavo concentrated on falconry, using hawks they bought on the then-open market in wild raptors.

"It was a great time for us," he recalled. "We just set our eyes on one thing and spent all our time doing it. When I was seventeen or so, everything I did was related to hawks. I even took them to school, and the teacher would let me bring them into the classroom."

One of the first hawks Ernesto trained was a roadside hawk, a tropical buteo that is closely related to the broadwing, and a rather sluggish hunter of small vertebrates and insects—hardly the best candidate for falconry. Or so one would think. The young men trained the bird to hunt out of a Volkswagen Beetle, with one person driving and Ernesto sitting ready to launch the hawk, spearlike, out the open window at startled flocks of great-tailed grackles. It was a long way from the medieval sport of chivalry, but the roadside hawk learned to excel at this bizarre style of hunting.

Ernesto has since moved away from falconry, but his friend Sergio still flies one of Ernesto's former birds, a male Harris's hawk with vibrant chestnut wings, long yellow legs, and a white tail tipped with a wide black band. Sergio would stop by the banding station occasionally when we were working, the hawk sitting quietly on his gloved fist or perched free on a fencepost while Sergio flushed cowbirds from the grass. This precipitated long, twisting aerial pursuits that usually ended with an exhausted, empty-handed hawk. Maybe they needed to try a Volkswagen.

Such a concerted interest in natural history as Ernesto's is still rare in Mexico, exacerbated by a lack of such bare necessities as decent field guides in Spanish; Ernesto and his friends used North American guides when they were growing up, books that include Arctic ducks but not the tropical parrots, trogons, and toucans they saw every day. Yet the *Rio de Rapaces* project has begun to tap an underlying curiosity among the local residents. On the soccer field at Chichicaxtle, a short distance from the local elementary school, an excited crowd of kids was gathered around Evodia Silva Rivera, the young woman heading up the project's educational effort that year. Borrowed binoculars were being passed back and forth, focused skyward by wobbling, uncertain hands. Other youngsters peered through spotting scopes or helped the counters record weather data and flight information. Flotillas of buteos and turkey vultures passed above us in endless lines.

"*Aura!*" one eight-year-old boy shouted, pointing to a string of birds sailing overhead. Turkey vultures!

His two companions looked disgusted. "*No, gavilanes,*" one said, jabbing a finger for emphasis on a drawing of a hawk in a brochure they had been given the day before in class, showing the flight profiles of the common migrants.

At first the counters had been regarded with watchful curiosity by the schoolchildren, but several weeks of daily contact had won the kids over. Now they were happy to tell me, with evident pride, about the importance of their community to these travelers. Did you know, *señor*, that the *gavilanes y halcones y auras*, the hawks and falcons and vultures, come from all over *el norte* to fly above our state? Did you know that there are more of them here than anywhere in the world, *señor*? And we must not hurt them, no, for they are good birds . . .

Ernesto was helping three boys line up a spotting scope on an approaching kettle of hawks. "We have three targets for our educational program, and I think they are equally important—hunters, farmers, and children," he said. "The farmers especially are very curious and very open, and they are a good source of information about things like roost sites and who's shooting birds illegally. The children, of course, will eventually grow up to be farmers and hunters themselves, so we need to start talking to them right now."

Since that first season in 1992, Ernesto and his colleagues have continued to demonstrate the astounding importance of Veracruz to migrant raptors. On a single day in the autumn of 1994, counters in Cardel recorded a staggering 925,000 hawks—more than double the number we counted on our best day. Over the years, the total count has averaged more than 3 million raptors each fall, including about 1.4 million broadwings, half a million Swainson's hawks, and 1.2 million vultures.

The scientific world agrees that Veracruz represents far and away the largest concentration of migrant raptors in the world, and one of the most critical migratory bottlenecks on the planet. For that reason, Hawk Mountain has made Veracruz a centerpiece of its new "Hawks Aloft Worldwide" initiative, which seeks to identify crucial raptor migration sites around the world and train local conservationists to protect them. Education programs aimed at both children and adults have been expanded. And as word has spread through the worldwide birding community, tour groups have begun trekking to Cardel each autumn, sparking civic pride and providing incentives for Mexicans to preserve the tremendous hawk migration that ebbs and flows across their land.

Not all the discoveries have been made in Mexico, however. In just the past few years, birders have discovered a hitherto-unknown raptor migration corridor around the western end of Lake Erie, where nearly a quarter million broad-winged hawks have been seen during one day—a dramatic reminder that we know less than we think we do about the natural world.

The day after my visit with the children of Chichicaxtle, I sat with Ernesto in the canvas blind at the banding station, looking out over the distant hills, ephemeral in the haze of dawn, my final day in Veracruz. Inca doves hooted, and a gentle wind stirred the mist nets, but nothing else moved until the sun came up like a yellow hammer.

As the heat grew, the air around us began to expand and rise. We couldn't see this, but the birds could sense it. Here and there columns of circling hawks rose out of the isolated patches of woodland, coalescing into kettles. Within minutes the sky, which had been completely empty, was swarming with thousands of Swainson's hawks and turkey vultures. We stood outside under the shade of the old acacia tree, watching the river of hawks flowing south above us. When after long minutes I glanced at Ernesto, he was still staring up through his binoculars, smiling.

Part Two

HIATUS

6

La Selva Maya

At the center of the earth, in Paradise (so the ancient Maya believed), grew *yaxche,* the great World Tree whose roots sank down into the Underworld and whose soaring branches brushed the heavens. In its green shade, the fortunate dead could rest forever, and at its top sat a bird.

Not just any bird, but Itzam-Yeh, the Cosmic Bird. "He was wealth incarnate—his eyes were bright with silver and jade, his teeth were blue with beautiful stones, and his nose glistened like a brilliant mirror. But he was boastful and prideful. Lost in his arrogance, he proclaimed himself to be the sun and the moon," according to one modern retelling of Maya mythology. The legend goes on to say that the Hero Twins taught Itzam-Yeh a lesson, knocking him out of the tree with a blowgun and ripping away his dazzling teeth and eyes. But to this day he sits atop the World Tree, carved in the stone of uncounted monuments that were left to the enveloping jungle a millennium ago.

I was taking my rest in the shade of the real *yaxche,* the sacred ceiba, or kapok tree, which grows throughout the lowland forests of Central America. No other tropical tree has the ceiba's majesty—a smooth, gray-barked trunk rises from buttress roots that flare out to the ground like the fins of a rocket, graceful and curving. The flat canopy of this specimen was a hundred feet up, vines dangling like guy wires; its branches, leafless for the dry season, were furred with mosses, bromeliads, and orchids, a garden in the sky. And at its top sat a bird.

It was not Itzam-Yeh, of course, but an eastern kingbird—black back, shining white belly and throat, a neat border of ivory along the tip of its flared tail. It gave a hoarse, twittering call and took off, joining more than thirty others that poured out of the nearby trees into a compact flock that wheeled once against the sun, then vanished from sight.

The homeland of the Maya civilization stretched from southern Mexico across Guatemala and Belize to Honduras and El Salvador, encompassing humid lowlands, the dry northern Yucatán, and cool cloud forests in the volcanic highlands. Here, they built great city-states like Tikal, Copán, and Palenque, shadowed by stepped pyramids and stuccoed, painted temples. It was a land rich with jade, gold, and copper, rich with cacao, turquoise, and pelts, with hardwoods, fruits, and honey, the bounty of *la selva Maya*—the forest of the Maya.

It also was—and still is—fabulously rich with birds. Belize alone has some 550 species, and Mexico and northern Central America together hold nearly 1,100. More to the point, nearly 160 species of migrants winter here, and for migratory songbirds, especially warblers, the forests that swallowed pyramids and temples a thousand years ago are the most important wintering region of all. But this is no idle winter's vacation for northern birds. They must recuperate from the stresses of the breeding season and the long journey, and prepare for the next round just a few months ahead. Why they stop where they do and how they fit into the intricate webs of tropical ecology are only now becoming clear to researchers.

Migrants come here for food, which is available at a time of year when insects and fruit are rare in many parts of North America. Songbirds from the United States and Canada winter most abundantly in the northern tropics, diminishing in numbers and variety the closer one gets to the heart of South America. In the 1970s this fact led ecologists like Robert MacArthur of Princeton to theorize that "ecological counterparts"—closely related species with similar habits and habitat preferences—kept them out of the luxuriant, lowland rain forests farther south. Today, that idea has lost support. "There is no obvious Chestnut-sided Warbler counterpart or Bay-breasted Warbler counterpart, for example, that prevents either of these species from expanding its range southward into South America," observed Smithsonian ornithologist John Rappole, noting that every tropical habitat has one or more species of migrant during the nonbreeding season. Avian ecologists now think the importance of northern Latin America may lie in its proximity to the breeding grounds—it is far enough south to offer wintering birds the benefits of the tropics, without the stress of a long flight to South America.

But our recent understanding of how migrants adapt to the tropics comes at a time when the most crucial of their habitats are being destroyed at a rate unseen since the collapse of the Classic-period Maya civilization, reminders of which still haunt the forests. The ceiba under which I rested grows in what had been the plaza of a Maya ceremonial center in the forests of northern Belize. Low mounds, the remains of old structures, rose fifteen or twenty feet on either side of me, their surfaces a jumble of weathered limestone blocks green with moss and the intertwining roots of vines and trees. The pale scar of a fresh trench stood out in stark contrast against the dark undergrowth; looters had been at work here not many months earlier, breaking into a tomb, searching for gold and other treasures.

I'd been in the woods before daybreak, when the air was cool and damp and the forest was alive with sound. A little too alive, at times—chachalacas, like slim, green-gray turkeys, argued among themselves from either side of the trail, an antiphonal squabble that made my ears ache. ChachaLAC! yelped one. ChachaLAC! screeched the other. ChachaLAC! ChachaLAC! CHACHA-LAC! Others in the flock joined in, until I could scarcely hear myself think. They were still belting it out when I moved, with relief, out of earshot.

Farther down the trail, in the branches of a tall, spreading tree, oropendolas had set up housekeeping. Relatives of the blackbirds and orioles, oropendolas are the size of crows, black with warm chestnut wings and bright yellow tails with black centers. Like orioles, oropendolas build woven bag nests, but theirs are large enough to hold a human infant, and they build them in colonies in the tops of trees, giving the appearance of strange, out-sized fruit. The males have a charming, absolutely ridiculous courtship display—falling forward so that they hang almost upside down from the branch, batlike, wings outstretched, making a loud, gurgling noise that sounds precisely like water glub-glubbing out of a large jug.

The Neotropics are a wonderland for a northern birder. Parrots and mot-mots flew through the canopy, maddeningly brief glimpses of color and sound. I watched a pair of red-legged honeycreepers, sparrow-sized but the color of lapis lazuli, taking turns bathing. Their tub was a bromeliad that grew on a low branch, a small pool of rainwater held in the cup of its pineapple-like leaves. One honeycreeper would perch nearby, watching for danger, while the other hopped down inside the plant, sending up a spray of droplets from its churning wings that sparkled in the shafts of sunlight. A jacamar—copper and green, with a long tail and a stiletto bill—hawked a blue morpho butter-fly from the air, landed on its favorite perch, and ripped away the iridescent wings, which floated to the ground to join dozens of others already there, a fabulous confetti.

I lunched in the shade of the ceiba, eating a mushy banana and sipping tepid water, while watching that flock of kingbirds—the "waxwings of the tropics," a regional nickname most North American birders would find puzzling. During the breeding season, kingbirds are ferociously territorial insect-eaters, tackling even dangerous prey like wasps and bees. But in migration, and on their wintering grounds, these flycatchers become gregarious, noisy fruit-eaters, like cedar waxwings in the north—"a behavioral about-face that would do justice to Jekyll and Hyde," wrote tropical ornithologist Steven Hilty.

On their way south through Central America in September and October, the wet season when there is relatively little fruit available, the kingbirds hurry along, filling up on insects when they have to. But once they reach South America, they find a bonanza—a widespread tree named *Didymopanax* that, along with dozens of other species, bears fruit throughout the northern winter months. These kingbirds were heading north again, this time during the dry season in March and April, when *Didymopanax* fruit was ripening in the Maya forests, too. Gone was the frenzied rush of autumn; now the kingbirds would be taking their time, gorging as they went. Yet as soon as they crossed the Gulf of Mexico, an eighteen-hour trip, they would switch once more to insects for the summer.

It may not be a coincidence that *Didymopanax* fruit ripens when the kingbirds arrive. Fruit is, after all, a bribe—a nutritious blandishment to convince a bird, a bat, a monkey, or some other animal to eat it and the seeds it contains, moving them away from the parent tree. As seed dispersers, migratory birds may be a better choice than residents, which don't travel as far. Botanists have noticed that the fruits of several species of trees ripen just as the waves of northbound migrants are passing through Central America, and some appear to be eaten almost exclusively by them.

After the kingbirds left, I started back along the trail for the long hike to camp. The afternoon breeze shifted, carrying with it the tang of smoke—it was the end of the dry season, when farmers clearing land torch the chopped or bulldozed vegetation, charring it to ash that fertilizes the ground for planting. Each afternoon, when the wind stirred, I'd been able to smell what I could not, thankfully, see—huge areas of former forest on the northern edge of the preserve, smoldering where a Mennonite community was clearing new farmland. The day before I'd driven for miles through what used to be wooded hills, now laid bare; the farmhouses sat on knolls among tiny groves of trees, shipwreck survivors clinging to each other for comfort.

When serious alarms were raised in the 1980s that migrant songbirds were declining, much of the blame was quickly laid on the tropics. It is easy to see why; the pace of tropical forest destruction has been unmatched in human history. We've all heard the figures and been numbed by their enormity. On average, 8 million acres of forest are lost in New World tropics each year, and some years are worse than others; in the 1994–95 burning season, more than 11,000 square miles of the Amazon, an area the size of New Jersey, went up in smoke. In the El Niño–induced droughts of 1997–98, another 6,800 square miles burned, including a good chunk of the northern Brazilian state of Roraima.

But not to minimize the loss, the Amazon is huge, and much of it is still left. Central America is considerably smaller—about the size of Utah and Nevada combined—and forest destruction there has been going on for much longer. As a result, only tattered fragments are still intact. Panama, which has precious little of its virgin forest left, loses another 160,000 acres every year. El Salvador is stripped virtually bare of native forest, and Costa Rica, except for its vaunted series of national parks, is going the same route. And the same El Niño drought in 1998 sparked thousands of fires that consumed a great deal of what was left in Mexico, Guatemala, Honduras, and Belize.

Ironically, many scientists now believe that habitat fragmentation and nest parasitism on their northern breeding grounds threaten songbirds at least as much as tropical habitat destruction, something we'll explore in more detail in later chapters. Moreover, tropical rain forests, which get most of the conservation attention, are not even the most important winter habitat for migratory birds in general. But there is no doubt that tropical deforestation is rampant and dangerous, not only to migrants but to the richest storehouse of biodiversity on the planet.

This is the second time that la selva Maya and its legions of birds have been down this road. The Classic-period Maya cleared enormous areas of northern Central America, though not to the extent we have today; in its heyday from A.D. 200 to 900, the city of Tikal in northern Guatemala supported more than a hundred thousand people, and such population centers were common then. To feed these metropolises, the Maya practiced an especially intensive form of agriculture—not the primitive slash-and-burn usually associated with the tropics (and still practiced by modern Maya) but a sophisticated mix of hill terracing, multiple crop plantings, fruit and nut orchards, even raised fields in swamps, with fish and turtles reared in the adjacent ditches and the muck used as fertilizer.

For reasons still unclear, the Classic Maya culture collapsed around A.D. 900; the cities were abandoned, although the Maya themselves never disappeared, remaining in the shadows of once-great temples. The forest reasserted itself, although the outlines of raised fields and canals are still visible from the air across wide areas of Belize and Guatemala. The fact that the forest recovered gives us hope, if the pace of destruction can be reversed in Central America. But even if that were to happen, the lesson of the Maya is a sobering one. In the more than one thousand years since Tikal was given back to the jungle, the forest has not recovered its original form; it is still demonstrably a "second-growth" woodland, still in the flux of ecological succession. Nor is there any way of gauging the impact of the Maya deforestation on wildlife, including birds. While many species obviously survived the period of greatest habitat loss, others with restricted ranges may have vanished before Western science ever encountered them.

By the time I got back to camp, I wasn't thinking about ancient civilizations or modern conservation; I was hot and tired and ready for a swim. Clad in a pair of old shorts and sandals, I lay stretched out on my back in a fast-flowing stream, clear and warm, feet braced against a smooth boulder and my head breaking the current. Late-afternoon sunlight trickled through the spreading, flower-spangled branches of a wide bribri tree. Its name sounds like something out of a Dr. Seuss book, and the flowers look like one of his creations, too. Each bloom was shaped like a snowball—a densely packed half-sphere of white filaments exploding out from the bud like a sunburst a few inches in diameter. The tree was bustling with movement, seething with bees and hummingbirds. A keel-billed toucan swooped across the river, a crow-sized banana with wings that disappeared into the bribri's leafy center. Near the top sat a scarlet-rumped tanager, black and shiny as obsidian, flaring his patch of crimson feathers, looking for all the world like a bird with a clown's nose glued to its rear end.

I closed my eyes, feeling with relief the sweat of the day wash away. A clattering of noise made me open them again, and I found that the tanager had been replaced by a half-dozen male Baltimore orioles, orange and black among the milky flowers. More flew in to join them, until several dozen, all males, were hopping through the canopy of the tree.

Then began a snowstorm, the slow, drifting descent of bribri flowers, which fell to the stream and floated past me like waterlilies in an Oriental painting. It was enchanting, but also puzzling. I swam quietly under the tree's nearest limbs, wishing now that I had my binoculars. An oriole landed on a low branch right above me, shuffling sideways to reach a flower. It grasped the blossom at its base, gave a twisting tug to pull it free, then plunged its beak into the flower's tubular corolla to drink the nectar inside. The flower gently parachuted down like dandelion fluff, landing soundlessly beside my shoulder.

I am used to seeing Baltimore orioles singing from the top of the red maple beside my home, or carrying beakfuls of insects to their nests along the quiet stream where I fly-fish for trout. As is so often the case when I see a migrant bird in the tropics, I was taken aback to see them sharing a tree with toucans and parakeets, although their vivid colors certainly fit in with the jungle motif.

That is, of course, precisely the point—Baltimore orioles really *are* tropical birds, as much at home in the wet forests of the Maya as they are in the oak woods of the Appalachians or Midwest. Maybe even more so, for of the twelve months of the year, an oriole spends only about four in the temperate zone of North America. The balance is spent migrating or in its nonbreeding range, which stretches from Mexico to northern Colombia—nearly seven months out of each year in the tropics. The ratio is even more lopsided for other songbirds; one study of warblers in Ontario showed that most spent less than three months there.

For most of the history of ornithology in the Western Hemisphere, no one spent much time thinking about what happened to migrant songbirds on their wintering grounds. Most ornithologists were North American, after all, and their interests tended toward the accessible breeding season, not what went on in the distant tropics. What little information was published covered the very basics, like relative abundance and distribution. It wasn't until the 1960s that a few pioneering scientists began to seriously examine how migratory birds fit into the tangled ecology of the Neotropics. Around that time, a consensus was developing among researchers that cast migrants in the role of vagrant opportunists—existing on the fringes, excluded from the best neighborhoods (wet tropical forests) by better-adapted resident species, pushed into marginal, successional habitats where they exploited patchy, ephemeral food resources. At a conference on migratory songbirds at the Smithsonian in 1966, ornithologists "almost universally espoused the view that tropical deforestation will not harm North American migrants and may, in fact, increase the amount of winter habitat available for them. There was essentially no discussion of the possibility that different species of migratory birds might vary in their ability to use disturbed habitats," one overview of migrant research noted.

But by 1989, ornithologists like Eugene Morton and Russell Greenberg had reached a different conclusion. "This idea of the adaptable migrant, forever skimming off the cream and moving on, was a notion that comforted avian ecologists in the 1960s and 1970s. It implies that migratory birds are immune to the future shock of deforestation. Unfortunately, field studies of migrants in the tropics show that the idea is incorrect," they warned. The picture that has emerged in the past fifteen or twenty years is much more complex and subtle than the simplistic "new kids on the block" theory once proposed, and its implications for migratory bird preservation are far-reaching, with elements both unsettling and encouraging.

In fact, research has confirmed that many migrant passerines do tend to show up more frequently in disturbed habitats, like agricultural areas and second-growth forests, than do tropical resident birds. Evergreen rain forest, the ecosystem most people associate with the tropics, and one being lost at a great rate to human development, is one of two habitats least-used by migrants (the other is open grassland). But that is a dangerous generality; a whole suite of boreal travelers—wood thrushes, Kentucky warblers, scarlet tanagers, yellow-bellied flycatchers, and many others—winter largely or exclusively in mature rain forests, one possible reason why some of these species are declining at such frightening rates.

Although there are exceptions, as a rough rule of thumb, migrants seem to pick the same kind of habitat on each end of their journey. Common yellowthroats, which breed in marshes, brushy roadsides, and overgrown meadows, winter in similar situations in Central America and the Caribbean, especially in abandoned agricultural land. Acadian flycatchers, which depend on mature deciduous forests in the eastern United States, winter in moist, closed-canopy lowland forests.

But that still leaves a complicated question: If migrants aren't wandering interlopers, excluded from the choicest locations, how do they interact with resident birds and with each other? Although the field is still in its infancy, studies are showing that migratory birds sometimes fill niches that tropical species don't; they hold and defend territories, often returning to the same place year after year; and habitat choices sometimes differ between the sexes of some species and between age classes.

Any discussion of ecology must start with the basic division of resources—who eats what, where, and how, divisions that ecologists refer to as niches. (Don't make the common mistake of confusing niche and habitat—to a scientist, a niche isn't a physical space. As thousands of ecology students have been told over the years, habitat is the address, while a niche is, in a way, an organism's job description.) Many species may occupy the same habitat, but direct competition for precisely the same niche is rare.

One way that migrants may avoid competition with resident birds is through diet. A recent study, by Brigitte Poulin and Gaëtan Lefebvre of the Smithsonian Tropical Research Institute, was the first to compare the diets of resident birds and migrants passing through the same habitat—in this case, a humid evergreen forest in Panama. They found that the migrants (thrushes, warblers, and flycatchers) focused their attention on two main kinds of food—small, hard-shelled insects like ants and beetles or invertebrates like millipedes and termites that produce toxic chemicals. These were two classes of food, presumably lower in nutritional value, that are rarely taken by the residents, which tended to eat higher-quality bugs like caterpillars and spiders. The thrushes were also found to take quite a bit of fruit, and their appearance in the study area seemed to coincide with the ripening of local fruit supplies—but it also corresponded with a seasonal abundance of millipedes. Poulin and Lefebvre found very little overlap between the diets of migrants and residents but quite a bit of competition between migrants of the same "guilds"—ecological shorthand for groups of species with similar feeding techniques, like foliage-gleaning insectivores.

I waded out of the creek, dried off, and gathered up my binoculars and daypack. Sunset was an hour away, and more and more birds were moving, freed of the midday heat. For long stretches, as I stood or shuffled quietly along the path, the forest would be eerily quiet. Then a flock of birds would roll into view, and suddenly I'd be surrounded by frenzied activity; for every bird I caught in my binoculars, a half-dozen others slipped by, barely noticed and unidentified. But of what I did see, there was little or no duplication—just one or two individuals of more than a dozen different species.

After one flock hustled past, I dug out my notebook and pen from the pack and scribbled a list of the birds I'd just seen:

Unident. woodcreeper (1)	Summer tanager (1)
Long-billed gnatwren (2)	White-bellied wren (2)
Blue-gray tanager (5–6)	Plain xenops (1–2)
Red-throated ant-tanager (1)	Sulphur-rumped flycatcher (1)
Red-crowned ant-tanager (2–3)	White-eyed vireo (1)
Unident. *Empid* (Least flyctr.?)	Blue-winged warbler (1–2)
Magnolia warbler (several)	Northern parula (1)

Mixed-species flocks of this kind are common among birds. In the temperate zone, they are mostly a winter phenomenon, after the hormonal aggression of breeding season has diminished and the territorial barriers have broken down. Chickadees, titmice, nuthatches, kinglets, brown creepers, and woodpeckers travel together in loose flocks through the woods near my home, and while the mix is different in the tropics, the underlying principles are the same. Ecologists who have studied mixed flocks believe they offer a number of advantages over single-species aggregations—more eyes for better detection of predators, for instance, combined with different foraging techniques that minimize competition.

Watch a mixed temperate flock and you'll see this last point quickly illustrated; every species has its own preferred food and specialized hunting method. The acrobatic chickadees hang upside down from branches, peering into cracks and crevices for hibernating arthropods or egg masses, while the tiny kinglets flit and hover at the very tips of the twigs, where the heavier birds cannot reach. Downy woodpeckers work their way up the trunk and nuthatches go down, headfirst, each finding food the other missed; the creepers also ascend the tree in a spiral, but they pay particular attention to the spaces between bark flakes, while the woodpeckers, with their sturdier bills and barbed, spearlike tongues, focus on bugs hidden deeper inside the trunk. The foods they seek are often the same, but the ways in which they look, and where—their niches—are different.

Mixed-species flocks reach their greatest expression in the tropics, especially in the Amazon basin, where as many as six dozen species have been seen in a single gathering. At first, biologists assumed these were almost accidental assemblies, but once they began to study the flocks in detail, a much more intricate pattern took shape. Each flock is made up of several subsets of birds, each with its own role to play. At the core is one or more "nucleus species," as they're known, usually a mated pair or a family group; by their behavior and calls they hold the flock together, setting the pace for the group, somehow providing cohesion to the enterprise. They are joined by what ecologists call attendant species, some of which stay with the flock almost fulltime, others that may follow for a few hours but then drop out, and still others that join only briefly, while the flock is transversing their territory. "Because flocks cross some boundary every few minutes, flocks have a consistently changing individual composition, even though species composition is relatively stable," writes one researcher.

The core species of tropical flocks are always resident birds, but migrants often join the melee; the summer tanager, white-eyed vireo, and warblers I saw are all regulars in such mobs. Rarely, however, will you see more than one or two individuals of any species in the flock; while they get along with other species in a collegial way, they tend to attack members of their own kind, driving them out. This is probably related to competition. Each species has a slightly different diet and hunting technique, and while they can (ecologically speaking) avoid stepping on the toes of other species, they would be directly competing with each other.

Animal behaviorists studying mixed flocks have found that there are often sentinels, birds with sharp eyesight and a hunting technique that lends itself to constantly scanning the surroundings—like flycatching or hawking. One squeal of alarm from the lookout and the flock scatters for cover—but the lookouts aren't above fibbing, as Hilty explains in *Birds of Tropical America*. The trouble arises when a sentinel and another member of the flock go after the same insect.

> The sentinel, in what appears to be a conscious attempt at deceit, may utter a false alarm call in an attempt to divert the attention of its competitor . . . Examples of such lying are rare in nature. Crying wolf too often, of course, could lead to a lack of response, but the penalty for ignoring an alarm, if it turned out to be true, is very high in a forest with seven or more kinds of bird-eating hawks. Consequently, mixed species flocks may be willing to pay the cost of an occasional missed prey item as part of the benefit of having a sentinel on hand.

Not all migrants join mixed flocks. Hummingbirds focus on patches of flowers, sometimes moving from one to another as if running a trapline, and fruit-eaters like eastern kingbirds tend to stick with their own species. This is a tendency also seen in tropical fruit specialists; because they rely on widely scattered but locally abundant foods, it may be worth their while to stick with their relatives, following others that have already located a fruiting tree. In fact, Gene Morton has proposed that the white tail tip on the eastern kingbird serves the same function as bright tail markings on waxwings and other fruit-eating birds—a visual signal to other kingbirds to come and get it.

While overwhelmingly insectivorous, most of the migrants in mixed flocks have different feeding niches—gleaning the outermost branches (like blue-gray gnatcatchers), staying in the highest part of the canopy (like black-throated green warblers), or working along the trunk (like black-and-white warblers). Some have more specialized techniques. Worm-eating warblers, which are only loose associates with mixed-species flocks, spend most of their time seeking out clusters of dead leaves caught in vines; the bird systematically rips and pokes its way through these, searching for the invertebrates hiding inside. Golden-winged and blue-winged warblers, which spend much of their time in mixed flocks, do the same thing with living leaves that are diseased or curled—a good bet for a meal, because many tropical insects roll up leaves and hide inside.

The best place to see mixed flocks, including migratory birds, is around an army ant swarm. Like piranhas, anacondas, and tarantulas, army ants are tropical creatures whose reputation far outpaces reality—what I think of as the "Green Hell Effect," the inescapable exaggeration of danger in the jungle. I remember being terrified when, as a kid, I saw a late-night rerun of the old Charlton Heston flick *The Naked Jungle*, in which Our Hero is a South American rancher who must battle hordes of flesh-devouring ants that strip livestock, people, and everything else to the bare bone.

Army ants are remarkable animals, and they are dangerous—if you happen to be a cockroach or a nestling bird unable to move out of their path. The soldier ants sport monstrously outsized, curving jaws, which Indians occasionally use as a crude but effective version of surgical staples. To suture a deep wound, the Indians carefully catch several army ants and hold them so that, as the ants bite, their gaping mandibles draw the lips of the cut closed. Then the Indians twist off the bodies of the ants, leaving their heads in place (the jaws locked shut by death) to hold the wound closed.

The ant swarm, which may number 1 or 2 million individuals, makes no permanent home but alternates between active and quiescent phases— "marching" and "bivouacking" are the militaristic terms used. While bivouacked, the colony forms a living globe the size of a basketball around the newly laid eggs, while sending raiding parties out in various directions. When the bivouac ends and the whole colony moves, it forms a seething ribbon through the forest, carrying the developing larvae along with them. Insects, spiders, scorpions, and small vertebrates scramble to get out of their way, and anything unlucky enough to be caught is simply overwhelmed, bitten to death, and dismembered.

The panic that army ants instill in otherwise hidden creatures is a boon for birds, and many species—antbirds, antthrushes, antshrikes, antpittas, and ant-tanagers, among others—make their living following the swarms, which act as unwitting beaters. (The birds rarely eat the ants themselves, because they pack a nasty dose of formic acid that renders them barely edible.) These ant followers form a distinct class of mixed-species flocks, one that northern migrants often join; as many as a dozen species have been seen trailing army ants in southern Mexico. Interestingly, not all migrants have this opportunity. Thrushes will follow ant swarms and feed on the flushed insects if they have a chance, but generally one or more species of antbirds—which, like the thrushes, tend to be largish, ground-foraging songbirds—will chase the thrushes away, excluding them from the smorgasbord.

As with the kingbirds, a migrant's behavior may change through the winter, depending on the local food supply. A study in Costa Rica showed that most Tennessee warblers joined flocks, either mixed-species assemblies or groups of other Tennessees. But should they find a tree in bloom, the birds become defensive homebodies, chasing away any other small birds that try to steal the nectar. A large tree like a bribri or a coral bean, researchers found, might hold four or five warblers, each staking out its own patch of canopy, battling hummingbirds and other nectar-eaters. This might seem like opportunism, but there is evidence that the link between Tennessee warblers and some tropical plants is as complicated as that mentioned some chapters earlier, between orchard orioles and the coral bean trees, in which the blossoms (as they are peeled open by the birds) mimic the color of the male orioles' plumage as a way of encouraging flocks to move on to another tree to cross-pollinate it.

In the case of the warbler, the evolutionary partner is a nectar-rich vine named *Combretum*, which bears plumes of showy red flowers. If it is socially dominant, a Tennessee warbler can stake out a territory around a vine, chasing away others and claiming all the nectar for itself. These successful birds pick up red pollen on their faces and heads—"war paint," as ornithologists have nicknamed it. Because Tennessees are rather drab, gray-green birds, the splash of color really stands out, and it seems to make the most dominant birds even more intimidating to others. When they move on to another *Combretum* vine, the pollen stain helps them displace any unpainted birds, making it more likely that they will earn a chance to feed from the new flowers—and thus transfer sticky pollen from the last vine they visited to the blooms of the next, completing pollination.

The idea of a migratory bird having set territories in the tropics once struck avian ecologists as absurd; everyone knew migrants simply roamed around the landscape, grabbed whatever food they could, and then moved on. But as interest in the neotropical connection grew, field-workers made the startling discovery that not only are many songbirds faithful to their winter homes, they sometimes show a higher degree of fidelity to them than to nesting sites, returning to exactly the same patch of forest year after year after year. Nor are their choices of winter territories homogeneous; it turns out that a growing list of species segregate themselves by sex and age, sometimes using radically different habitats.

One of the best examples came from a study of hooded warblers that James Lynch, Eugene Morton, and Martha Van der Voot conducted in the 1980s. These yellow songbirds are common in hardwood forests in the East, the males with black, monkish cowls on their heads, the females with faint, incomplete hoods, or (especially in very young females) none at all. In winter, the researchers found, males and females stay in different parts of the Yucatán Peninsula—males predominantly in the eastern Mexican state of Quintana Roo, where the forest is wetter, taller, and evergreen, the females on the western side of the peninsula in the state of Yucatán, where the drier climate and two millennia of Maya agriculture have produced a lower, scrubbier, deciduous woodland. Noting that many migrants tend to choose winter habitats similar to their summer breeding range, and that wet forests are presumably more fertile than dry second-growth, the team wondered if males weren't somehow excluding the females from the better territories—relegating them, in effect, to the poorer side of town.

Besides their plumage, male and female hooded warblers show differences in physique; the males are somewhat heavier, with longer wings that may give them an advantage when foraging in flight. Might these slight variations be significant when it comes to claiming and holding territories? Lynch and his colleagues suspected they were. "One working hypothesis is that males, perhaps by virtue of their greater size, are behaviorally dominant over females, and that overwintering males tend to exclude females from deep forest," they wrote. If so, it is not a watertight system—they saw one assertive female chasing males out of her territory—but most male warblers seemed to be able to displace most females.

But remember, while adult males have a heavy black hood, the females have a fainter pattern, and young females usually show no black at all. Lynch and his team observed that females with darker, more malelike faces—presumably those that are older and more experienced—held territories that were more like a male's. That led the researchers to suggest that males (and masculine-appearing females) might simply be excluding lower-ranking birds from prime real estate through social dominance. This seems to be the case with black-throated blue warblers on some Caribbean islands, where males chase females out of the lusher lowland sites, banishing them to scrubbier forests at higher elevations.

Nor are they alone; segregation by gender has now been documented in almost a dozen other species of warblers, with males generally occupying the taller, wetter forests. But perhaps male warblers aren't the chauvinists they first appear. When Gene Morton and several coworkers conducted follow-up investigations on hooded warblers, they found that females seem to actively choose the shrubbier habitat, even when given a chance to occupy vacant male territories. And captive-raised warblers seemed to gravitate to their respective gender's combination of plant height and structure. This was apparently an innate choice, not the result of a pecking order; it may well be that, in ways we cannot yet discern, the sexes are best adapted to the habitats they use.

At one time, migrants were thought of as a fairly monolithic bunch. But as scientists look ever more closely at the wintering ecology of migratory birds, especially the question of habitat selection and use, they're finding all sorts of intricacies they never imagined. Here are three examples:

• Ornithologists once thought that young migrant birds, encountering the tropics for the first time, wandered around more or less randomly looking for suitable habitat, a behavior called exploratory assessment. But research by Kevin Winker of the Smithsonian Institution indicates that habitat choice may be innate among even the most inexperienced migrants. Winker worked in Sierra de los Tuxtlas, a volcanic mountain range in southern Veracruz and the northernmost rain forest in the Western Hemisphere; first-year migrants arriving there in autumn would never have seen that ecosystem before. Yet in species after species—Swainson's thrush, ovenbird, yellow-breasted chat, Wilson's warbler, and a dozen others—young birds showed an almost uncanny ability to find their species-specific habitat with no discernible fumbling around.

• American redstarts winter in different habitats in Belize, depending on gender. Males, which are black with orange markings, are most common in forest edges and interiors, and least common in scrub; females, which are grayish-olive with yellow wing and tail patches, are most common in open pine savannas and scrub, and least common in forest edges and interiors. What's interesting is that immature males, which look like females, choose habitat intermediate between the two adult genders. The youngsters most often use pine savannas, switching to deep forest only after they have acquired black adult plumage in their third year.

• The last example involves habitat segregation again, but this time by subspecies rather than by age or gender—a very unusual situation. The yellow-rumped warbler, one of the most widespread and abundant members of its family, comes in two distinct forms, which once were considered separate species—"Audubon's" warbler of the West, which has a yellow throat, and the "myrtle" warbler of the East, Canada, and Alaska, which has a white throat. Not only do the two forms occupy different summer ranges, overlapping only in the northern Rockies, but they winter in different habitats in the tropics. Audubon's migrate to high-elevation oak and conifer forests from central Mexico south to Honduras, while the myrtle race congregates in lowland areas in the Yucatán (especially in grassy dunes and heavily altered landscapes like roadsides and pastures), along the eastern Central American coast, and on Caribbean islands.

Why do closely related forms of the same species winter in such radically different places? Why would male warblers of many species prefer forest and females brush? How do young, inexperienced birds find and then recognize their species' winter habitat without ever having seen it before? We don't know. The questions may seem esoteric, but the answers have obvious implications for bird conservation. For instance, is it essential, for the future of hooded warblers, that we save large tracts of wet forest in Central America? Because the males seem to prefer that habitat—and because you need males to keep a population viable—the answer would seem to be yes. But let's not jump to conclusions. Male redstarts also prefer wet forest in Central America—yet in Jamaica, where most of the mature forest was cut long ago, both sexes of redstarts seem to flourish in the kind of scrubby habitat only females prefer in Belize. Are the redstarts just making the best of a bad situation, or can they do well in what appears, on the surface, to be marginal habitat? And if they can, how many other species of migrants have this same flexibility?

These are not idle musings. Marginal habitat is the wave of the future, as the last pristine forests disappear. The chainsaw and the machete are always busy, the smoke is hanging in the air, and *la selva Maya* is a little smaller now than it was when you started reading this chapter.

Of the ten countries with the highest deforestation rates in the world, half are in northern Latin America and the Caribbean—Mexico, Guatemala, Costa

Rica, Jamaica, and Haiti. Mexico alone cuts an additional million acres of trees each year—not just rain forest but tropical dry forest, cloud, and pine-oak forest—with a similar amount lost annually in the rest of Central America.

While the impact of deforestation hits resident tropical birds hardest, since many have small, ecologically or geographically restricted ranges, migrant birds obviously suffer, too, despite their penchant for disturbed habitats. "The task for students of migratory bird ecology is to ascertain the level (and form) of disturbance tropical forests can withstand and still be suitable as winter habitat for Neotropical migrants," concluded the authors of a recent study into just this issue. "Clearly, however, wildlife conservation plans must also consider resident tropical birds (and other taxa) which often are more adversely affected by alteration of natural forests."

Complicating the issue is the fact that many tropical resident birds migrate, a fact still little appreciated even by many ornithologists. Exactly to what extent this intratropical migration occurs is hard to say, since so little in-depth fieldwork has been done on it, especially in South America. Many intratropical migrants eat fruit or nectar and travel between highlands, where their food is abundant only part of the year, and warmer, wetter lowlands. Others pursue widely scattered resources, like the seeds of erratically flowering bamboo, or are wading birds that move in response to local water levels.

Some years ago, on my first trip to the Peruvian Amazon, I steered a dugout canoe through the tree trunks of the varzea, the seasonally flooded forest, and out into an oxbow lake covered with hundreds of acres of floating grasses and water hyacinth, the latter crowned with masses of blue flowers that looked like another pool of water hovering above the lake. Picking its way surreptitiously among the hyacinths was an azure gallinule, a rail related to the coots—a pale bluish creature that matched the flowers around it, with long yellow legs and a short bill. Like many Amazonian birds, it was always considered sedentary—a belief that changed, oddly enough, in part because of a discovery made in a Long Island, New York, back yard in December 1986.

That winter, Angela Wright found a strange, chickenlike bird with gangly legs lying dead by her feeder, and rather than toss it in the garbage, she called the local bird sanctuary. It was the first time an azure gallinule had been seen north of Venezuela, and most experts assumed it was either an escaped cage bird or had hitchhiked on a ship, although they acknowledged that neither alternative was very likely. Because everyone knew azure gallinules aren't migratory, the obvious alternative—that it traveled the 2,000 miles to Long Island on its own—wasn't seriously considered.

Spurred by the New York specimen, however, J. V. "Van" Remsen, Jr., and the late Ted Parker at Louisiana State University analyzed specimen records and field observations of azure gallinules in South America and came to the conclusion that these lovely marsh birds do make seasonal migrations across the northern half of the continent. Admitting that their evidence was thin, they nevertheless found a consistent pattern of movement, especially toward the northern and southern edges of the gallinule's range. Along the upper Amazon of Peru, where I was paddling, the birds appear in greatest numbers from January through July, during the peak of the flooding, then disappear from July through December for parts unknown.

Although they are considered nonmigratory, other wading birds like capped herons seem to come and go seasonally in parts of the Amazon, as do some of the large storks like the jabiru and the maguari, which local Indians say gather at shrinking ponds during the dry season to catch the fish trapped within them. There's little about this in the scientific literature; the Remsen and Parker paper is one of only a handful of journal articles ever printed on intratropical migrants in the Amazon—and the gallinules are one species out of nearly nine hundred, many of which may undertake seasonal movements.

Much better understood are the altitudinal migrants that make regular movements down to the lowlands. They are especially common in the mountains of Central America, which experience pronounced seasonal changes in temperature and rainfall. The most stunning example is the resplendent quetzal, which moves to the warmer, drier lowlands after breeding, when the rainy season begins. Male quetzals have iridescent green plumage, a crimson belly, and flexible tail plumes that may reach two feet in length; these were once reserved, on threat of death, for Maya nobility. But three-wattled bellbirds in southern Central America, which are said to have one of the loudest voices in the bird world, perform similar short-distance migrations, as do more than seventy-five species in the highlands of Costa Rica, most of them fruit- or nectar-eaters. By some estimates, a fifth of all tropical bird species in the Central American highlands are altitudinal migrants.

Intratropical migration adds another reason for concern about the loss of so many Latin American ecosystems. Most of the attention in the neotropics has been on lowland rain forests, often to the exclusion of other, more critically imperiled habitats such as mangrove stands, fire-dependent grasslands (in South America), and high-altitude cloud forests. Migrant songbirds, because they are colorful, make up the great majority of northern birds using the tropics, and are relatively well known to the average North American, have often been drafted as the standard bearers for rain-forest preservation. While this may seem like savvy public relations, not everyone is convinced that it is good conservation. In one of the most comprehensive reviews of neotropical bird conservation, Douglas Stotz and Debra K. Moskovits of the Field Museum of Natural History, John W. Fitzpatrick of the Cornell Laboratory of Ornithology, and Ted Parker at Louisiana State warned that the myopic focus on lowland forest may ultimately backfire against migrants and against many species of tropical residents.

"The great importance of pine and pine-oak forests as a wintering habitat for northern-breeding birds has gone almost unnoticed, or at least unheralded," they caution, noting that these woodlands are the primary winter habitat for more than half of all migrant species that require undisturbed forests. "From the standpoint of migrants alone, pine and pine-oak habitats of Mexico and Middle America warrant more attention than any other type of forest . . .

"It is clear, then, that although lowland forests are important for some migrants, they do not warrant the disproportionate attention they have received as the winter homes of northern migrants. By placing overwhelming stress on lowland forests to the exclusion of other habitats, conservationists ignore the fact that a large proportion of both resident and migrant bird species depend on nonforest habitats. Undue concentration of conservation efforts on lowland forests could result in a major *loss* in diversity among both migrants and residents as the other habitats quietly disappear."

The degree to which some migratory songbirds are in trouble, and why, is an enormously complicated puzzle, one that we'll get to in due time. For now, suffice it to say that while the signals are mixed for some species, they are unequivocal for others. Birds that winter primarily in the interior of lowland evergreen forests in Central America also tend to be the migrants in deepest trouble. The wood thrush is perhaps the best (and best-loved) example, the bird of whose liquid song Thoreau wrote: "Whenever a man hears it he is young, and Nature is in her spring; wherever he hears it, it is a new world and a free country, and the gates of heaven are not shut against him." Wood thrushes do not sing while in *la selva Maya* (song is a breeding activity only), but you can hear their clucking call notes as they hop through the understory, and see a flash of bright rust as they pass through shafts of sunlight.

According to an analysis of results from the North American Breeding Bird Survey, a continent-wide census of nesting birds conducted at thousands of sites each year since 1966, wood thrush populations were fairly stable through the 1970s. But starting in the 1980s, the thrushes began declining drastically in almost every region of North America. "Population declines are prevalent throughout the Wood Thrush's breeding range. Areas with increases tend to be small and isolated," a summary of thirty years' worth of BBS data concluded.

The reasons may be legion. Wood thrushes nest in hardwood forests, and as North America's woodlands have been chopped into smaller and smaller fragments by logging, roads, and development, the forest birds nesting within them have become increasingly vulnerable to predators like raccoons, opossums, crows, and jays that prowl the edges of forests. Wood thrushes are also susceptible to nest parasitism by brown-headed cowbirds, which lay their eggs in the thrush's nest at the expense of the thrush chicks.

It isn't that wood thrushes use only virgin rain forest. In southern Mexico, more wood thrushes were actually caught in mist nets set in neighboring second-growth forest, but by marking the birds, the researchers realized there was a tremendous rate of turnover among those thrushes—second-growth wasn't good enough to support a thrush through the entire winter. In contrast, nearby areas of untouched forest had a lower density of thrushes, but the birds maintained long-term territories, probably an indication of better-quality habitat. Biologists also found, using tiny radio transmitters, that wood thrushes in the second-growth spent a lot of time each day probing the rain forest, looking for vacant territories they could usurp. Second-growth was a life raft, not a preference.

This may be a moot point. The area where that study was conducted, the Los Tuxtlas Mountains of southern Veracruz, has been so relentlessly cut and burned that less than 5 percent of its forest remains, and wood thrushes—once abundant there, along with sixteen other species of migrants—are close to being eliminated.

"I am not suggesting that wholesale extinctions are in the offing," Duke University biologist John Terborgh told a symposium crowd in 1989, speaking about the impact of tropical destruction, "but rather that many formerly common species will become rare and local in their breeding distributions. You can imagine having to travel hundreds of miles to see a Hooded Warbler or Canada Warbler, just as we have always had to do to see a Kirtland's Warbler."

The governments of the region have made praiseworthy attempts to preserve native habitat, a task made all the more difficult by economic hardship and, in some areas, decades of civil war. National parks, forest reserves, and protected areas have been set aside in Mexico and all the countries of Central

America, although many are little more than "paper parks," lacking even rudimentary, on-the-ground protection. Even more heroic have been the efforts of mostly small, chronically underfunded, nonprofit conservation groups, which have lobbied for land preservation, wildlife protection laws, and stronger enforcement of existing statutes.

One of the largest protected areas is the Maya Biosphere Reserve, a four-million-acre swath of the Petén region of northern Guatemala that is one and a half times the size of Yellowstone and was declared a World Heritage Site by UNESCO. It is linked to the east with the Rio Bravo Conservation Area, a privately owned 152,000-acre tract in Belize, and with several forest reserves in southern Belize; and to the north and west with the Calakmul Biosphere Reserve and other preserves in Mexico, forming one of the largest unified conservation areas in the northern tropics.

But here again, the difference between a real preserve and a paper park can be damning. Since the mid-1990s, the Belize government has sold timber concessions totaling more than a half-million acres, including tracts within the Columbia River and Maya Mountain forest reserves—areas Conservation International has said hold the nation's greatest biological wealth. And across the border in Guatemala, deforestation within the Maya Biosphere was running at 120,000 acres a year in the early 1990s, even within the supposedly sacrosanct core of the reserve (outer portions are designated as buffer zones, where some extractive industries are allowed).

Thousands of immigrants from the highlands of Guatemala were already moving each year to the Petén, but the peace agreement that ended a long-running civil war in 1997 set off a flood of new squatters, convinced (incorrectly) that the government would be parceling out land for farms within the biosphere reserve. Conflicts with conservation groups intensified—there were kidnappings, beatings, and their facilities within the reserve were looted and burned. The Guatemalan government, instead of confronting the issue, has allowed the squatters to stay. "It seems clear that if present rates of human immigration into the region are not reduced, it will be difficult for the conservation potential of this reserve complex to be realized," a team of U.S. and Guatemalan conservationists warned.

So many ecosystems are under assault in the tropics, so many seemingly inexorable pressures are working against conservation, that it is easy to despair. But one of the more intriguing ways to save migratory songbirds, and many of the tropical plants and animals with which they coexist, may also be the simplest: Have a cup of coffee. Strangely enough, this global addiction is both responsible for considerable environmental destruction and capable of reversing some of the damage.

Originally from the forests of Ethiopia, coffee moved along Arab trade routes to the Europeans, who in turn brought it to the New World in the 1720s. There, a kind of cultural cross-pollination occurred. While white settlers first grew the coffee bushes in open, sunny plantations, native workers hit on a better method. The Maya and other Mesoamerican societies had grown cacao (chocolate) in plantations shaded by native trees, and they discovered that coffee thrived in the same conditions. The most basic form of this technique, known as rustic cultivation, entails little more than planting the coffee shrubs within an existing forest, although more often the farmers remove some trees and replace them with species that produce fruit, timber, or other products, a method called traditional polyculture. "To the passing traveler with an untrained eye, a mature coffee or cacao plantation can easily be mistaken for a natural forest," notes John Terborgh, who studied birds on such farms in the Dominican Republic.

Shade-grown coffee was the norm in Latin America for hundreds of years, and the industry had a profound impact on the landscape and the economy. In Mexico, Central America, the Caribbean, and Colombia—the most critical area of the tropics for migratory birds—coffee covers almost 6.7 million acres, roughly half of all the cropland in the region. In some badly deforested countries, like El Salvador, coffee plantations account for nearly two-thirds of the surviving forest land. The traditional forms of coffee-growing, researchers have found, are surprisingly compatible with wildlife, especially birds. Russell Greenberg, the director of the Smithsonian Migratory Bird Center, recorded at least 180 species of birds using shade-coffee farms in Chiapas, Mexico, with migrants especially abundant in this slightly disturbed habitat. What's more, overall biological richness on traditional plantations—everything from mammals to snakes to beetles to orchids—is second only to undisturbed forest. The same holds true for cacao plantations.

"Our research has shown that the bigger and more diverse the trees, the greater the diversity of forest birds," Greenberg writes. "Although some farms are dominated by only one species of highly groomed trees, many are more diverse. We commonly find 40 to 50 species of trees on a few hectares of coffee lands." Greenberg and other scientists found that, while virtually all neotropical migrants used shade-coffee farms from time to time, some species were especially dependent upon them, with the exact mix depending on the region.

"As migrant habitat, the man-made vegetation served almost as well as the native forest," Duke University's Terborgh wrote. "I say almost because there were two common migrants to the Dominican Republic that I never found outside of primary forest: the worm-eating warbler and the Bicknell's (gray-cheeked) thrush. Thus, even the best-case scenario for artificial vegetation falls short of perfection." His findings have been echoed by others, who note that coffee plantations are less valuable to resident tropical birds, many of which seem to require mature, undisturbed forest. The plantations are also of secondary importance to migrants like wood thrushes and olive-sided fly-catchers that winter in moist, old-growth forests.

If nothing had changed in the world of coffee, all this would be an interesting ecological sidelight and nothing more. But coffee farming has undergone a revolution in the last twenty years. It started with the discovery in 1970 in Brazil of an infamous fungal blight, known as coffee leaf rust, which had already devastated Asian coffee farms. Because the blight spreads best in damp, shady conditions, panicked farmers were encouraged (through government tax incentives and international aid) to switch from their shade-tolerant varieties of coffee to a dwarf shrub known as *caturra,* which grows well in full sun.

Compared to the oasis of a shade plantation, sun-coffee farms (also known as intensified or technified farms) are biological deserts; Greenberg's team found 90 percent fewer bird species on those they studied. The coffee bushes grow in neat, orderly rows, packed close together and devoid of tree cover. Deprived of the companion plants and organic mulch that foster soil fertility and prevent erosion, the farms must be augmented with synthetic fertilizers, and the coffee shrubs—growing in a monoculture that encourages pests—must be soaked with liberal applications of insecticides, herbicides, and fungicides, usually applied by workers with little training or protective clothing.

But with the chemical life support, technified farms produce up to 30 percent more coffee, fueling a stampede to convert from older, more benign shade techniques, even though full-sun cultivation costs are substantially higher. Brazil, which leads the world in coffee production, has replaced almost all its shade farms with sun, and more than a third of the coffee land in Mexico and Central America has been converted, especially in countries like Costa Rica. Low coffee prices in the 1980s also prompted landowners to convert farms to sugarcane, pasture, or other agricultural uses with little wildlife value. (Ironically, the rust blight that started this debacle has turned out to be less of a problem than originally feared. Apparently the higher elevation of most Latin American coffee farms, and the sharp dry season most areas experience, have curbed its virulence.)

Just as market forces have driven the change to sun coffee, so scientists and conservation groups have tried to harness the market as a way to save the remaining shade plantations. A number of organizations have launched twin efforts to educate consumers in North America about the ecological benefits of buying shade coffee, thus encouraging them to buy it, and to certify coffee grown under conditions that promote biodiversity. It is too soon to tell what effect the nascent campaign is having. At the moment, only a handful of outlets carry certified shade-grown coffee, and interested buyers who have read about the programs may be unable to locate suppliers. The vast majority of commercial coffee—the mass-produced stuff sold in cans on grocery store shelves—comes from technified farms in Brazil and Colombia.

"Sipping a cup of coffee is a ritual that is played out millions of times a day throughout the world. It is an act that ties together consumer, retailer, roaster, broker, producer and farm laborer in complex relationships about which we rarely ponder," writes Greenberg, a prominent campaigner for shade-coffee initiatives. "Migratory birds are emblematic of the importance of coffee to both the local environments where it is grown and the global environment that we share." In *la selva Maya,* where ecological contradictions hide in the forest like the ruins of old pyramids, the unlikely intersection between conservation and commerce is simply one paradox more.

7

Hopping Dick and Betsy Kick-up

The land around Jamaica's Portland Bight is about as far as you can get, physically and economically, from the splashy, North Coast tourist havens like Montego Bay and Ocho Rios. The mountainous midriff of the island wrings most of the moisture from the trade winds, so that the climate here on the south coast, in the rain shadow, is hot and dry, scorched by an annual six-month drought. The survival of sugarcane fields, which stretch for miles, depends entirely on irrigation water pumped up from the Salt River. The few small hamlets like Lionel Town and Mitchell Town sit astride crumbling roads that wind through cane fields and stingy forests of thorn trees and cactus, the only plants that can survive the brutal climate and centuries of grazing, woodchopping, and burning.

A few kilometers outside Mitchell Town, Leo Douglas pulls off the hardtop road and onto a dirt track that runs north into the thorn-scrub; ironically, the dirt is smoother under the tires than the potholed asphalt was. A twenty-five-year-old graduate student at the University of the West Indies in Kingston, Leo is studying the impact of human disturbance on the birds that inhabit the dry forests of southern Jamaica.

It is barely seven o'clock on an early March day, but the air already has a warm crackle to it; Leo takes off his baseball cap, lays an open bandanna over his head, and pulls the hat back on, so that the handkerchief shades his ears and neck. I slather myself with sunblock, as my friend Sharon Gaughan, an environmental educator who's joined us on the trip, shoulders her pack. Moving in a defensive half-crouch, we start threading our way through the thorns, following a straight course set by Leo's compass.

When Columbus first bumped into Jamaica in 1494, the landscape looked nothing like this. The native dry forest was composed of evergreen hardwoods, palms, vines, fleshy agaves—but like the gentle Arawak people that inhabited it and who are now extinct, that woodland all but vanished under waves of white immigrants from Europe and their slaves from Africa. The land was cleared, burned, planted, cleared again, and what finally grew was this unforgiving, hardscrabble excuse for a forest. There are just two species of trees here now, honey acacia and mesquite, neither of them apparently native to the island. Both have finely divided, fernlike leaves that filter the harsh sunlight, giving the interior of the woods a translucent green light, and both are armored to the teeth with long, needle-sharp thorns that score and scratch us as we try to maneuver around them while keeping to the compass's transect line.

There is little birdlife evident. Loggerhead kingbirds, gray with outsized black heads and sturdy bills, chatter from the tops of the trees. A couple of smooth-billed anis, black and glossy as warm tar, slouch from branch to branch in an acacia, making petulant, whining calls. Anis, which are a peculiar sort of cuckoo, are bigger than grackles, with long, round-tipped tails that wag and flop when they move, as though held on by only a thread. Their beaks are enormous—high and bulging in profile, but flattened like plates; they look like prizefighters that have taken a few too many jabs on the nose.

Each tree is a chaotic thicket of small, coppiced saplings, crowding up a few inches from the ground from a thicker base that has obviously been cut many, many times, like a Hydra whose heads keep multiplying. We pass a shallow, room-sized depression, surrounded by newly chopped trees, where charcoal burners have been at work, converting the wood into easily transported fuel—one of the main causes of forest destruction in Jamaica. Everywhere, the soil is parched and dusty, strewn with dried goat and cow dung, grazed to within a hair of its life, the pathetic result of centuries of abuse. The thorn trees alone can stand the relentless browsing, and the goats and cows distribute the trees' seeds within their droppings, making for a self-reinforcing cycle. This is habitat hell, an impoverished, completely alien ecosystem, imposed on the land like some organic form of demonic possession.

"First stop," Leo says, unslinging his pack and shoulder bag. As Sharon assembles a tape player and portable amplifier, Leo uses a handheld GPS unit to get a precise fix on our position, and I mark a tree with a strip of orange surveyor's tape that identifies this as point count No. 195. When all is ready, Sharon hits the Play button and the woods are filled with the rolling *ka-ka-ka-ka-ka-ka-ka* of a Jamaican lizard-cuckoo, one of the island's endemic birds—a species found nowhere else on earth. After a minute, the cuckoo call is replaced by the blue jay–like squeals of a rufous-tailed flycatcher, and eventually by the calls of four other resident species.

"By the way," Leo whispers, "if you see a bird, don't point it out to me. That itself would be a bias in the study. Today we have six eyes, but most times I only have my two."

There is no danger of this in any case, because there is little obvious reaction to the recording. Through the six-minute tape, and for long minutes after it finishes, Leo stands with his metal clipboard poised, looking and listening for any bird that responds, keeping track of the elapsed time with a stopwatch dangling from his neck. I hear a few distant birds, which I cannot identify, but by looking over Leo's shoulder as he makes cryptic notes, I learn their identities after a moment's confusion—YFGR, I realize, are yellow-faced grassquits, and STFL are stolid flycatchers. But the forest is otherwise empty.

Sharon has switched tapes, and at Leo's signal, a new set of songs and calls pours from the speaker. These I recognize immediately—the buzzy song of a black-throated blue warbler, the rollicking *tee-chur, tee-CHUR, TEE-CHUR!* of an ovenbird, the calls of American redstarts and Swainson's warblers and several other northern migrants. Leo starts pishing loudly, and the results are as immediate as they are electrifying. Out of nowhere, we are charged by a mob of agitated, scolding warblers, which surround us like Indians circling a wagon train, each new song from the tape bringing in more recruits from the pale green depths of the thorn-scrub. There are dozens of western palm warblers, yellow-rumpeds, prairie warblers, yellow warblers, American redstarts, black-and-white warblers, parulas, all mad as hell at us, thinking that the taped calls are intruders in their winter territories.

For the next three hours, at five more stops deeper and deeper into the scrub, the story is roughly the same—a near-absence of resident birds, but hordes of migrant warblers, including many, like ovenbirds, that I would not have associated with such a dry, desiccated landscape. At our final stop, beside another wide charcoal burn, Leo points to a small shape in a tangle of spiny branches—a northern waterthrush, a species that even in winter is supposed to choose moist, lush forests and mangrove swamps. But here it is, bobbing its tail in the white midday sun, its black shadow puddled in the dust between its feet.

This is the paradox that Leo keeps finding in his research—while even the slightest degree of human disturbance apparently renders a forest unsuitable for some Jamaican endemics, many of the northern migrants that come to the island each winter thrive in grossly altered habitats. Nor is thorn-scrub the only haven; traveling around the island in subsequent days, we saw Cape May warblers feeding among mango blossoms in back yards, black-throated blues poking their bills into red hibiscus flowers beside hummingbirds, and common yellowthroats and palm warblers darting among the long stalks of sugarcane. By contrast, the island's remaining areas of natural vegetation had noticeably fewer migrants, but a much richer cast of resident birds.

It would be tempting to say that this proves the reassuring adaptability of migrants—and to a degree it does, at least the adaptability of some migrants, under some circumstances. But that is scant comfort to conservationists. Not every species of bird that travels to the Caribbean does well in dry acacia scrub; others require high-altitude rain forests that have been cleared for coffee plantations or coastal mangrove forests that are being chewed up for resorts or industrial sites. At least one migrant to the Caribbean has become extinct, in part, it is believed, from the loss of winter habitat; others are endangered, or nearly so. The islands of the Caribbean—the Bahamas, the Greater Antilles (Cuba, Jamaica, Hispaniola, and Puerto Rico), and the Lesser Antilles—together comprise only 90,000 square miles of dry land, an area smaller than Oregon. But like the narrow funnel of Central America, they are crucial to migrant songbirds out of all proportion to their size. Nearly three dozen species, a great many of them wood warblers, winter on the islands, particularly the Greater Antilles—islands that have been subjected to a degree of human pressure and habitat destruction that is unequaled almost anywhere else in the hemisphere.

No one knows precisely how such changes affect the birds, both residents and migrants. The picture that has emerged from studies over the past twenty years in the Caribbean is incomplete and, at times, contradictory; some researchers have seen long-term declines in wintering migrants, even in undisturbed forest, while others have detected no downward trends. The same species seem to respond to similar forms of habitat change differently on different islands, for reasons that are unclear.

Although ornithologists have been working in the Caribbean since the eighteenth century, relatively little serious modern research has been conducted here, and what has been done is mostly the work of a handful of determined individuals, both professional scientists and skilled amateurs. While ornithological interest in the Caribbean has increased in the past two decades, much of the research has been conducted by scientists from North

America and has focused on northern migrants. That leads to some under-standable frustration among Caribbean conservationists. While acknowl-edging the need to study migrants, they point to the lack of the most fundamental information concerning the islands' resident birds, many of which are rare or local in their distribution and threatened by continuing habitat loss.

It was a point brought home to me long before I went to Jamaica. I'd con-tacted Catherine Levy, vice president of the island's Gosse Bird Club and a past president of the Society for Caribbean Ornithology, to ask some ques-tions about bird conservation—migratory birds, that is. Despite her polite replies, it was clear that she was a bit exasperated by my single-minded inter-est in migrants. When I arrived in Jamaica some months later and had a chance to chat with Mrs. Levy, a charming woman who has made Jamaica's birds her life's work, she elaborated on her concerns.

"We know almost nothing about Jamaica's twenty-seven endemic birds," she said. "We don't know the breeding seasons of many of these birds. In many cases we don't even know what the nests look like—it is very hard to find nests in a place like Jamaica—so we don't know what they feed their young, we don't know if they breed in immature plumages. We know very, very little about them."

Part of the problem is a simple lack of observers. "I could count the number of serious birders in Jamaica on two hands," Mrs. Levy said—a statement I took to be an exaggeration, but one that turned out not to be. Unlike the booming pastime it has become in the States, birding is still very much an oddity here, pursued mostly by a dozen or so affluent white islanders.

The Jamaican educational system has largely ignored environmental mat-ters, although the Gosse Bird Club, with funding from the U.S. Fish and Wildlife Service and the National Fish and Wildlife Foundation, among others, has started producing educational materials and conducting teachers' workshops. Private conservation organizations are small and chronically short of funds. But the real difficulty lies much deeper, Mrs. Levy and oth-ers believe, rooted in the fact that Jamaica has no aboriginal connection to this land.

"We devalue our native vegetation by calling it 'bush,' and bush is to be destroyed, because in a sense, the history of Jamaica is one of exploitation," Mrs. Levy said, sitting with me in a shady nook on the university campus, as tiny vervain hummingbirds jousted with each other for control of a flowering tree. "There are no native people left, and therefore their relationship with the land has not formed part of the culture. Most of our folktales are based on Africa or India—we speak of Brother Tiger and this and that, animals that we've never known in Jamaica. The reaction to lizards is very African, and the reaction to snakes. Whereas the Jamaican boa is not a seriously dangerous snake, it is treated as if it were poisonous and a danger. So you see, an imported culture has been imposed upon our wildlife, and those are aspects that we as a society haven't recognized."

All of which makes someone like Leo Douglas, a young black Jamaican who wants to make ornithology his career, the rarest avis of all. Leo started out, as so many naturalists do, as a kid with lots of wild animal pets; although he grew up in Kingston, his father owns a farm on the North Coast, so Leo spent a lot of his time outdoors. After university and degrees in zoology and terrestrial ecology, he took a temporary job with the Gosse Bird Club and worked on a couple of summer projects studying birds. He was hooked. Leo is now pursuing a master's degree under the tutelage of Peter Vogel, a respected terrestrial ecologist at the University of the West Indies, and Thomas Sherry, an ecologist from Tulane University in Louisiana. Sherry, with his colleague Richard Holmes from Dartmouth, put Jamaica on the map from a migratory-bird perspective, studying the winter ecology of American redstarts and black-throated blue warblers for nearly fifteen years. For more than thirty years, they have studied the same species during the summer at the Hubbard Brook Experimental Forest in New Hampshire. Together, the twin sets of research provide an unprecedented degree of detail on the population dynamics of songbirds, and the researchers have produced a long series of scientific papers that explore everything from sexual segregation to overwinter mortality.

That first evening in Jamaica, Leo led Sharon and me up through the streets of Kingston and into the foothills of the Blue Mountains, past gated, fenced homes built by the island's increasingly prosperous upper-middle class. Soon we were among half-completed houses in a tony new development, following a freshly paved road that dead-ended by a limestone cliff where bulldozers had recently been working, clawing their way even farther up the hill.

As we turned the car carefully around and parked it, several teenagers rode slowly by on battered bicycles, staring at us, then bounced down a path that dropped into the heavily forested ravine below. "They're suspicious of us," Leo said. "Their families live back in there as squatters, and whenever they see someone they don't know, they think they're coming to drive them out." He went over and talked to the boys, slipping from formal English into the liquid Jamaican patois I could scarcely follow, explaining that we were just going to watch birds. One could almost see the reassessment going on behind the boys' eyes: Oh, these people aren't dangerous, just crazy. As we began to hike back up the dry streambed that ran through the heart of the gorge, I could see two or three shanties perched on the hillsides, barely visible through the dense foliage.

The trees were awash with birds, active in the evening cool, many of them Jamaican endemics I had never seen before. A male orangequit, a tanagerlike bird that was deep blue except for a patch of chestnut at its throat, fed among the blossoms of a flowering tree. A Jamaican woodpecker, whose hammering we could hear long before we figured out where it was, poked its white face out of a new hole it was digging in a dead snag. A Jamaican mango, a large bronze-black hummingbird whose wings clattered in the still air, joined the orangequit in feeding among the flowers. Several white-chinned thrushes, their orange legs stark against the dim shadows, loped ahead of us as we walked up the trail.

Jamaicans have a flair for naming wild birds. The turkey vulture is known as the "John Crow" (for which a rugged mountain range is named), while several species of flycatchers are "Sarahbird," "Willie Pee," and "Little Tom Fool." The crested quail-dove, a forest species with a haunting call and a way of staying out of sight, is called the "mountain witch," while the Caribbean dove is the "white-belly maid." And the white-chinned thrushes, which bobbed up the trail ahead of us like the robins to which they are related, are known as "hopping Dick." Interestingly, rural Jamaicans tend to lump all the migrant warblers together as "Christmas birds," because of their winter arrival, or as "chip-chips," a reference to the simple call notes they make while on the island. Only a few merit special notice: the male American redstart is called the "butterfly bird," both for its orange-and-black colors and its habit of constantly flashing its wing and tail feathers. But my favorite Jamaican name of all is that given to the ovenbird, which spends the winter scuffing through the leaves of the forest for food. To the islanders, an ovenbird is "Betsy kick-up."

Although the hills outside Kingston are covered with second-growth forest, the woods are thick and relatively undisturbed, a fact reflected by the mix of species that use it. There were quite a few ovenbirds, lots of redstarts, and a number of worm-eating warblers, a species that in winter is almost never found outside mature forest. Biologists have found that the degree to which migratory warblers can adapt to human disturbance varies considerably, with some species showing remarkable flexibility and others restricted to only a few plant communities. Redstarts are a good example of the former, common on Jamaica almost everywhere there is forest, from mangroves to dry thorn-scrub. Swainson's warbler, on the other hand, a quiet, ground-haunting bird the color of dead leaves, is almost never found away from damp, shady stands of timber.

Undisturbed forest is a scarce commodity in the Caribbean. By one esti-mate, barely 20 percent of the region's forests still survive, much of it badly degraded second-growth instead of mature, virgin woodland. (As early as 1912, only 1 percent of Puerto Rico's virgin forest was left, although second-growth has since returned to about 40 percent of the island.)

> The once extensive forests of the Antilles ... are a thing of the past [Duke ecologist John Terborgh told a symposium on bird conservation some years ago]. When [E. E.] Smith wrote his treatise on the forests of Cuba in 1954, he estimated that only five percent of the forest estate in that country remained intact. The situation today can hardly be any bet-ter. Alexander Wetmore wrote of traveling for days through unbroken forest between Cotui and La Vega when he explored the Dominican Republic in the early 1900s. I traveled the same route in 1970, and hardly a tree was to be seen. The entire landscape, as far as the eye could see, had been converted to sugarcane. Every one of us knows that Black-throated Blue Warblers, American Redstarts, and Ovenbirds do not winter in sugarcane fields.

Jamaica is somewhat better off than Haiti or the Dominican Republic; it retains significant tracts of montane rain forest in the Blue and John Crow Mountains, although these areas have been logged off and on since the first Europeans arrived, cleared for coffee over the past two centuries, or, more recently, planted with non-native Caribbean pine. Part of these adjoining ranges were given at least nominal protection in 1993, when they were named national parks, although enforcement is scant and pirate logging, charcoal-burning, and farming continue to eat away at the reserve's cool, fern-carpeted forests.

In the early 1990s, Jamaica was still reported to have the highest rate of deforestation in the world. Most of the island's center has been cleared for agriculture, but in the northwest, the deeply eroded landscape known as Cockpit Country still holds on to its forest, the largest contiguous tract left in

Jamaica. The woods have survived because the terrain is so damnably difficult to cross; the limestone bedrock has eroded into what is known as karst, a landform marked by steep, closely packed conical hills that look like egg cartons set upside down, with deep hollows known as cockpits in between. Level ground is almost nonexistent. The Cockpit Country's wet limestone forests are among the last refuge for several of the island's most imperiled native birds, including the black-billed parrot.

The area where Leo conducts his fieldwork, on the south coast of the island, has only a few areas of undisturbed vegetation left. One is in the appropriately named Hellshire Hills west of Kingston, protected by the extremely hot, dry climate and by the eroded, honeycombed limestone bedrock, which looks like a cross between Swiss cheese and concrete, and is as sharp underfoot as knives.

"The first time I hiked back in there, it just sliced through the soles of my tennis shoes," Leo told me. "Cut them to ribbons. Fortunately, I found an old tire, cut it into pieces, and tied them around my feet like sandals. I couldn't have gotten out otherwise." Now he wears heavy boots, which nonetheless wear out with distressing speed.

Thanks to this natural barrier, few woodcutters penetrate the Hellshire Hills—nor does much of anyone else, including scientists. In 1990, the Jamaican iguana, a five-foot-long lizard believed to be extinct, was rediscovered there, and Leo thinks the John Crow Mountains, which are equally difficult to travel through, may still harbor a few Jamaican petrels.

This chocolate-brown seabird once nested in burrows in the mountains, moving to and from the ocean under cover of darkness, but it is thought to have vanished after predatory Indian mongooses were introduced to the island in the 1800s. Yet Leo and a few other Jamaican conservationists hold out hope that the petrel survived in its mountain fastness, and on a recent expedition to the John Crows, he and two colleagues saw an unidentified bird one night—a silhouette that was not an owl or a potoo, the only other large nocturnal birds on the island. They found burrows, but they may have been only the diggings of hutias, endemic Jamaican rodents like long-tailed guinea pigs, known locally as coneys. "Maybe they were just coney burrows," Leo said, wistfully, "but now I really wish we'd dug them up to be sure."

Since European colonization, there have been more extinctions in the Antilles than anywhere else in the Western Hemisphere, the inevitable result of rampant habitat loss and killing combined with the tiny ranges of island species. Many of the losses have been mammals, including hutias, native rice rats, and six species of nesophonts, large, shrewlike insectivores. But birds also took an enormous hit; at least eight species of macaws became extinct, two of them from Jamaica, which also lost a pauraque and an endemic sub-species of a small rail, the uniform crake. But these were resident birds. It's unclear whether any migrants vanished from the rapid deforestation of the Caribbean starting in the 1500s. Early American ornithologists, working in the East in the late 1700s and early 1800s, described several birds that haven't been seen since, like John James Audubon's small-headed flycatcher and Blue Mountain warbler, or Townsend's bunting, only one of which was ever col-lected, in Pennsylvania in 1833. They may represent nothing more than hybrids or specimens with aberrant plumage, or they may have been legiti-mate species that were already on the brink of oblivion, glimpsed briefly by science before toppling into the void.

There are, however, two North American migrants to the Caribbean that we know have had a rough time of it—one that is now probably extinct and another that is within a whisker of it.

In the spring of 1832, the Reverend John Bachman—pastor of St. John's Lutheran Church in Charleston, South Carolina, but at least as avid a natu-ralist as he was a man of the cloth—collected a songbird he did not recognize. It was a warbler, obviously, greenish on the back and yellow below, with a spotty black bib and a touch of a dark hood; the beak was narrow and curved down a bit. He sent the skin to Boston, to his close friend Audubon, who expressed some doubts about the find, wondering if it was simply a female or an immature of a better-known species. (This was, after all, a century before the earliest field guides, and Audubon had been burned several times in announcing "new" species that weren't.) Some weeks later, though, Bachman shot another specimen, a male this time, bright yellow with a smooth black cap and bib, and clearly different from anything previously described. Audubon named it "Bachman's Swamp-warbler" for his colleague, and painted the pair perching among the large white blossoms of a Southern tree.

There is a poignancy and an eerie prescience in Audubon's selection of the background for that painting. The tree he chose was *Franklinia*, a species dis-covered in Georgia but not seen in the wild since 1790. Yet while *Franklinia* at least survives in cultivation, Bachman's warbler is almost certainly extinct, and the reasons for its demise, while unclear, seem tied to habitat losses on both ends of its migration.

After its discovery by Bachman, the warbler remained infrequently observed for another fifty years. Very little is known about its ecology or life history; it was found across a wide area of the South, from the Gulf as far north as Missouri, Kentucky, and Virginia. In winter, the entire species withdrew to Cuba and (possibly) the adjacent Isle of Pines. Its nest was found only forty times, thirty-five of them by the same dogged collector, so very little is known about its breeding behavior. Although initially rare, at least judging from the paucity of reports in the mid-1800s, Bachman's warbler became relatively common in the Southeast during the last two decades of the nineteenth century; in northern Florida in 1891, the eminent ornithologists Frank Chapman and William Brewster noted that it was more abundant in migration than red-eyed vireos and orange-crowned warblers, and on one day that spring they saw almost thirty.

Then, around 1910, Bachman's warbler started inexplicably fading from the scene, and by 1920 it was considered quite rare. Its downfall was short and spectacular; the species was virtually gone after 1930, although a few scattered individuals were still reported through the middle of this century, mostly from the I'On Swamp in South Carolina. The last incontrovertible report was of a male photographed in the I'On in 1962; there have been a few sightings since then, including an inconclusive photograph of what may have been a female from Florida in 1980, but most experts believe "Bachman's Swamp-warbler" is extinct.

What happened? It is assumed the warbler was rare to start with, but may have benefited, somehow, from the small-scale forest clearing that occurred in the South after the Civil War, accounting for its sudden increase in abundance prior to 1900. But there is another, more intriguing theory, one that has gained acceptance in recent years. Van Remsen at Louisiana State University has speculated that Bachman's warbler was a bamboo specialist, both during the breeding season on the mainland and in winter on Cuba. Its extinction may well be tied to the loss of a unique habitat, the Southern canebrake.

In the American South, the most abundant native bamboo was *Arundinaria gigantea*, known as "cane," and the almost impassable stands that blanketed riverbanks and lowlands were known as canebrakes. "The Canes are ten to twelve feet in height, and as thick as an ordinary walking-staff; they grow so close together, there is no penetrating them without previously cutting a road," wrote naturalist William Bartram of a Florida canebrake in the 1770s. Near Mobile Bay he found cane "thirty or forty feet high, and as thick as a man's arm, or three or four inches in diameter," which he took to be "proof of the extraordinary fertility of the soil."

Apparently Bachman's warbler nested in these canebrakes, building its deep cup of dead leaves and grass between stalks growing amid pools of standing water. In winter it may have sought out the same kind of habitat, for large stands of bamboo grew in the hot Cuban lowlands. ("May have," "apparently," "it is believed"—there is so little concrete information about this lovely bird that almost everything one writes about it must be hedged round with disclaimers.) One explanation for the otherwise inexplicable boom in Bachman's warbler sightings from 1880 to 1900 is that low-intensity logging in the mid-nineteenth century may have created small clearings that were ideal for cane. Another possibility, one proposed by Remsen, is that the upsurge in warbler numbers may have been linked to the irruptive nature of bamboo, which grows and dies en masse in cycles spanning twenty years or more.

Finally, by the turn of the century, industrial logging, with its vast clearcuts, coupled with the almost total destruction of canebrakes for farmland, eliminated much of the warbler's breeding habitat. And John Terborgh has pointed out that canebrakes in Cuba, where the bird presumably spent the winter, were vanishing at about the same time, as almost all the island's lowland vegetation was cleared for sugar production. The remaining scraps of Cuban forest and bamboo could support relatively few warblers, and when the birds spread out in spring across their vast, fragmented American breeding range, he suspects they simply couldn't find each other. (Songbirds do not travel in mated pairs from the wintering grounds but pair up anew each spring, the males arriving first to set up territories, the females coming days or weeks later and picking their mates.)

It was this "gross imbalance between the area available for wintering and the areas available for breeding," Terborgh concludes, that doomed Bachman's warbler. "I imagine that each spring a tiny remnant of birds crossed the Gulf of Mexico and fanned out into a huge area in the Southeast, where they became, so to speak, needles in a haystack. Toward the end, it is likely that most of the males in the population . . . were never discovered by females. Once this situation developed, there could be no possible salvation for the species in the wild."

It is ironic that Kirtland's warbler, a species that was probably never as common as Bachman's, and certainly never as widespread, is still with us, albeit just barely. First described from a specimen shot during migration in Ohio in 1851, its wintering grounds in the Bahamas were known as early as the 1870s, but its breeding range remained a mystery until 1903, when a nest was finally found on Michigan's lower peninsula. That area, centered on just two counties, remains the only place this slim, yellow-breasted warbler breeds, nesting in thickets of young jack pine that sprout up after wildfires.

By 1971, the annual breeding-grounds survey showed only 201 singing male Kirtland's warblers, the nadir in a long decline caused by wildfire suppression, which eliminated new breeding habitat in young pines, and nest parasitism by cowbirds. An aggressive cowbird-control program was instituted the next year, and controlled burns and pine plantings have increased the nesting area somewhat, but the number of singing males (the easiest part of the population to monitor) remains at fewer than eight hundred.

Kirtland's warbler is known to winter only in the Bahamas, an archipelago with barely 5,000 square miles of land, giving it both the smallest winter and summer ranges of any North American bird—a recipe for disaster if ever there was one. So why did Kirtland's warbler survive when Bachman's warbler, which wintered on an island with almost ten times the land area, and which nested across nine or ten states instead of a small part of one, become extinct?

For one thing, the Bahamas—despite coastal development on the larger islands—still have a great deal of suitable habitat for Kirtland's warbler, which likes low, scrubby brush in winter. "Today, one can fly the length of the Bahama chain and see little in the *interiors* of the islands that has changed since Columbus landed here in 1492," wrote Harold Mayfield, who has studied the species since the 1940s. "Large-scale agriculture and lumbering have been attempted more than once in the last two centuries but have been discouraged by the shallow soil and scarcity of fresh water. Indeed, periodic clearings may have been beneficial to the habitat favored by Kirtland's Warblers."

Kirtland's behavior may also have saved it from extinction. The very aspects of its ecology that place it at risk—the highly restricted size of its breeding range and its tendency to nest in semicolonial groups—prevent it from scattering to the four winds, diluting itself over the vastness of the eastern United States, as probably happened to Bachman's warbler. But this is a two-edged sword: A natural catastrophe, like a powerful hurricane sweeping up the Bahamas, could wipe out many wintering birds in a single blow; in fact, a severe drought in the Bahamas during the winter of 1970–71, when "bushel baskets" of dead birds of many species were seen each day, may have caused the sharp drop in the number of Kirtland's warblers recorded in Michigan the following summer.

Just as Kirtland's warbler migrates only to the Bahamas, Bicknell's thrush winters only in the Greater Antilles, particularly on Hispaniola, but also on

Jamaica; hiking through the rain forest in the Blue Mountains, I kept hoping to see one of these retiring mouse-brown birds hopping among the ferns and orchids. In the words of one researcher, Bicknell's thrush is utterly dependent on "islands, both real and figurative"—the Caribbean in winter, and alpine islands in summer, high mountaintops in New England and eastern Canada where it nests. Here again is a bird with a restricted range at both ends of its migration, causing concern among conservationists—although it must be admitted that, until recently, few people spared this somber-colored thrush much thought at all.

That's because Bicknell's thrush was always considered merely a subspecies of the widespread gray-cheeked thrush, which nests across most of Canada, Alaska, and northeastern Siberia. Smaller and a warmer brown than the other gray-cheeked races, Bicknell's was discovered in 1881 on Slide Mountain, in the New York Catskills, and was named for its discoverer, Eugene Bicknell, when it was officially described for science a year later.

Along the way, Bicknell's has been caught in one of the great ebbs and flows of biology, the tug-of-war between the splitters and the lumpers. The time of its discovery in the 1880s was an era of field biology when even the slightest differences in size, shape, or coloration were enough to cause a specimen to be labeled a new subspecies, or even a completely new species. It is ironic, given what we now know about Bicknell's thrush, that it escaped this latter fate, because in those days the splitters—scientists who claimed to find new species behind every bush—held sway. This tendency reached ridiculous extremes, as with mammalogist C. Hart Merriam, who recognized a total of eighty-seven full species of North American grizzly bears and eleven species of coyotes.

Through most of the twentieth century, however, the pendulum swung in the other direction, away from the splitters and toward the lumpers, scientists who believed that even profound differences in size or plumage signify little, so long as there is any degree of interbreeding between two groups of organisms. This school of thought reached its apogee in 1973, with the release of "The Thirty-second Supplement to the American Ornithologists' Union Checklist of North American Birds," a biennial AOU publication that outlines changes in the taxonomy of the continent's birds.

Usually the release of a new checklist supplement is of barely academic interest, but the 1973 edition was incendiary, lumping and renaming species left and right, many of them beloved by the general public. To pick one contentious example, the Baltimore oriole of the East, Bullock's oriole of the

West, and the black-backed oriole of Mexico were lumped, on the basis of two thin zones of hybridization, into a new entity known as the northern oriole. Yellow-shafted, red-shafted, and gilded flickers, which also hybridize in places, were lumped as the northern flicker. Northern oriole, northern flicker: the utter lack of poetry in the new names only added to the sting. The checklist lumped two species of towhees, two of titmice, and four species of juncos. Later checklist revisions lumped rosy-finches, herons, and many others. The result was a firestorm of criticism, among both birders (many of whom saw their carefully accumulated life lists tumble) and average folks like the citizens of Baltimore, Maryland, who lost their baseball team's namesake overnight.

Through most of this century, biologists have defined a species as a group of organisms that successfully interbreed with one another, but not with other organisms, something called the "biological species concept," or BSC. By this definition, the kind of rampant hybridization that occurs between flickers in the Great Plains or titmice in the Southwest means they are not separate species. But about a decade after the 1973 supplement was published, the pendulum started back the other way, and it continues to gain momentum. There are two reasons. In some cases, like that of the orioles, biologists realized the hybridization wasn't as widespread as first thought. Also, as they reexamined many of the bird subspecies around the continent, weighing details of their breeding ecology, songs, and behavior, biologists realized that a good many are, in fact, reproductively isolated from each other under the tenets of the BSC. And the use of DNA analysis has become widespread, revealing that many "subspecies" that look quite similar are, in fact, more different from each other than they are from clearly unrelated species of birds. What's more, the BSC, taxonomy's guiding principle for more than fifty years, is being replaced with what's known as the "phylogenetic species concept," or PSC, the notion that any genetically, physically, or geographically distinct population—including many of those now considered subspecies—warrants recognition as a full species.

Since 1985, more than twenty North American bird species have been split—some, like the orioles and the rosy-finches, rectifying what now appears to have been excessive zeal; others completely new, like the Pacific loon and the saltmarsh sharp-tailed sparrow. The solitary vireo has been split into three species, the scrub-jay into another three, the brown towhee into two. Rumors swirl that the red crossbill comprises no less than *eight* species, none of them distinguishable in the field except by voice. If those advocating

PSC have their way, North America's avifauna could jump from its current eight hundred or so species to several thousand.

In 1993, a Canadian taxonomist named Henri Ouellet published a study that carefully made the case for elevating Bicknell's thrush to full species status. Ouellet had examined scores of museum specimens, noting coloration and body measurements; Bicknell's tended to be smaller and a warm shade of cocoa, although the color was highly variable. He studied the kinds of summer habitat that each type chooses—gray-cheeked in a variety of coniferous woodlands, but Bicknell's almost exclusively in high-altitude stands of conifers, especially on steep mountainsides, which accounts for its spotty distribution. Ouellet recorded the songs of both thrushes and put them through computerized analysis, revealing marked acoustic differences, and he played graycheek songs for wild Bicknell's, getting no reaction—a strong suggestion that the two are separate creatures, indifferent to each other's courtship signals. He and another scientist compared the DNA of graycheeks and Bicknell's and found they diverged as much as a million years ago. Finally, Ouellet noted that while graycheeks winter in mainland South America, Bicknell's migrate to only a few islands in the Caribbean. The AOU followed his recommendation, and in 1995 the organization formally split Bicknell's into a new species.

Does any of this make the slightest bit of difference to the birds themselves? Of course not; taxonomy is the system of pigeonholes into which we humans try to shoehorn nature, and nature doesn't give a fig. But it makes a great deal of difference to people, apparently. Birding magazines were suddenly full of articles on how to identify Bicknell's thrush in the field (a tricky task), or where to find it during the breeding season. Species status had given Bicknell's a certain cachet it lacked before.

With the stroke of the AOU's pen, Bicknell's thrush was also pushed into the political spotlight. As a species with a restricted breeding and wintering range, it was almost immediately proposed for protection under the federal Endangered Species Act—the first of what promises to be a flood of newly minted endangered species, brought into being by taxonomic splits. It isn't that the thrush did not qualify for ESA protection before the split; the law provides full legal protection for subspecies, and in fact, most of the birds covered by the law are subspecies or regional populations. But there's that cachet again, an additional, almost mystical weight to an endangered *species*, as compared to a mere geographic race—a weight that prompts greater scientific, popular, and legal attention, and more funding for research and conservation.

Indeed, even before publication of Ouellet's research, as word of his preliminary findings spread, there was new interest in studying Bicknell's thrush.
The most ambitious project was organized by Chris Rimmer of the Vermont
Institute of Natural Science, and Jon Atwood of the Manomet Center for
Conservation Sciences, who with a large corps of volunteers set out to survey
more than seventy mountain peaks in New England where the bird was
thought to breed. Rimmer also visited the Caribbean, starting in 1994, to
comb the high-altitude rain forests of Hispaniola and other islands, looking
for signs of Bicknell's thrush.

There was, the scientists soon learned, a good deal to be worried about.
The species had disappeared from parts of its breeding range, including Mt.
Greylock in Massachusetts (its only nesting locale in that state) and a number of sites in the Canadian Maritimes. Elsewhere in the Northeast, the wind-
stunted tracts of red spruce and balsam fir in which the thrush nests have
been savaged by airborne pollution, especially acid precipitation. Even before
Rimmer went to the Dominican Republic, which occupies the eastern half of
the island of Hispaniola, it was clear that severe deforestation had robbed
Bicknell's of much of its wintering habitat there. (Haiti, in the island's west,
was already a lost cause.) Little wonder that Partners in Flight, a hemispheric
umbrella group of conservation agencies and organizations, has ranked Bicknell's thrush as one of the highest priorities for research and protection. By
one recent estimate, there may be no more than 5,000 of them in existence—
mostly on Hispaniola, but also on Puerto Rico, St. Croix, and in the Blue
Mountains of Jamaica. Yet in areas where conditions would seem to be ideal,
the thrushes are not found—perhaps a question of elevation, or the moisture
level, or the degree of disturbance.

Those sorts of questions keep Leo Douglas busy, trying to tease out the
reasons why some migrants thrive in the modern Caribbean and others keep
fading away. After working all day in the acacia forest, we drove to the Salt
River, which despite its name is fresh water and which is pumped out to
irrigate sugarcane fields in enormous, curving fountains of spray. We pulled
up to a cavernous, corrugated metal barn beside the road, out of which an
ear-splitting din emerged; this is where the sugar company's gargantuan irrigation pump is housed and where Leo has his field station. Inside the dim,
cool barn were what looked like three boxcars without wheels, two back-
to-back and one perched on top, a wooden staircase climbing to it. They were
old steel cargo containers, the kind used on ships, and the upper one had
been converted to a rustic apartment, complete with a refrigerator, a toilet,

and, in lieu of a shower, a pipe that sprayed river water and the occasional aquatic plant.

My sleep was fitful, and I tossed and turned on my bedroll; the pump, which fell silent at dusk, roared back to life at 1 a.m., easily overwhelming the foam plugs I had stuffed in my ears, and mosquitoes buzzed through the open door, which couldn't be closed because of the heat. Leo, on the other hand, slept straight through the night, apparently used to the pandemonium, and when the alarm went off at four-thirty, he was the only one of the three of us with much pep.

Now it is dawn, and we are in the small market village of Lionel Town, in the back yard of a home owned by a gracious older couple of East Indian descent. The property is heavily planted with fruit trees, as are most Jamaican yards, and it is neat and lovingly maintained, which many are not.

The last notes have spilled out of the tape player, and the fruit trees are full of birds, although not the overwhelming response we'd experienced in the thorn-scrub the previous day. One of the mangos, which is in full bloom, has been staked out by several male Cape May warblers, each busily defending the brownish, steeple-shaped flower clusters against one another. Cape Mays and black-throated blue warblers, which eat little but insects during the northern breeding season, switch to nectar while in the Caribbean; the black-throated blues, Leo tells me, often claim hummingbird feeders for their own, chasing away even the big streamertail hummers that are Jamaica's national symbol. So reliant are Cape May warblers on nectar that they have specially adapted, fringed tongues to help them lap up the sweet liquid, even though this is probably of no use in the summer, when they eat insects.

Through the morning we visit four yards, doing bird surveys, then back-track, conducting vegetation analyses in each so Leo and his advisors can search for links between the plants and the birds that use them. Every tree and shrub on the property must be identified, the diameter of the trunk measured (a joy when working with acacias or mesquite, their multiple trunks bristling with spines), and its height determined. This last job is mine; I have a heavy telescoping pole of yellow plastic pipe that extends more than fifty feet, and I must thread this flexible wand up through the tallest branches of these old mangos and breadfruits and almond trees, until Sharon, standing some distance away with a clipboard, tells me I'm on the mark. We work our way down fencelines choked with shrubs and saplings, and through overgrown corners. The sun is high and the work tedious.

I'd love to know exactly what the local homeowners make of all this. The previous evening Leo had gone from door to door, handing out letters of introduction on university stationery, asking permission to come back the next morning to do his surveys. He's been working here for weeks, and many of the people who pass him on the street wave and call his name, but I'm surprised to find that one aspect of his census technique causes a uniquely Jamaican problem.

"It's pishing," Leo explains. "Going *psst, psst, psst* like that is a Jamaican way of calling a girl—and a very, very rude way. A nice girl, a properly raised girl, wouldn't even look over if she heard that. So people see me here and say, 'There's Leo, the guy who goes *psst, psst, psst* in our back yards.'"

The next day, our last in the field with Leo, he has a special treat in store— a visit to one of the last remaining parcels of native dry limestone forest on the south coast. For the better part of a week, we've been working in places that bear the heavy mark of human disturbance, but as we thump and bounce up a ragged dirt road that climbs Portland Ridge, we finally glimpse the island that the long-dead Arawaks knew.

Curving out like a fishhook from the center of the south coast, Portland Ridge is a long peninsula that rises several hundred feet in elevation and stretches for miles out toward the sea. The only entry is through a high, locked iron gate, for almost the entire point is owned by two well-to-do gun clubs, whose members use it for pigeon hunting. The difference is stark—just outside the fence is the same depressing thorn-scrub we've been seeing all week, with the usual signs of woodcutting and grazing. But on the gun-club grounds, the forest is completely transformed. There isn't an acacia in sight; the canopy trees are tall, broadleafed species, with a thick understory of thatch palms and enormous agaves with fleshy leaves five or six feet long. Epiphytic cacti, long and snakelike, spraddle across the ground and hitch themselves up the trunks of trees, held in place by delicate root hairs.

Hermit crabs the size of plums, their claws red and smoky blue, lug snail shells across the trail as we hike back into the woods, weaving around deep sinkholes that drop like mineshafts into the ground. I peer carefully down one and cannot see the bottom for the shadows, but it is at least twenty or thirty feet deep, narrow as a well. The sinkholes, Leo explains, are a rich source of fossils, animals that tumbled in and couldn't get out; we take the warning to heart, placing each step with care, not only to avoid the pits, but because the eroded limestone over which we walk would slice us savagely should we trip and fall.

The drill here is the same as in the acacia woods—every 250 meters along the transect we stop, mark the site with a ribbon and a numbered metal tag, and play the tape of resident bird calls, followed by that of migrants. But for once, it is the residents who come flocking to the recording—big lizard-cuckoos with spectacles of bright crimson skin around each eye, Caribbean doves, endemic Jamaican vireos singing their *see-weet, see-weet* songs end-lessly, gray Bahama mockingbirds, the strangely named sad flycatcher, black-whiskered vireos, and many others.

There are northern migrants, too—ovenbirds, parulas, worm-eating war-blers—forest specialists for the most part, birds that do badly in the more degraded habitats that have become the norm in the Caribbean. Notable by their near-absence are species like palm and prairie warblers, which were so abundant on Leo's point counts in disturbed habitats.

It is easy to slip into complacency and oversimplification about the natural world and humanity's impact on it. What is good or bad? Where do you draw the lines? Cutting the native forest is a disaster for lizard-cuckoos and Swain-son's warblers, a boon for palm warblers and yellowthroats, and a wash for many other birds, residents and migrants alike. By lucky chance, some find themselves well adapted to the new reality of the Caribbean, and some don't; some will thrive, and some will fail.

I am a North American birder, to whom orangequits and black-billed par-rots are once-in-a-lifetime exotics, but to whom prairie warblers and redstarts are the stuff of everyday life; their colors, their songs, the timetable of their comings and goings are part of the fabric of home. My natural allegiance, I suppose, is with the migrants, and it is hard not to feel a sense of satisfaction when I encounter a squabbling flock of warblers down in the thorn-scrub or black-throated blues dancing around the flowers of a back-yard mango tree.

Some small, unworthy part of my mind says, "Well, certainly it's a shame that the resident birds are disappearing, but at least *my* birds are doing all right." But that's not just craven, it's untrue; ask a Bicknell's thrush how things are going these days, while the charcoal burners and the farmers tear at the last mountain forests in the Caribbean. As scientists like Leo Douglas, Tom Sherry, Richard Holmes, Peter Vogel, and others dissect the intricate relationships between bird and habitat, we are coming to realize the future is a tangle of possibilities, endlessly complicated and impossible to boil down into neat, simple predictions. What does the future hold for migrants in the Caribbean? No one can yet say. One thing, though, is clear—on these tiny islands that sparkle like jewels, there is no room for complacency.

8

Aguilucheros

The ribbons of migration can be short or long. When I was a kid, the phrase "flying south for the winter" conjured images of robins forsaking our snowy yard for some Spanish moss–hung antebellum estate in Dixie. Now, of course, I realize that "south" covers a lot of territory.

Gather up all the migratory birds in North America along an imaginary line, hundreds and hundreds of species, and fire the starter's pistol. Some hardly budge—Clark's nutcrackers that trade high country for lower valleys; the panhandling, park-pond mallards that keep just one lazy, bread-fattened step ahead of the frozen water. Others make a respectable hop but stay largely north of the Mexican border, like rusty blackbirds traveling from northern Canada to the Southeast states.

Beyond the Rio Grande it becomes a marathoner's race. Birds begin dropping out wholesale in Mexico and Central America—more than two hundred species winter in the coastal forests and the Yucatán, and almost as many in the Caribbean, a palette of colorful warblers, vireos, tanagers, thrushes, and orioles. Fewer (between one and two hundred species) make it as far as northern South America—Acadian flycatchers and cerulean warblers in humid forests of the northern Andes, yellow-throated vireos along the edges of woodlands and plantations in Colombia, dickcissels in insectlike swarms on the grass-filled *llanos* of Venezuela. Even fewer, roughly three dozen species, winter in the Amazon.

Yet a handful of birds push on even farther—past the Amazon, past the endless marshes of the Pantanal in Brazil, almost to the ends of the earth, where the tapered cone of South America points like a finger to Antarctica.

These birds are the long-distance champs, mostly shorebirds and swallows with their scythe-shaped wings, built for speed and endurance over the long haul.

Many stop, at last, in the pampas of Argentina—that great sheet of grassy land that stretches from the Rio de la Plata west to the Andes. Here, they are known by Spanish names. The barn swallow is *golondrina tijerita*, the "little scissor-tailed swallow"; the upland sandpiper, which holds its wings open for a moment when it lands on a fencepost, is *batitú*, "the flapper." And with them come the *aguiluchos langosteros*, the grasshopper hawks, which North Americans call Swainson's hawks—long-limbed and graceful, traveling in flocks thousands of birds strong, chasing the sun across the equator in a journey that may carry them 7,000 miles in three months.

But the land they find is not the one that sheltered their forebears; the plains of waving bunchgrass and wind-sculpted groves of acacias now are yellow with sunflower fields that stretch to the horizon and busy with the drone of tractors and crop dusters. In recent years, the only thing waiting for the *aguiluchos* at journey's end has been death on a frightening scale. In some respects, it is the same tragic tale of environmental contamination and chemical irresponsibility played out around the world. Yet theirs is also a story of unprecedented international cooperation, and although it is too early to say for sure, it may have that rarest of endings—a happy one.

A big storm is coming, an anvil-shaped cloud piled high on the northern horizon and painted with the pink-orange of early evening. It's been hot and muggy, a typical midsummer day on the pampas, but as the sun slides down and the shadows lengthen, the *estancia* is once again coming to life. Siesta on the ranch is over, and horsemen move out into the pastures where the herds of black cattle and dusty sheep are stirring. From inside the cool, dark ranch house, with its red tile roof, arched colonnade, and white stucco walls, I can hear the clatter of pots and Silvina calling to the kids.

Estancia La Chanilao, about 350 miles west of Buenos Aires, is a typical pampas farm. Its 1,200 hectares (about 3,000 acres) are a mix of field and pasture, dominated by a *monte*, a planted forest of eucalyptus trees that surrounds and shelters the main house, workers' quarters, and other buildings. In the past, Chanilao was solely a livestock operation—black Angus cattle, sheep, and horses—but today much of the land is given over to a variety of row crops, especially sunflowers, sorghum, and wheat, a change mirrored across the pampas in the past fifteen years.

This shift from grazing to farming has brought with it a growing reliance on potent agrochemicals, many deadly to wildlife. The previous January, the ornithological world had been shocked by the discovery of thousands of dead Swainson's hawks in this small part of the pampas. Now, Estancia La Chanilao is serving as a field station for a team of American and Argentine scientists trying to determine exactly what was happening, and how to stop it.

We rose at four this morning to set live-traps, and snared a dozen Swainson's hawks for blood samples and banding, a job that took until after noon to complete. Following the midday meal, I spent several more hours sitting in the shade of a thatched pavilion, tediously retying the small, monofilament nooses on the traps, readying them for the next day's use.

I get up to stretch around six o'clock and walk out into the low, slanting sun. To the naked eye, the sky overhead looks empty, but when I raise a pair of ten-power binoculars, I can just make out tiny black flecks, like pepper suspended in the pastel blue. The specks grow and enlarge—becoming birds, becoming hawks, drifting down languidly from beyond the limits of vision.

At first there are hundreds of Swainson's hawks, then thousands, arrayed across the sky in meandering layers. A cool breeze from the west begins to ruffle the long grass of the lawn, blowing through the deep grove of tall eucalyptus, carrying their spiced scent. The hawks have dropped lower still, close enough for me to see the light flashing off their pale undersides as they begin to circle, coalescing into a single, swirling kettle, a galaxy of birds. The show is starting, and I shout for the others.

No other bird soars with the unmatched grace and utter physical beauty of a Swainson's hawk—long, reaching wings of charcoal with knife-blade tips, a subtly proportioned tail, and a kaleidoscope of colors in every flock: streaky juveniles, pale-bodied adults with their brown hangman's hoods, rufous birds, cocoa birds, hawks the color of ebony wood.

As the hawks gather, so do we, a forest of tripods on the lawn and the constant click of camera shutters like busy crickets. Forgotten, at least for a time, is any thought of science, of transects and samples, of toxicology and behavioral ecology. Scientist and layman, Argentine and American, we are all equally lost in the spectacle. Now, at least 8,000 hawks are suspended in the air over the *monte*, preparing to roost there for the night but not yet willing to give up the open sky and freshening wind. At times the flock looks like a river; at others, a tornado. Frequently, several thousand will settle into the trees, but then a *gaucho* on horseback trots beneath them on his way to the far pastures and the hawks roil up again in a frenzied mass.

As the sun begins to set, the *aguiluchos* form a great sheet that runs the length of the narrow, kilometer-long forest, hanging almost motionless in the wind. Then the hawks at the front of the pack wheel back, soaring in a tight circle, spinning down the line toward us, gathering more and more hawks into the maelstrom, as though someone is rolling them into a ball. Within minutes, the whole chaotic mass is directly overhead, a riotous ballet of light and movement against the darkening storm.

Swainson's hawks are members of a select group. Of the more than 500 species of migratory birds that breed in North America, fewer than three dozen travel as far as Argentina's pampas. Roughly twenty of those are shore-birds: Hudsonian godwits, American golden-plovers, a variety of sandpipers, Wilson's phalaropes. The Eskimo curlew, once among the most abundant birds in the Western Hemisphere, but driven to virtual extinction in the nineteenth century by market hunting, wintered in the pampas; no one even knows if the curlew still exists. Blue-winged teal, common and roseate terns make the journey, as do barn, cliff, and bank swallows, bobolinks, and a few peregrine falcons of the Arctic race.

No other raptor in this hemisphere is as clannish and social as the Swain-son's hawk. Only for a few months during the breeding season do they split into pairs; the rest of the time they feed and roost and travel in the kind of immense flocks that appear each night at Chanilao—riding the thermals south through Veracruz, down the Panamanian isthmus, and across the heart of South America, a trek that crosses thirteen countries on two continents. "The migrations of the Swainson's Hawk are the longest and, because of the immense flocks in which they sometimes travel, the most spectacular of any North American hawk," wrote ornithologists Leslie Brown and Dean Amadon. "They may be the most impressive avian gatherings in North America since the demise of the Passenger Pigeon."

Swainson's hawks are birds of the wide horizon and empty prairies. In the northern summer, they breed from Texas to the Dakotas, north into the Canadian prairies and west to California and eastern Washington—roughly the same part of the continent covered by grassland. They once nested in the tallgrass prairies of the Midwest, but as the sod was plowed under for crops, the hawks fled west, like the vanished buffalo, and today only scattered pairs still hang on in states like Illinois and Missouri. Elsewhere they have made their peace with agriculture, hunting for rodents in alfalfa or harvested wheat fields, nesting in small copses of aspens and willows that line creekbeds, or in the shelterbelts planted around farms. Some now even nest in suburban neighborhoods.

For generations, almost everything known about the Swainson's hawk came from the northern breeding grounds, and what happened to it the rest of the year was something of a cipher. In the 1960s, Brown and Amadon could say only that "in Argentina it is said to winter in flocks, and is evidently local judging from the number of observers who have not seen it." The 1988 *Handbook of North American Birds*, perhaps the most exhaustive reference work of its day, gave only this vague statement: "East of the [Panama] Canal Zone and in all directions beyond n. S. Am., the whereabouts of any number of Swainson's hawks is a mystery . . . It has long been supposed that the major 'wintering' area was the pampas grasslands of Argentina, where the birds moved about in groups searching for food and roosting on the ground, but some who have sought them there have not found them."

Biologists could do little more than shrug when asked where the hawks went in the northern winter. Which is why this story about death and hope on the Argentine pampas begins, strangely enough, in northern California.

Klamath National Forest, up toward the Oregon line, is big-woods country, home to old-growth trees, salmon-rich rivers, spotted owls, and bitter controversy between environmentalists and loggers. But the million-acre Klamath also spreads out of the mountains and into the more arid lowlands. Butte Valley National Grassland, part of the Klamath, grows alfalfa where the land is irrigated, and sagebrush and grass where it is not. It also supports astounding numbers of Belding's ground squirrels, at a rate of roughly 16,000 per *acre* in places. Farmers in the region hate the squirrels, which can devastate their crops. But the squirrels in Butte Valley (along with equally abundant grasshoppers) are food for more than sixty pairs of Swainson's hawks, which a Forest Service biologist named Brian Woodbridge began studying in the early 1980s.

Swainson's hawks are rare in California, and becoming rarer; by some estimates, their numbers have dropped 90 percent in the state since the 1940s. In fact, the hawk seems to be in trouble in many parts of its wide breeding range—

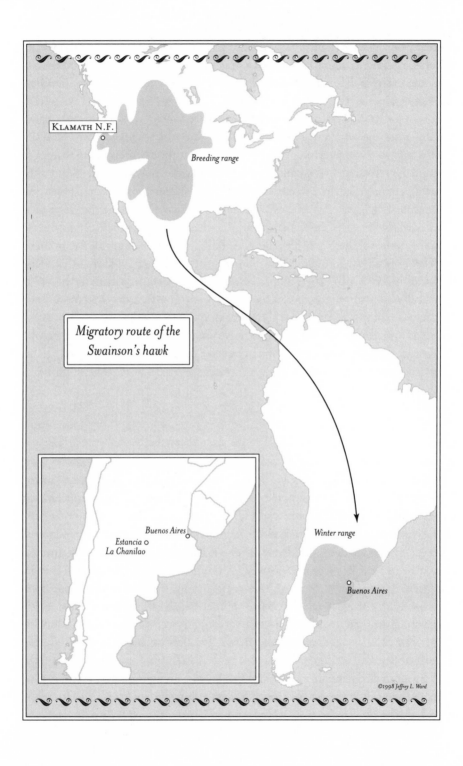

KLAMATH N.F.

Breeding range

Migratory route of the
Swainson's hawk

Winter range

Buenos Aires

Buenos Aires
Estancia
La Chanilao

©1998 Jeffrey L. Ward

in Oregon and Nevada, in Montana and the Canadian prairies, in the eastern Plains and elsewhere. Alarms have been sounded for decades, with increasing urgency through the 1980s and '90s, although there was little hard evidence and no consensus on a cause. Habitat loss, that perennial Grim Reaper of wildlife, was often cited, as was chemical contamination. "Both grasshoppers and ground squirrels . . . are considered detrimental to agriculture and have been targets of widespread poisoning campaigns in the w. states during the general period when the hawk is said to have declined. Coincidence of timing suggests a possible connection," noted the *Handbook of North American Birds*. But then the author made a statement that later seemed prophetic: "Whether biocides on austral summer range in S. Am. may be factor is unknown."

Thirty-nine years old, Brian Woodbridge has dark hair, thinning in front and worn long over his collar, a mustache and dark-rimmed glasses. Running several miles a day keeps him trim, as does a year-round cycle of fieldwork that ranges across the biological spectrum, from spotted owls and goshawks to salamanders. Summer, though, is hawk season. Each year for the past fifteen years, Woodbridge and his team have captured Swainson's hawks in the Butte Valley, color-marking them with large-numbered plastic bands that can be read with binoculars. They watch the hawks mature, pair up, raise families, and disappear south, season after season.

Not surprisingly, there was some degree of turnover among the nesting pairs in their study area each year, a certain percentage of the marked birds that would fail to show up in the spring. This is normal; birds get old, they get hurt on migration, they die. But Woodbridge also noticed something odd. Every so often, he would see a dramatic spike in the turnover—a year when the rate would double or triple, when maybe 40 percent of the breeders didn't return to Butte Valley.

Woodbridge assumed, naturally enough, that there was trouble somewhere along the migration route or on the wintering grounds, but he had no idea how to find out for sure. Biologists knew the hawks swung down through eastern Mexico, because the new hawkwatch in Veracruz was reporting hundreds of thousands of them, as were observers in Panama. It was known that they crossed Colombia, where they were shot in large numbers each year, and it was assumed that they wintered in Argentina. Everything else about the six- or seven-thousand-mile journey was a mystery, open to educated guesses. Brian and his friends would brainstorm wild schemes about attaching a tiny radio transmitter to a hawk, then following it south in a light airplane, but in reality they were stuck in northern California, wondering and worrying.

That changed in 1993. Engineers created a tiny transmitter that broadcast its signal, not to a handheld radio receiver, but to weather satellites owned by the National Oceanic and Atmospheric Administration (NOAA), eliminating the need for biologists to physically follow along behind. Biologists had been using larger, clumsier models for years on big creatures like caribou and sea turtles, but these new satellite transmitters weighed just 27 grams, about an ounce—small enough to fit on a mid-sized bird like a peregrine falcon without hampering it. And that's exactly what scientists started doing left and right, slapping the $3,000 units on peregrines, by far the most glamorous of all raptors. It was a mad rush that Brian Woodbridge ignored. "Everyone jumped on peregrine falcons, but the first thing that came to my mind was 'I have to find enough money to put some on some Swainson's hawks.'"

The following year, the Forest Service allowed Woodbridge to buy two transmitters, and gave him thousands more to pay NOAA to download the weekly locations. He attached the units to two breeding females, and when they left on their fall migration, Brian was tagging along—not in a plane, as he'd always dreamed, but at a computer terminal.

At first, as the two hawks migrated down through California and into the Southwest, everything worked as promised. Then one of the transmitters gave out in Arizona, leaving all Woodbridge's hopes riding on a single bird. The hawk coursed down Central America and South America, leaving footprints across a map in the form of signal locations. For a time it vanished due to poor satellite coverage around the equator, but its signal reappeared in southwest Brazil, and by November it had entered Argentina. In December 1994, the signal "localized," as biologists say, in eastern La Pampa province, about 300 miles west of Buenos Aires. The hawk had settled in for the winter.

The next month, January 1995, Woodbridge and two former assistants, Trent Seager and Karen Finley, flew to Argentina. "We thought, We'll rent a car, drive around, and find the bird, which was incredibly naïve," Brian now admits. The pampas are huge, and they simply swallowed the tagged bird. Fortunately, through a bit of almost accidental networking among the English-speaking community in Buenos Aires, they were introduced to a family with an interest in ornithology and a ranch in La Pampa province. The *estancia* became their base of operations, and the owners' son told them about a huge flock of hawks he'd seen while flying a crop duster a day or two before. They followed his directions to a nearby ranch, and their jaws dropped. Over a small shelterbelt of trees, a cloud of hawks swirled like a waterspout.

"There were two thousand birds, almost exactly," Brian recalled. "We were whooping and shouting and hugging each other, couldn't believe what we were seeing. I remember having this sense of—well, at the time it felt like closure, little did I know. Feeling like, okay, now we've got the full cycle, now we can begin to understand the ecology down here."

There was a lot to learn. Argentina has a long tradition of ornithologists and naturalists, dating back at least to W. H. Hudson, the author of *Green Mansions,* who grew up on the pampas in the mid-nineteenth century. But the pampas were, and in a sense still are, the frontier—the boondocks, the sticks, a place that most educated folk simply didn't bother to visit. Until quite recently, even some Argentine ornithologists questioned the assumption that Swainson's hawks occurred there regularly at all, despite the recovery of dozens of banded hawks in the pampas since the 1940s.

"With all the band recoveries, and all of the people who have built their careers on Swainson's hawks, no one had ever come down here. All the excellent ornithologists in Argentina—no one had ever bothered to come and record this phenomenon. If I wanted to, I could cleverly write a scientific article that almost mocks the people who have claimed to be interested in Swainson's hawk migration," Brian said. "But I'd have to include myself, too. One of the most important parts of this is how tenuous, how incredibly tenuous, the whole thing was. I'd like to say I came down here and used my skill, all my resources to find this problem. But it was sheer luck, the most incredible and fortuitous kind of luck."

With the discovery of the first large flocks, Woodbridge and his team decided to set up a huge study plot, taking advantage of the grid of dirt roads that checker the pampas at five-kilometer intervals. The Americans methodically drove their rental car up and down the almost endless number of roads, looking for concentrations of hawks. One morning at the western edge of the 6,400-square-kilometer area, just as the Swainson's hawks were leaving their roosts, they happened upon Estancia La Chanilao.

You can see Chanilao from a long way off—not the *estancia* itself, but the great grove of trees at its center, massive and sloped like Gibraltar, visible for miles in any direction. Every farm has one of these *montes* of Australian eucalyptus, but Chanilao's is huge, a kilometer long and half a kilometer wide, made up of straight trees sixty or seventy feet high. In the flat, open world of the pampas, Chanilao's grove is a local landmark, *monte grande,* as reassuring to a traveler as the sight of a familiar mountain.

Woodbridge spotted a huge mass of Swainson's there, perhaps 7,000 over the *monte* in the morning. "We just couldn't believe what we were witnessing, having been watching pairs intensely for twelve years. It was just phenomenal. We thought, We need to meet this *estanciero* and ask him if we can hang out and watch the birds."

End to end down the center of the forest, deep in the green shade, runs a packed-dirt driveway, with the smooth-barked trees standing like columns on either side. Brian, Karen, and Trent turned down it, out of the bright morning sun. Brian's voice fades for a few moments, lost in a difficult memory. "The road itself was littered with dead birds. It wasn't just one here and one there—they were touching each other all the way down. We'd seen a couple of road-kills [since coming to Argentina], but nothing like this. There were these flattened carcasses *everywhere,* and quite a few of them had been run over repeatedly, pounded down into the dirt, and in some places there were enough of them that they sort of interleaved with each other.

"We were just horrified. At first we thought somebody'd been driving down the road at night shooting them, but then we noticed other birds lying off to the side of the road, and although we couldn't tell at the time what the numbers were, it was clear there were quite a few."

They went first to the workers' quarters of the ranch, where the wife of one of the *gauchos* told them the birds hadn't been shot—they'd flown in one evening the previous month and simply died. But she knew little and said they should speak with Señor Lanusse, the owner.

The next day, Agustín Lanusse—a tall, spare man nearing forty, with penetrating eyes, a jutting black beard, and a reserved, almost formal manner—invited them into the house for beer and the bitter tea called *yerba mate* that is sipped from small gourds through ornate silver straws. An avid birder, Agustín had already begun to suspect a link between the dead hawks and pesticides he and other farmers were being encouraged to use to control grasshoppers—the mainstay of the Swainson's diet on the pampas. The day before the hawks had died, he said, one of the nearby fields had been sprayed with a chemical called monocrotophos, an organophosphate no longer used in North America.

With Agustín's permission, Brian and his friends started a macabre survey of the *monte,* trying to determine how many hawks had been killed. Slowly, painstakingly, they combed the forest, walking parallel transect lines within an arm's reach of each other.

"My first walk, from one end to the other, I had almost 200 dead birds," Brian said. "We were trying to find both legs, looking for bands, but the armadillos had really done a good mix-and-match on a lot of the carcasses. We tried to identify age and anything else like that, but mostly we just counted bodies."

Besides the old carcasses, there were a number of live hawks that had been injured in a violent hailstorm a day or two earlier. Some of them the scientists put up in trees, out of the reach of predators, in the hope that they might recover, but many more were beyond redemption, with smashed heads, broken wings and legs. These they euthanized.

"It was a gut-wrenching situation. It helped that I was with Karen and Trent, who were very close friends, so it was easy to feel what we were feeling and not play the scientist and have a stiff upper lip. But there were times we just quit, just had to get away from it."

After two days of laborious work, they had counted 714 dead Swainson's hawks. In fact, as they realized even then, the total kill was much worse, since they couldn't survey the dense fields of sunflowers on either side of the *monte*. The next year, when Brian returned, the fields had been plowed, and there were feathers and bone fragments everywhere. He now suspects the true figure at Chanilao was at least a thousand dead hawks, and probably much higher, plus another thousand that died on a neighboring farm.

As they picked through the stinking remains, Trent and Karen each found hawks that carried numbered leg bands. (Later, they learned, one bird had been banded in Colorado, the other in Saskatchewan.) But Brian made a far more personal and troubling discovery: White 05 Left, a female Swainson's hawk he had banded as a chick in the Butte Valley. A few years later, when the mature bird had settled down to breed, he'd captured her again, and around her left leg he'd placed a white plastic band, marked in black with the numbers 05. Now, far from the sagebrush of northern California, White 05 Left was dead.

"And that really hit home, really highlighted the link. This wasn't just a conceptual thing—these were birds from around my house," Brian said.

Back in the States, Woodbridge, Finley, and Seager wrote a paper for the *Journal of Raptor Research* about their discoveries, focusing on confirmation that a large number of Swainson's hawks do indeed winter in Argentina. Following advice from those wise to the politics of conservation, they said little about the poisoning deaths, hoping to return the next year for iron-clad proof. While they blamed the 700 dead hawks on an "unknown pesticide," and said only that the "description of the birds' symptoms suggested organophosphate or carbamate insecticide poisoning," they were already fairly certain that monocrotophos was the culprit.

Monocrotophos is an organophosphate, or OP—a family of potent chemicals derived from phosphorus acids and similar in structure to nerve gases like sarin. OPs do not linger in the environment, and they came into wide use after persistent chemicals like DDT were banned in the early 1970s. But they are highly toxic to vertebrates, especially birds, and monocrotophos is considered among the most virulent of all avian toxins. It was withdrawn in 1988 from use in the United States, where it was sold under the trade name Azodrin, but sales continued in many other countries, Argentina among them.

In January 1996, Woodbridge went back to Argentina, this time with toxicologist Michael I. Goldstein of Clemson University's Institute of Wildlife and Environmental Toxicology and Marc Bechard of Boise State University, an expert on Swainson's hawks. Based this time at Estancia La Chanilao, they spent a fruitless couple of weeks looking for signs of any more dead hawks. But the pampas, as I said, is a big landscape, almost all of it privately owned; the way to find something is to ask around, not to look. They asked, but turned up nothing before it was time to go home. Maybe, Woodbridge thought, the dead hawks he'd documented the year before had been an anomaly.

No such luck. That austral summer of 1996 was a drought year; by January crops were stunted and sparse, and the dry conditions favored swarms of grasshoppers. Farmers—feeling they couldn't afford to lose even a few hectares to insects—reached for the strongest weapon they could find: monocrotophos. The reports didn't start coming in until a few days after the team had returned home. People were calling Agustín, and he in turn was frantically calling the United States, looking for guidance. How should they take samples? How should they store them? What should they do?

Woodbridge and Goldstein hurried back to La Pampa, joining Argentine biologist Sonia Beatriz Canavelli. Once more, they began the grisly task of surveying poisoning sites, but this time on a much larger scale. One location alone contained nearly 3,000 dead hawks—the largest single pesticide kill of raptors anywhere in the world. Many of the dead hawks still had grasshoppers in their mouths and throats, so quickly had the poison struck them down.

In all, Woodbridge and the others conservatively estimated, between fifteen and twenty thousand Swainson's hawks were killed that winter just in the small area of La Pampa where they were working; heaven knows what the mortality rate was for the entire pampas. Almost all the dead hawks were adults, the core of the breeding population; at that rate, the scientists warned, the species' population would crash in just a few years.

Usually, when a wildlife tragedy like this comes to light, there's lots of hand-wringing and finger-pointing, but not a lot of immediate action. A few cynics predicted that nothing significant would be done until the Swainson's hawk was placed on the U.S. Endangered Species list, a process that can take decades. This time, though, scientists, government agencies, and private conservation groups moved with remarkable speed on several fronts. As soon as the 1996 kills were located, word spread across the Internet, generating tremendous grassroots concern. The U.S. Fish and Wildlife Service, the U.S. Department of Agriculture, the Canadian Wildlife Service, and the Argentine government cooperated on training and field surveys; the Argentines even released a gutsy report that showed how poorly their own institutions were regulating pesticides. The Argentine agricultural agencies cracked down on the use of monocrotophos and launched an aggressive education campaign aimed at farmers, while U.S. toxicological experts began training their Argentine counterparts in the exacting lab procedures required to monitor pesticides in the environment.

A few months before my arrival in Argentina, the American Bird Conservancy, an umbrella group of more than fifty conservation organizations, brokered an agreement with the chemical giant Ciba-Geigy and three other manufacturers, who agreed to pull monocrotophos off the market in certain areas of the pampas, add warning labels, and help educate farmers on safer ways to control insects. (Ciba, now known as Novartis, later agreed to entirely phase out production and sale of monocrotophos and five other OP pesticides, although others still make the chemical.) All in all, the international effort was one of the most rapid-fire, concentrated attempts to head off an emerging environmental crisis. People said it might serve as a blueprint for similar action in the future.

There was only one question, which we were hoping to answer: Would it really work?

At 3:45 A.M., the Argentine sky is awash with stars, the Southern Milky Way like confectioner's sugar splashed across the middle, and Orion standing on his head in the north. But I am blind to the spectacle, barely able to drag myself out of my sleeping bag and dress. We are caught in a collision between biology and culture. Agustín and his family, like most Argentines, keep a very European schedule, and the evening meal isn't served until ten-thirty or eleven o'clock at night; no one gets to bed before midnight. Yet the hawks rise before the sun, and at this time of year, the sun rises very early indeed. We Americans are falling further and further behind in our sleep each night.

In a cramped room beside the old ranch bunkhouse, a bleary-eyed Mike Goldstein snaps on a battery-powered lantern and growls, "Let's get moving. It'll be light in an hour." Twenty-nine years old, bearded, with a shock of dark hair framing a high forehead, Goldstein is a doctoral candidate at Clemson, and his job at Chanilao is to document whether the hawks are being exposed to chemicals—not only monocrotophos, but a range of potential toxins.

Mike grabs two brown lab mice by the tail and pops them into a flat, wire-mesh cage about a foot square and two inches deep, covered on top with dozens of small nooses made of fishing line. This is a *bal-chatri* trap, an ancient Asian falconer's device once made of loosely woven baskets and horsehair loops. If a hawk tries to grab a mouse, its toes will become harmlessly ensnared in the nooses. We will place a dozen of them in a field of newly sprouted sorghum alongside the *monte*, but with a thoroughly modern twist. Beside each, we will bury a small, cigarette-pack-sized radio transmitter, its whip antenna poking up through the dusty soil and clipped to the trap by a thin cord. Any movement of a captured hawk will activate the transmitter, allowing us to monitor many traps over a wide area of the ranch.

It's cool, with a chilly breeze. I walk through the dark *monte*, listening to the carpet-slap wing beats of frightened hawks overhead. With me are two of Agustín's teenage daughters from his first marriage—Candé, fourteen, and Agustina, seventeen, slim and dark like their father, both very pretty and both very capable hawk catchers. They show me how to arrange the traps in a huge grid across the field, mounding up the soil to make a little platform for each, burying the transmitter and a small, dumbbell-shaped weight nearby. As we finish, the sky is turning purple and the Milky Way fades to nothing.

Then there is a little time for *mate* or coffee back at the bunkhouse, listening to the throaty coos of spot-winged pigeons. Raptors start flying out of the woods in the half-light, but they are chimangos, not Swainson's hawks. Chimangos are caracaras, a group of scavengers related to falcons, although they look nothing like those sleek, aerial hunters. The size of crows, with long, naked legs, the dingy brown chimangos are ubiquitous in this part of the pampas; it is hard to look in any direction without seeing a few sitting on fenceposts, flapping across the sky, or squabbling over a road-killed lizard.

Shortly after daybreak, the first black silhouettes of circling hawks rise above the trees, gliding off toward the flat, bare earth of the newly plowed fields. For reasons that remain unclear, Swainson's hawks spend the first hours of daylight simply sitting on the ground, sometimes feeding on grasshoppers, but mostly just standing around. Perhaps they are waiting for the sun to warm the air and produce powerful thermals, so they can depart together to look for concentrations of grasshoppers. Maybe they just aren't morning people.

Goldstein refills his mug of coffee, flicks on a walkie-talkie–sized radio receiver, and begins to punch in a series of frequency numbers while waving the large, H-shaped antenna over his head. Immediately the air is filled with electronic chirps. "Okay, we've got birds in 6425, 6450, and 6475 in the south field, and"—he hits another sequence of frequencies—"6325 and 6375 in the east pasture." He grabs a couple of old T-shirts from a heap in the workroom and jumps in his pickup truck; Candé and I pile in excitedly as he skids around in the dirt driveway toward the field, sloshing coffee over everything. We vault the fence, flushing hundreds of hawks from the ground as we pound a quarter mile across the deep, loose soil, my breath coming in ragged rasps by the time I reach one of the hawks.

An *aguilucho* is snagged by two toes, sitting on its rump with its wings thrown wide and its free foot ready to lash out—trying to make itself look bigger and more threatening as I trot toward it. The flight feathers are silvery gray, marked with darker bars, while the wing linings and breast are creamy white. The head and chest are brown, with a white throat like a spotlight and eyes the color of strong tea. I gently lower a fold of the T-shirt over its head and swiftly grab its legs while it can't see. Once it is untangled, I wrap the bird securely in the cloth, then one-handedly replace the trap, fixing the matted and tangled nooses as best I can. I hurry, because still more hawks are gliding out of the trees, passing just over my head toward the far end of the field.

Meanwhile, back at the ranch (I've always wanted to write that), an assembly line is in operation, processing the dozen and a half birds we catch through the morning. While one person holds a hawk, another carefully washes its feet and legs with ethanol and catches the rinse, which will be analyzed back at Clemson for contaminants that might have been picked up walking in sprayed fields; the freshly cleaned toes are as yellow as dandelions.

Tiny snippets of feathers are clipped from the belly, wings, and back, and slipped into plastic specimen bags; these, too, will be tested for chemicals. After banding, another two-person team draws two small vials of blood from each hawk for analysis; then the bird is handed to another set of researchers, who weigh, age, and measure it. Finally, an hour after it was captured—and no doubt feeling as poked and violated as an alien abductee—the hawk is released in the yard of the ranch. As a final indignity, the pugnacious flycatchers that nest in the tall tree by the bunkhouse harass each hawk out of sight, strafing it until it clears the edge of the woods.

Goldstein's toxicological work is just one facet of the research going on at Chanilao. Every few days, the team receives, via computer from Idaho, the latest locations of two dozen Swainson's hawks fitted with satellite transmitters; the coordinates show that many of them stopped well to the north, where no one has been looking for hawks. (Several members of the team later drive into the northern pampas, confirming the presence of large flocks there.) Because even the simplest questions about how the hawks live on the pampas remain unanswered, another twenty have been fitted with standard radio transmitters and are being shadowed by a team of Argentines headed by Sonia Canavelli. What kind of habitat do they choose for roosting? When and where do they eat? Are the huge flocks an accident—a result of patchy food supplies, perhaps—or are the hawks somehow dependent on dense aggregations for their survival, as were the passenger pigeons to which they are often compared? How sensitive are they to human disturbance? Do the flocks wander widely, or do they remain in the same general area for long periods of time? Is there segregation by age, since most of the birds seen in La Pampa are adults, while flocks of mostly juveniles have been seen hundreds of kilometers to the east? There are far more questions than answers at this point, and any one of them could be crucial to the long-term survival of the Swainson's hawk.

In that sense, the *aguiluchos* are perfect examples of the challenges facing conservation biologists when dealing with migratory birds, for the most fundamental information is often lacking. Take food. There are mountains of studies on what Swainson's hawks eat in North America, but only guesswork and limited observation in Argentina. It is assumed they feed heavily on grasshoppers; after all, that's what got them in trouble with pesticides in the first place. But there is one report, from the eastern pampas south of Buenos Aires, of flocks of juvenile birds feeding heavily on swarms of migratory dragonflies. It may be that the hawks here are catching dragonflies as well, on the wing during the hot midday hours when they rise beyond the limits of vision.

If true, that would be reassuring, since the grasshoppers are under fierce assault by farmers.*

To answer the question of diet, we must collect food pellets, or castings, the indigestible remains of the bird's meal that are regurgitated once or twice a day. The ground beneath the roost trees is a mess—splashed with white guano and littered with freshly molted wing and tail feathers. Scattered liberally beneath the trees are dry, pinkish lumps the size of walnuts, many of them disintegrating into granular piles. Most of the hawk pellets I've seen over the years were made up of packed fur and bone fragments, but from what I can see, these are comprised entirely of the chitonous bits of grasshoppers—legs, mandibles, and pieces of the outer shell, graphic evidence of the *aguilucho's* insectivorous appetite. Someone—not me, thank God—will have the thankless job of picking the pellets apart under a dissecting microscope and identifying all the fragments, one by brain-numbing one.

On this hot midday, there is little bird life within the *monte*—little life of any sort, since these Australian trees give off a toxin that kills plants growing beneath them, so the forests are empty of any sort of understory except for young eucalypts. Most of the birds prefer the wide-open grasslands or the lower, scrubbier native acacias called *caldén*. But there is some activity. Chimangos scold and squeal, and spot-winged pigeons—bulky, gunmetal-blue birds—flap explosively out of the branches. Tropical kingbirds machine-gun twittering calls as they zip from tree to tree, flashing saffron bellies and wing linings. The kingbirds are migrants, but their pattern is the reverse of the Swainson's hawk's. During the austral summer the robin-sized flycatchers breed across Argentina, but with the coming of autumn they migrate north, spending the midwinter months of June and July foraging along rivers in the Amazon basin. Likewise with the fork-tailed flycatchers, which would be show-stoppers were they not so common on the pampas—slim birds the size of thrushes, black above and white below, with scissorlike tail feathers that

---

* In fact, the hawks have already made one shift in diet. Until the 1940s, the Swainson's probably fed most heavily on a species of migratory locust, which the Argentines call the *langosta*. This insect bred in relatively small areas of subtropical northern Argentina, then swarmed south by the billions across the pampas in hordes that looked like black thunderclouds—a bonanza of food for bug-eating hawks but a nightmare for farmers. With the introduction of DDT and other powerful pesticides after World War II, the Argentines were able to virtually exterminate the *langostas* by spraying their egg-laying grounds. It may not be coincidental that scientists in the United States and Canada first noticed an enormous crash in Swainson's hawk numbers around this same time. The hawk population never recovered to its former levels, and the survivors shifted from locusts to smaller, nonmigratory grasshoppers known locally as *tucuras*.

may exceed ten inches in length. Forktails are among the most conspicuous birds in Latin America, found from southern Mexico to Patagonia, and both the northern and southern populations are migratory, using northern South America as their non-breeding grounds.

As I walk down the rows of trees, a few Swainson's hawks lumber into the air, quickly catching thermals and riding high into the blue. I squat and shuffle along on my knees, searching for pellets, and my hand uncovers several old, matted feathers, still attached to a bit of dried bone. Gingerly, I brush away the covering of curled leaves and find the complete skeleton of a hawk, partially mummified by the dry climate, its feet still clenched in the paralyzing grip of organophosphate poisoning. Looking more carefully now, I quickly discover that I am kneeling in a boneyard—there are dozens, perhaps hundreds of skeletons in this corner of the *monte*, buried beneath just an inch or two of leaf litter, most still whole and articulated, held together by desiccated tendons. This is the same patch of woodland where Brian made his first gruesome tally two years before, where he found White 05 Left and more than 700 other dead Swainson's. I had seen the photographs of dead hawks, talked to Brian and others about their experiences, read the reports and news accounts. But only now, confronted with the weathered bones, does it really hit me just what is at stake.

Every day we wait in mingled hope and fear, carrying on the research while dreading the first call reporting a major kill. But as the weeks pass, Agustín's phone remains silent, perhaps for several reasons. Unlike the drought of the previous year, this season is bountiful and wet; the sunflowers are eight or nine feet tall, capped with platter-sized seed heads, and the soybeans and sorghum grow almost while you watch. Farmers are more forgiving of insect damage when there is so much to go around, and the wet weather has both reduced the number of grasshoppers and provided enough lush grass in road margins and waste places to keep them out of the fields. Few people, this year, are spraying for *tucuras*. By the end of the winter, in fact, the research team will have heard of only one small incident, involving just two dozen hawks— a dramatic change from the tens of thousands killed the year before.

It may also be that the new regulations restricting the use of monocrotophos and strenuous efforts to educate the farmers are paying off. Just as the growing season began, the government announced a partial ban on monocrotophos, which could no longer be used on alfalfa (a favorite grasshopper food) or sprayed to directly control *tucuras*. All well and good, but enforcement is tough in the empty pampas, and the chemicals being recommended by the government as replacements, including other organophosphates like dimethoate, are almost as toxic.

Many of the farmers had grown uneasy about monocrotophos even before news broke of the dead hawks; they blamed the chemical for killing their dogs, cats, and chickens. A publicity onslaught aims to galvanize that concern into action. Newspapers have been full of the story, flashy buttons with a conservation message about hawks have been handed out by the boxful, and videos about the dead birds have been played repeatedly on local TV stations. Everywhere I go in La Pampa, I see posters and signs about the hawks imploring farmers to use care with their chemicals. "Help an ally: Don't use Monocrotophos," says one, produced by a private ornithological group. "The grasshopper hawk ingests more than 100 *tucuras* per day," explains another, created by a coalition of pesticide makers. "It's an important ally in the control of the pest. Do not harm it." C. Hart Merriam, the nineteenth-century biologist, made pretty much the same argument. Merriam, who championed birds of prey at a time when it was fashionable—even among ornithologists— to condemn them as wanton killers, tried to quantify the agricultural benefit of a flock of Swainson's hawks he'd seen in Oregon. "Assuming that each hawk caught 200 grasshoppers a day and that there were 200 hawks, the daily catch would be 40,000 grasshoppers. At this rate these hawks would destroy 280,000 grasshoppers in a week and 1,200,000 in a month."

Everyone in La Pampa, it seems, knows about the hawks, and we're minor celebrities wherever we go. When we stop in town to buy more ethanol for the toxicology tests, the pharmacist drags us (and a fast-growing train of onlookers) through his adjoining house to see his hunting trophies, curling boar's tusks and stuffed red deer heads. An elderly woman running a tiny soft-drink kiosk in the equally small town of Alta Italia gasps when she finds out we are the *aguilucheros,* the famous hawkers, and rounds up the family to meet us. Out at the airfield, as we are trying to book a small plane for a survey flight, a van painted with the call letters of the local cable TV station billows up in a cloud of hurried dust. The interviewer is wearing an S.O.S. AGUILUCHOS button with a photograph of a hawk on it; so is his cameraman, a grim mountain of a fellow forested in wiry black facial hair, the video camera a matchbox on his beefy shoulders.

They arrange Mike Goldstein and me in the blistering sun next to a plane and start the interview, even though I'm not really part of the research team and my baby-talk Spanish is no match for this situation. I catch perhaps one word in fifteen as Mike fields the questions, but I laugh when everyone else laughs and try to look like something other than a dumb-ass *Norte Americano* who's going to keel over from the heat. Finally, a question for me: How do you like La Pampa? "A lovely land," I say in English, counting on Mike's fluent Spanish for translation, "a place of remarkable natural beauty, full of warm, generous, hospitable people who have shown that they have a deep concern about their environment." Mike, who is tired, hot, and grumpy, barks maybe four words in Spanish—but the right four, I guess, since everyone smiles in delight.

The next morning I head out alone, driving aimlessly down the dirt roads looking for birds, a train of dust in my wake. It is hard to remember where I am, hard not to merge the pampas with similar backroad drives in eastern Montana or the Dakotas. Since the sixth grade I'd carried a mind's-eye image of the wild pampas, gleaned from a dog-eared geography textbook—mustached *gauchos* on horseback, rock-tipped *bolas* twirling, ready for the throw, careening across the plains chasing flocks of ostrichlike rheas. But my first day in the region, driving west from Buenos Aires, hour after endless hour of fields of sunflower and sorghum spiked with windmills, I found it hard to imagine this as ever being wild land—as hard as it is to picture wolves in England or Iowa blanketed in bison. But sometimes my eye would slide past the regimented fields, past the squared-off groves of trees scattered like children's blocks, out to a horizon ruled like a straight-edge—a horizon blurred only by the flowing humps of bunchgrass, beneath a sky as wide and expansive as the laws of geometry allow. In that instant I caught a glimpse of the old pampas, of a time of Indians and rheas. But then the image evaporated into the present, and the only rheas I saw that first day were standing in the shade of a roadside Coca-Cola billboard.

This morning, I drive past close-cropped pastures, where families of burrowing owls sit on fenceposts like little gnomes and pump their heads in agitation as I clatter by. A tinamou, brown and grouse-like, scuttles across the road with its head down, a spy hunched inside his trenchcoat trying to be inconspicuous. In a newly plowed field, hundreds of American golden-plovers move desultorily, gray birds with pale eyebrows, a far cry from the brassy plumage they wore a few months ago in the Arctic. Eventually I come upon a complex of small lakes and marshes, dotted with flocks of white-faced whistling-ducks and stately Chilean flamingos. On a wireless old utility pole sits an aplomado falcon, which appears to be asleep. A tight knot of white-faced ibises jostle among themselves, big, black birds with a maroon sheen in the sun, long curved bills plunging into the bright green marsh grass—a confusion of angular legs and knobby knees.

Clustered at one end of a shallow pond are thirty or forty greater yel-lowlegs, slim, gray shorebirds. Reaching delicately into the water with their slender, knitting-needle beaks, they pause for a fraction midway through each stride, like a snapshot—weight on one straight leg, the other cocked up and back; poised, self-contained, trim. I see yellowlegs during the spring and fall migration in Pennsylvania, where the old duck hunters called them "tell-tales," because of their suspicious nature and piercing alarm cries that alert other birds. When the falcon rouses itself, shaking its feathers and looking around, the yellowlegs bunch up nervously, calling *dear! dear! dear!* The ibises raise their heads with a snap, like impala that have scented a cheetah; they look like a frieze from an Egyptian tomb, a two-dimensional arrangement of nervously bobbing heads and drooped bills. Then the falcon smoothes its feathers and raises one foot to its belly, blinking in the warm sunlight, and the lesser birds relax as well.

Besides all the resident birds milling about, like South American stilts and white-winged coots, there are six or seven species of migrant shorebirds in the marsh, including white-rumped sandpipers, scurrying by the dozens between the legs of the flamingos, and upland sandpipers sitting on a slumping fence that edges a neighboring pasture. Shorebirds travel so far, and spend so much of their lives on the wing, that they were dubbed the "wind birds" by writer Peter Matthiessen. "The white-rumped sandpiper, which flies 9,000 miles twice each year in pursuit of summer, is only exceeded in the distance of its north-south migration by the Arctic tern, and the golden plover far exceeds the tern in the distances covered in a single flight; it is thought to travel well over 2,000 miles nonstop on both its Atlantic and Pacific migra-tions," he wrote.

While Argentina's grasslands are vital for many species, its seacoasts attract still others. The race of red knots that nest in the islands of the Canadian Low Arctic winters along the coast of Patagonia and Tierra del Fuego, joining sanderlings, semipalmated sandpipers, whimbrels, long-billed dowitchers, ruddy turnstones, and several other species. And because I like to believe in miracles and lost causes, I am keeping my eyes open down here for an Eskimo curlew, although my chances of seeing a band of wild pampas Indians may be greater.

The Eskimo curlew was, by all accounts, one of the most abundant land birds in the Western Hemisphere, numbering in the millions through the middle of the nineteenth century; people called it the "prairie pigeon" because of its sky-blackening flocks. Along with the American golden-plovers, with which it associated, the curlew made an elliptical migration each year—down the western Atlantic from eastern Canada to northeastern South America and on to Argentina, then up the middle of the continent in spring, across the Gulf of Mexico, and back to its breeding grounds in the western Arctic. Curlews were fat and guileless, circling back time and again when the flock was fired upon, and market hunting virtually wiped them out by the 1890s. "The gentle birds ran the gauntlet all along the line, and no one lifted a finger to protect them until it was too late," wrote Arthur Cleveland Bent in 1927. "They were so gentle, so confiding, so full of sympathy for their fallen companions, that in closely packed ranks they fell, easy victims of the carnage."

The last specimen collected in the United States was a single bird killed in Nebraska in the spring of 1915. In fact, the species has been declared extinct several times, but in 1962 one was photographed on the Texas coast, and the next year one was shot in Barbados. Since then, presumably reliable reports have dribbled in, and the hopes of birders and conservationists have been periodically raised, then dashed. The most hopeful report was a flock of twenty-three in Texas in 1981. In 1987, Canadian Wildlife Service personnel reported seeing two Eskimo curlews in the Arctic, although the sighting was officially dismissed as probably being whimbrels, a slightly larger but almost identical species of curlew. In 1993, birders were thrilled to hear that someone had seen an Eskimo curlew in Argentina; but no, an investigation later concluded, the report was an error based on a hypothetical record from several years earlier.

And so it goes. Maybe a handful of curlews still make their brave flight between the poles every year—one expert recently put the global population at fifty, a figure that surely ranks as one of the wildest of all wild guesses. I spotted no curlews during my weeks in the pampas, but I did finally get to see one after I got home. In the Academy of Natural Sciences of Philadelphia, a few floors above the museum's display areas, I followed David Agro through a maze that smelled of mothballs. The ornithology department is a rim of offices surrounding a large, central space filled with floor-to-ceiling storage cabinets. The cabinets, in turn, are filled with 165,000 carefully preserved bird skins, the result of two centuries of scientific collection and study.

Agro, who manages the bird collection, unlocked a cabinet at the end. "This is where we keep the endangered and extinct birds, and all the type specimens," he explained. Type specimens are the Adams or Eves of taxonomy, the first of their species to be meticulously described for science—the individuals that serve, in a sense, as the official representatives of their kind. Agro pulled out a wide drawer; I saw minute hummingbirds, parrots, and parakeets, tropical barbets. Wrong drawer. The next slid out, and I was confronted with row upon row of passenger pigeons, blue-gray bodies with a wash of wine across the breasts, their eyes vacant with wisps of escaping cotton. Most of the labels, written in spidery script, were dated from the 1870s, forty years before the pigeons' extinction.

A third drawer rolled silently open. There, with extinct Carolina parakeets shot and stuffed by John James Audubon, lay a row of Eskimo curlews. Agro picked one up, holding it delicately between his thumb and middle finger, a bird not much larger than a jay, with a curving bill, its legs neatly crossed and tied together with the thread of the specimen tag. "This was the last Eskimo curlew ever collected, the one that was shot in 1963 in Barbados," he said. "Shot by a hunter who didn't know what it was, and somebody salvaged the specimen." When I was finished looking, he replaced the bird in its row, next to the green-and-orange parakeets, and rolled the drawers back into the darkness. I found myself wondering how unintentionally symbolic his actions might be. Was the curlew, like the pigeon and the parakeet, a thing of darkening memory? Or were there still a few out there, somewhere, still making the great arc from tundra to pampas each year? And if we cannot stem the chemical tide on the pampas, will someone else one day look with grief and longing at the dry skin of a Swainson's hawk in just such a drawer, the last remnant of lost multitudes?

On one of my last days in Argentina, several of us jam in the truck during a predawn thunderstorm and drive north a couple of hours, to a part of the pampas that still retains much of its original character, a mixture of bunch-grass prairie and native *caldén* acacia forest.

We slew our way out a muddy track into the savanna, through scattered groves of humped, wind-shaped trees and faded, waist-high grass. The difference between this and the farmland around Chanilao is striking, not only in appearance, but also in the much greater diversity of bird life here. Flocks of monk parakeets chatter past us, green, gray, and hyperactive, and monjitas—white, thrushlike birds with black wings and tails—sit on the tops of trees like ivory carvings. This, I find myself thinking, is the pampas of my childhood imaginings—and as we round a bend in the nearly invisible track, we find ourselves face-to-face with a flock of rheas.

Rheas are ratites, the same primitive, flightless order of birds to which ostriches belong, and they share with their more famous cousins the same long-necked, long-legged body plan. They say that birds are kin to dinosaurs, and after watching rheas, it is hard to disagree. They move with a saurian grace, heads rising on supple necks like periscopes, their bodies gray and shaggy, as though wrapped in German-shepherd fur instead of feathers. There are eight in the flock, one a male with a black neck, all moving slowly into the wind-rippled grass like swimmers breasting waves.

There are few places left in the pampas where such timeless scenes exist, the animals and habitat still in relative harmony. If the *aguiluchos*, the rheas, and all the other creatures of the pampas, migrants or resident alike, are to survive into the next century, then a much broader effort will be needed. The steps taken so far—banning a single chemical in a small area for a particular purpose, educating people about the requirements of an individual species of hawk—are fine beginnings, but they are insufficient of themselves.

I said at the beginning that this story was one of hope. The hope, however slight and tenuous, is that the plight of the *aguiluchos* seems to have galvanized not only North Americans but the Argentines as well. They are slowly awakening to what they have lost in the pampas, the land that is their own mythic frontier, their own Wild West.

The only way to safeguard part of the pampas—the migratory birds—is to preserve the functioning whole. It is a lesson that has been learned at great cost in North America, and the Argentines are coming to understand it, too—along with the difficulties it presents. There is almost no public land in the pampas, no huge national parks and preserves that can serve as a protected core, so the onus is on private landowners. They are taking the first, halting steps toward a new relationship with the land—exploring approaches like integrated pest management, which relies less on chemicals and more on biological controls, and on land-use planning that looks beyond crop yields to issues like biodiversity and habitat restoration.

The rheas are more nervous now, the male glaring balefully at us while the hens drift toward a thicket of *caldén*. He fans open his short, useless wings, lowers his head, and charges toward us a step or two in threat, then wheels and trots after his harem. The acacia branches close behind them, and we are left alone in the sea of grass.

9

When Anywhere Is Better than Home

This was not the New England of travel posters. There were no pastoral summer hills, lush with wildflowers and rustic farms, no snow-wrapped mountainsides exploding with photogenic skiers. There were no smiling couples at quaint country inns or weathered sugar houses tucked in groves of ancient maples.

This Vermont is rarely acknowledged by tourism bureaus—griddle-flat, wind-raked, achingly cold. The sky was thick with overcast; it was midday, but not much brighter than twilight. Fields rolled away to every side, a thin skiff of old snow corrugated with endless lines of bare, frozen furrows, which bled brown dust to stain the white. Gap-toothed rows of elm trees marked the edges of the fields—many of them dead, Dutch-elm disease giving them an amputated look, reduced to their heaviest limbs; others were simply bare with the dormancy of winter. A barbed-wire fence, its posts uneven, lurched out of sight into a small draw, and the wind made the rusted strands hum. Even the sound they made was thin and miserly.

I blinked away tears from my stinging eyes and bent over to peer through my spotting scope. What had looked like an old squirrel's nest in a distant tree was a hawk perched on a thin branch that bent beneath its weight. Aside from a tiny spot of yellow at the base of its beak, the raptor was the color of burnt wood, even its legs, which were feathered in black right down to the orangish toes that clamped hard on the branch.

The hawk hadn't moved in the ten minutes I had been watching, although the cold had me shuffling from foot to foot in a futile attempt to stay warm. It was staring down at the stubble of wild grass at the edge of the field, a narrow swath missed by the summer tractors. It stared with patience born of instinct and an intensity born of hunger, as if it could conjure something edible from the empty grass by force of will.

And it did. The hawk dropped without preamble, falling most of the way before it opened its wings, flaring in the last moment as the tangerine feet shot down and out. Wings and tail fanned on the ground, forming a semicircle of silver-and-soot barring; the hawk lowered its head, biting, then relaxed, standing upright and rattling its feathers back into place. It returned to its perch, the limp form of a vole in its delicate feet, and began to feed.

If you have wings, it makes sense to use them—to reach if not a warm tropical haven, then at least the temperate climate of the Gulf Coast or the Southwest. The damp, bone-deep cold and snow of the Lake Champlain basin in western Vermont is nobody's idea of a pleasant winter getaway. Yet for this rough-legged hawk, probably born on the Ungava Peninsula or Baffin Island more than a thousand miles to the north, the desolate fields and wind-scraped pastures were a great improvement. At least the sun rises here in the morning, which is more than you can say about the tundra. When you're facing the long, perpetual darkness of an Arctic winter, anywhere's better than home.

In summer, the Champlain Valley is rich farmland, fields of corn and pastures full of Vermont's trademark Holsteins. But winter seals it in snow, a flat-bottomed channel for the north winds that burst out of Quebec, funneled between the Adirondacks and the Green Mountains. With nowhere else to go, the winds scour the hundred-mile-long valley, sending ice crystals hissing along the drifts like a low, painful fog.

Roughlegs aren't the only birds that come here each winter. Over the next few days, as I drove the icy roads that web the valley on the Vermont side of Lake Champlain and hiked along frozen streambeds, I found many refugees from the north—snowy owls that sat on fenceposts, rotund snowmen with eyes of glowing yellow instead of coal; flocks of Lapland longspurs, snow buntings, and horned larks, scuffing undigested seeds from strips of cow manure spread like dirty paint on the fields. When the small birds took flight, they looked like leaves blown tumbling on the wind. In the woods along nearby Dead Creek, I heard the musical chatter of purple finches and watched tiny pine siskins rise and swoop on wings that flashed yellow with every beat. Up in the hills, where the forests were thicker, there were flocks of evening grosbeaks, crossbills, and red-breasted nuthatches, natives of the vast spruce belt that girdles Canada.

A surprising number of migrants travel only to southern Canada or the northern United States, regions that receive the full blast of winter. Some are regular migrants, such as the rough-legged hawk, which completely vacates its Arctic breeding range each year. The hawks winter in a wide band across the middle latitudes of the United States, sometimes reaching south of Kentucky in the East and into the deserts of the Southwest. Most, though, hang north, from the tidal marshes of the Chesapeake to the grasslands of the Dakotas, Montana, and Wyoming. Exactly where they fetch up depends on food supplies and snow depth, which helps conceal the small rodents upon which they feed almost exclusively.

Other winter birds are much more fickle, as those who enjoy watching their back-yard feeders know only too well. Many of the common feeder species—titmice, downy woodpeckers, chickadees, cardinals—are permanent residents, living out their lives within the same patch of forest and yard. Others are a mix of resident and migratory populations, like blue jays and song sparrows, whose numbers rise and fall through the seasons but which are usually present year-round. But the birds that really brighten the winter are the northern finches, which follow an erratic, wholly unpredictable pattern of migration known as irruption, which makes their appearance something of an event. (This is not a misspelling, incidentally. The departure of the birds from their normal range is an *eruption,* but their arrival somewhere else is the reverse—an *irruption,* a bursting in.)

The phone rang the other evening, just as I was sitting down to dinner. My friend Tom was on the line, his voice excited. "Have you heard? There are grosbeaks everywhere. I had sixteen at my feeder today, and there are reports from all over the county of grosbeaks, purple finches, siskins, even redpolls." This is not as odd a conversation as it might seem. Tom's the local compiler for the state ornithological journal and the guy everyone calls when they see a notable bird. He's quick to share the news, so I'm used to receiving such breathless updates.

I was jealous; I had seen no evening grosbeaks at my feeder, but I would gladly have traded a sack of my drab house finches and juncos for a mob of them. Evening grosbeaks are dramatic, the kind of creatures that make non-birders stop in their tracks and rub their eyes in disbelief. A bit smaller and pudgier than robins, the males have bright yellow bodies that shade to dark brown on the head, except for golden eyebrows. The pale yellow beak is heavy and conical, the wings black with large patches of white. The females look as though they were painted with watercolors, wet-on-wet, so the grays and buffs and lemons melt together. They are noisy, quarrelsome, and eat like pigs. Birders love them.

Unfortunately, grosbeaks do not appear predictably each winter. Some years they are content to stay on their breeding grounds, which stretch through New England and southern Canada to British Columbia and down through the Rockies. These winters are also usually bereft of other northern birds—pine siskins, purple finches, pine grosbeaks, white-winged and red crossbills, common and hoary redpolls, red-breasted nuthatches, boreal chickadees, Bohemian waxwings, and, in the West, Clark's nutcrackers. (Black-capped chickadees also irrupt south, but their arrival is usually masked by resident populations of the same species.)

The sudden influx of northern birds is often considered a sign of a bad winter ahead, but their wanderings are tied to food, not weather. Most of these birds depend for part or all of their diet on the seeds of conifers, especially pines and spruces; given plenty of food, they are quite capable of surviving both cold and snow. The trouble is, they must bank on a very unreliable food supply. Some years, tree branches sag beneath the weight of their cones, but in other years the crop is poor, and occasionally it is all but nonexistent. The good years and the lean ones often occur synchronously across vast expanses of northern forests.

This capriciousness is called *masting*, and while it is probably linked to how good or bad the previous growing season was, it is also an elegant strategy by the conifers to increase the survival of their seeds. (Many other trees also mast, including oaks.) The lean years reduce, by starvation and lowered reproduction, the number of so-called seed predators—birds, mammals, and insects that feed upon the seeds. During a heavy masting year, the trees flood the landscape with nuts and seeds, far more than the animals can possibly find and eat. Well fed and healthy, the animals prosper for one year, producing an abundance of young—but then the trees usually pull the rug out from under them with several poor cone crops in a row, as they recoup the energy reserves needed for another round of fruiting.

It is during these bad mast years, often exacerbated by high populations from the preceding good season, that seed-eating birds come south. It is not a migration in the usual sense of the word; unlike most migratory songbirds, which have traditional wintering ranges—even individual wintering territories—irruptive species do not have a specific destination in mind. They simply wander the landscape, looking for pockets of food. In the past that would have meant stands of more southerly conifers like white pine and hemlock, or hardwoods like tuliptrees that bear nutritious seeds. Today, a great many of them wind up at back-yard bird feeders, methodically cracking sunflower hulls in their massive beaks; no wonder the evening grosbeak's genus name is *Coccothraustes*, "seed-shatterer."

While this is delightful for bird and birder both, their appetites can be prodigious, and the expense can be considerable. Some years back, a friend of mine was fortunate to have more than 150 grosbeaks at his feeders all winter. It was a stunning sight, a seething mass of cocoa and yellow, so raucous that from inside the house you could hear the birds calling and screeching, like parrots. When a hawk cruised by, they would explode in every direction; the effect was vertiginous, like toppling into a field of sunflowers. But this specta- cle came at a price. My friend had to shell out for a couple of fifty-pound bags of seed each week to keep them fed, and by the end of the winter—having purchased something like one and a half *tons* of food—he was more than happy to see them go. "It's fine for you," he grumped during a visit, when I was enthusing over the birds. "You don't have to feed the damned things. They're not grosbeaks, they're gross-pigs."

Evening grosbeaks are something of a new development in the East, where I live. Early naturalists along the Atlantic seaboard did not know the bird, which wasn't discovered by scientists until 1823, northwest of Lake Superior. But as whites were moving west, the grosbeaks began moving east. By the winter of 1854 they were spotted near Toronto and began nesting in Ontario by 1920 and Newfoundland by 1961; they have also spread as far south as northern New York. Every few winters, they would irrupt south and east, flooding the back yards of America.

The reason for this eastward shift may lie, in part, in the settlement of the Canadian prairies, once a natural barrier to the forest-dwelling grosbeak. In the late nineteenth century, farmers there began planting millions of trees as windbreaks and fencerows, and they were especially fond of using ashleaf maple, a Midwestern species also known as box elder, whose seeds and buds are favorite foods of the grosbeaks. The result: stepping-stones across the prairies, allowing the grosbeaks to infiltrate the land of milk and honey. As further incentives, the eastern boreal forests are subject to massive infestations of a caterpillar known as the spruce budworm, a perfect summer food for grosbeaks and their chicks; also, conifer forests that had been clear-cut grew thick with pin cherry and chokecherry, whose pits are another good grosbeak food.

The first major grosbeak irruption into the East occurred during the winter of 1889–90, with smaller flights during the first half of the twentieth century.

Then, during the winter of 1962, grosbeaks irrupted in enormous numbers, and from then through the late 1970s they invaded the eastern United States every couple of years. Interestingly, the migrants did not necessarily come from the newly established eastern population; banding studies show that grosbeaks and purple finches from northern Michigan and beyond often migrated to the Atlantic coast—one of the few cases of east-west migration in North America. Nor did they show much year-to-year consistency in where they pass the winter. In one case, grosbeaks banded in a single location turned up, in subsequent years, in seventeen states and four provinces. Another bander ringed more than 14,000 grosbeaks and other winter finches at his feeders over an eighteen-year period, and not a single one ever returned.

In a big "finch year," as birders call these irruptions, the grosbeaks do not arrive alone; the same food shortage often sends the rest of the cast of northerners along, too. Among the most common are purple finches, the males looking as if they'd been dipped in red wine, the females striped with brown and white; and the delicate pine siskins, a kind of forest goldfinch covered with dark streaks. But sometimes the occupying army includes common redpolls, pale, straw-blond finches with red caps; and red-breasted nuthatches, diminutive, tree-creeping birds with orangish undersides and black-and-white-striped heads. There might be pine grosbeaks, chunky and red-orange, or white-winged and red crossbills, two species with the oddest beaks in the bird world, their tips twisted like fingers crossed for luck, an adaptation to opening conifer cones.

This twenty-five-year period of regular grosbeak invasions coincided with a boom in the popularity of birdwatching, and I have no doubt that the explosion of interest in feeding birds was fueled, at least in part, by the regular presence of such vividly beautiful creatures just outside the kitchen windows of America. People became a little bit spoiled—and we've been complaining bitterly in recent years, because the well eventually ran dry. Starting in the early 1980s, grosbeaks once again became scarce winter birds south of the Canadian border, a dramatic change from just a decade before.

American Birds, the journal of the National Audubon Society (and the magazine's successor, *Field Notes*), tracks the continent's bird life with the help of thousands of experienced volunteers who compile their observations on a seasonal basis. Through the 1980s and '90s, the litany about winter finches was consistently gloomy: "another poor showing this year" (1984); "generally another poor year for winter finches" (1985); "below average numbers" (1987); "poorest year in a decade for winter finches" (1988); "strikingly conspicuous in [their] absence" (1990). In 1991 the journal noted with an air of resignation, "We'll have to re-think what constitutes a 'good season' for these birds."

The reasons for the prolonged absence of major irruptions aren't clear; some people have speculated that as spruce budworm outbreaks declined in eastern Canada, many of the grosbeaks there either died off or shifted west again. But that doesn't explain the persistent dearth in invasions of other northern seed-eaters. Purple finches may have been elbowed out by the explosive spread of their western counterpart, the house finch, which was accidentally introduced to the East. But the real reason may be too much of a good thing. To borrow a term used by waterfowl biologists, the birds may be shortstopping in southern Canada and the northern United States, where back-yard bird-feeding is every bit as popular as it is to the south. "A flock of finches that comes out of the boreal forest, looking for food, can hardly travel any distance . . . without being intercepted by a kind of beneficent Maginot Line of feeders," wrote Kenn Kaufman, one of the continent's top birders, in 1992. "And if the first feeder is crowded, or the first ten feeders, more will be waiting just down the road."

Irruptions are among the most dramatic of wildlife events, and they occur around the world, not always with boreal or Arctic species. Among the most famous were the repeated invasions of Europe by Pallas' sandgrouse, a bird normally found on the arid steppes of central Asia. Between 1863 and 1908, the colorful sandgrouse repeatedly irrupted west as far as Britain, and even stayed to nest there. Sandgrouse irruptions still occur in Europe from time to time, although at less frequent intervals and involving far fewer birds, probably because their steppe habitat has been greatly altered and their populations are reduced.

In the past, tracking irruptions has been difficult, given the widespread, ephemeral nature of such events. But Project FeederWatch, a cooperative study run since 1987 in the United States by the Cornell Laboratory of Ornithology, and in Canada by the Long Point Bird Observatory, has given ornithologists a much clearer picture of what happens during irruptions. Every two weeks from November through April, nearly 7,000 volunteers across North America record the species of birds visiting their feeders, and the number of individuals of each type. Among other benefits of this grassroots effort, one may look at irruptions on a continental rather than a local scale.

For instance, while it was long recognized that pine siskin invasions occurred roughly every other year, the FeederWatch data clearly showed a similar biennial cycle in common redpoll migration, a pattern probably linked to the every-other-year failure of birch seeds in the Arctic and subarctic. The irruption of 1993–94 was unusually large, and the network of observers gave biologists a rare peek at its timing and nuances. The first waves of redpolls moved south in November of that winter, with more arriving each week through January, followed by a period when flocks grew larger as the redpolls massed together in places—such as bird feeders—with a good supply of food. At first, the redpolls focused on what researchers considered high-quality sites, feeding stations in rural locations with trees for cover and little disturbance by humans. As the invasion continued and redpoll numbers swelled, however, competition forced some into less desirable locations, urban and suburban yards with lots of people, pets, and fewer trees.

FeederWatch observations also showed a curious linkage between purple finches and American goldfinches, the former a forest dweller and a classic irruptive finch, and the latter an open-country bird that is nomadic in its winter wanderings. No one ever suspected these two dissimilar species might be connected. Yet the study showed that, year after year, purple finches and goldfinches are most common in the same parts of the continent at the same time—most abundant in the Southeast one year, say, and in the Northeast the next. The reasons for this synchrony are unclear, but may be tied to wild food crops—a mystery to be solved in winters yet to come.

Some years there are lots of siskins, other years lots of nuthatches; many years there isn't much of anything coming down from the north, leaving birders with nothing to combat post-holiday depression. But once in a very great while, the stars align and the boreal floodgates open. In the fall of 1997, birders in the Northeast and the upper Midwest began to suspect great things were afoot; not only were large flocks of grosbeaks, siskins, and purple finches being reported, but also almost unprecedented numbers of red and white-winged crossbills, two species rarely seen south of Canada. By October of that year, it was clear that something quite remarkable was under way—a "super-flight," as such mega-irruptions are known, the first in the East in almost thirty years. Experts suspect it was caused by an unusual coincidence—a failure of not only the spruce cone crop but also pine cones and birch seeds, the latter the favorite food of redpolls.

The true dimensions of the last superflight, in 1972, became clear only months after it ended, when observers submitted their reports to their local and regional bird journals. This time, the Internet made it possible for scientists and birders to track the irruption as it occurred. BirdSource, a cooperative online project of the Cornell Lab of Ornithology and the National Audubon Society, encouraged birders across the country to log on to its Web page and type in their reports—species, number of birds seen, nearest ZIP code. Thousands did just that, adding reports at the rate of twenty or more per hour, providing priceless information on the whereabouts and abundance of each of the nine species taking part in the superflight. The BirdSource project was a perfect collaboration between ornithologists and "citizen scientists," as Cornell likes to call enthusiastic birders who take part in research projects like this. With the irruption data, BirdSource researchers were able to create animated maps, updated every week, showing the movement of the phenomenon—spilling south out of Canada, rippling quickly across the Northeast and Midwest, down into Kentucky and Virginia, finally reaching its high-water mark in the Deep South, in places like Alabama, where boreal birds almost never go.

When the first big flocks of northern finches started pouring through the mountains near my home, a buddy called me one night, excited because he'd seen more than a hundred white-winged crossbills that day at a local hawk-watch. "Lay in extra wood," he said with a laugh, "I think it's gonna be a long, cold winter."

But the birds proved once again that they move according to the dictates of their stomachs, not the weather. That was the mildest winter ever recorded in the East.

I was cruising slowly down the street of a large new subdivision, recently carved out of fertile Amish country farmland in southeastern Pennsylvania. The directions I'd been given seemed clear the previous night on the telephone, but now, in the maze of identical, cookie-cutter houses, I was hopelessly confused and growing frustrated. Bundled against the cold, a woman was walking her terrier down the street, so I coasted to a stop beside her and rolled down my window.

"Excuse me, ma'am, but have you heard about the snowy owl that's hanging around here somewhere?" I asked, smiling as cheerfully as I could so she would assume, correctly, that I was a harmless kook, rather than a dangerous kook.

"Oh yes! I just saw it a few minutes ago—over there on Chestnut." She pointed to another section of the development. "It's so beautiful, and it let us walk right up to it! It's sitting on the chimney of one of the houses. You can't miss it."

She was right. There *was* an owl on a chimney on Chestnut Street—and no wonder it let her and Fifi walk right up to it. It was a plastic owl decoy, the kind people use to scare birds out of the strawberry patch; it wasn't even painted white. But when I turned back out of the street, the woman saw me, grinned, and waved, so I gave her a big thumbs-up; no sense in bursting her bubble.

A short time later, on the barn roof of a neighboring farm, I finally found the real owl—a white ghost squinting in the morning sun, with just a faint barring of black on its wings and breast. It sat unmoving, like an ice sculpture, for more than an hour, but I soon had to share the view with others, as dozens of birders arrived, peering from their cars with spotting scopes and binoculars. When the television news crew arrived, I decided it was time to leave, but the owl was still perched in what looked like quiet meditation, ignoring the growing gaggle of humans just fifty yards away.

If finch irruptions are a pleasant winter diversion, an invasion of snowy owls—huge, beautiful, and shockingly tame around people—can be a real event, one that often makes the front page and the top of the local evening news. Snowies have always been the textbook example of an irruptive species—content most years to remain on their breeding range, which rims the Arctic Ocean, but moving south of the Canadian border in large numbers every three to five years. And every so often, perhaps once in a decade or two, something much more remarkable occurs—a true invasion, with these spectral owls showing up as far south as Texas and Georgia.

In fact, several species of northern owls and hawks make winter incursions to the south. Early on, naturalists noticed that these movements had a curiously cyclical nature—that large numbers of goshawks, for instance, moved south every nine to eleven years. In New England, goshawk irruptions occurred in 1886, 1896, and 1906, and in 1926, 1935, 1945, 1954, 1962, and 1972. Snowy owl invasions coincided with goshawks in some years (1926 and 1945, for instance) but not in others, but they always lined up with heavy flights of rough-legged hawks.

Seeds aren't the only food that waxes and wanes in the Far North; a number of small mammals experience population cycles that ripple up through the food chain to predators like owls and hawks. Every three or four years in the Arctic and subarctic, the number of lemmings—small, short-tailed rodents—reaches plague proportions. During such "lemming years," there may be fifty or more per acre of tundra; then the population crashes, and it becomes hard to find even one. It isn't (popular myth aside) the result of migration ending in mass suicide by jumping into the ocean; no one knows what causes the booms and busts, although predation, overcrowding, reduced food, disease, or some internal reproductive trigger within the lemmings themselves have all been suggested.

Whatever the cause, the lemming cycle has a profound effect on all the predators of the Arctic, especially the birds. Snowy owls feed heavily, at times almost exclusively, on lemmings, and when the rodent population takes a nosedive, the owls must turn to alternate foods. Adults with well-established territories and more experience hunting may be able to get by without leaving, shifting to alternate prey species like ptarmigan, but younger birds will be forced south of their usual range. In especially bad years, most of the owls, young and old, will have to leave the Arctic, but even so, researchers have found that the different age and sex classes don't all end up in the same places. Adult females, which are socially the most dominant, are likely to stay the farthest north, followed by adult males a bit farther south, then immature females south of them. Immature males, at the bottom of the pecking order, are the real vagabonds, the group most likely to show up farthest from the Arctic.

Rough-legged hawks are also rodent specialists; although they are the size of red-tailed hawks, their feet are positively tiny, perfect for snatching voles and lemmings. This penchant for small mammals explains the synchronized nature of snowy owl and rough-legged hawk irruptions, timed with the rise and fall of the lemming cycle. That seems simple enough—but as scientists have studied snowy owls more closely in recent years, they've realized that the picture is far more complicated than once believed. The idea that rodent cycles rise and fall in lockstep over vast areas of the tundra now appears to be wrong. Instead, lemming populations form a mosaic—some areas in a peak, others in a trough, with the owls moving nomadically across the Arctic, breeding wherever their food is abundant.

What's more, banding studies in the northern Plains showed that many snowy owls migrate there predictably each winter, often returning to the same place year after year. In the East and West, however, the owls display the classic irruption pattern, which makes little sense if lemming populations aren't going through ecosystem-wide booms and busts. Perhaps, some biologists argue, the trigger is weather, especially snow depth that shields the burrowing mammals from hunting owls—although *that* doesn't explain the metronomic regularity of the owl irruptions, since weather is far less cyclical.

Once a tidy natural history tale, the linkage between snowy owls and lem-mings is growing increasingly messy, and increasingly interesting.

The reasons behind other raptor irruptions are equally murky. Boreal owls, northern hawk-owls, and great gray owls also periodically invade southward, but much less frequently, and not with the predictability seen in snowy owls. Great gray owls, which live like phantoms in the conifer forests of Canada and the Rockies, made limited, irregular irruptions from the early nineteenth century through the 1960s, then began a series of spectacular invasions over the next several decades.

During the winter of 1978–79, great grays burst into southern Canada and the Great Lakes states, setting off a stampede of birders to see this huge, mys-terious species. Then, five years later, an even heavier irruption of the north-ern owls occurred, focused east of the Great Lakes; southern Ontario alone had more than 400, considerably more than had been seen in all of eastern North America in the previous invasion. Two hundred fifty were recorded in Quebec, 140 in the Great Lakes states, and one even made it as far south as Long Island—and those, of course, are merely the ones recorded by birders. Most of the owls had probably never seen humans before, and they were utterly guileless and tame, often snatching pet-store mice from people's hands or plucking them from atop their hats. In 1991–92, great grays irrupted once more, this time concentrating in the Great Lakes region. Then, in 1995–96, great grays moved south across almost all of North America, in record or near-record numbers, from Alberta to New England.

Why the great grays irrupted when they did is uncertain. It may have been the result of a prey crash following several good breeding seasons, when owl populations were unusually high, or it may have been due to exceptionally cold, snowy weather during those winters. Even though they are the largest (though not the heaviest) owls in North America, with wingspans of more than four and a half feet, great grays feed on small rodents, especially voles—animals that thrive in deep, insulating, protective snow. Small rodents are the main prey as well of boreal owls, a species that also moved south in record numbers those same winters.

Northern goshawks are the largest and boldest of the accipiters, the clan of short-winged, long-tailed forest hawks. Birders call the adults "gray ghosts," for their pale bluish upperparts and silvery barred chests. I catch a few goshawks each fall at my hawk-banding station in the Appalachians; the species is a partial migrant, with some individuals staying on their breeding grounds in the north or the mountains and others traveling south as far as the

Carolinas. As a rule, goshawks are almost fearless; I have had them try to take from my hand smaller hawks I was banding, and even after capture, banding, and a tedious series of measurements, the goshawks we catch sometimes fly straight back into the nets, still intent on catching the pigeon lure safeguarded within: an appetite with wings.

But the number of goshawks counted in the Appalachians does not remain fairly stable from year to year, unlike the number of most other migrant hawks. At Hawk Mountain Sanctuary, just a few miles from my banding station, biologists have been counting migrant raptors since the early 1930s, and their goshawk tallies show a persistent three- to four-year cycle, with especially large spikes every ten years or so.

Here, at least, the reason seems pretty straightforward. Goshawks in New England and Canada prey heavily on snowshoe hares, which undergo a famous ten-year population cycle similar to that of the lemmings—a fact noted as early as the seventeenth century by Hudson Bay Company fur traders, who saw its effects on lynx, another hare predator. In some areas, goshawks also depend on ruffed grouse, which are also subject to ten-year cycles. When the hare and grouse populations crash, the goshawks stage major invasions, usually dominated by adults, perhaps because the poor food supply snuffed out any breeding attempts that season.

Irruptive finches, as I said earlier, usually move south in response to bad seed crops and stop whenever they hit back-yard feeders. But for one unusual species of songbird, those well-stocked feeders mean food of an entirely different sort.

I was peeling Thanksgiving potatoes at the kitchen sink, keeping half an eye on the busy feeders a few yards away through the window, which had already resulted in skinned knuckles. The usual mob of house finches was crowding the plastic tubes and scuffing through the empty sunflower seed shells on the ground, when, with a whir of wings I could hear through the glass, the flock shattered in all directions.

This happens several times a day at my house, the result of one of the local sharp-shinned or Cooper's hawks making a strafing run on the feeders, hoping to pick off a straggler. This time, though, I was mildly surprised to see a mockingbird hopping among the thick boughs of the spruce tree, gray against the dull green. Odd, I thought; mockers can be bullies at a feeder, but the smaller birds don't usually react with such fuss.

Then the bird jumped into the open, and I understood the reason for the panic. It was not a mockingbird at all but a northern shrike, a burly Arctic songbird that acts as though it's a hawk. The size of a robin, but with a thick neck and blocky head, the shrike was a pale misty shade of gray tinged with buff; its wings and tail were black, and there was a smudged black mask behind each eye. What was most arresting, though, was its beak, heavy and hooked like a raptor's.

Shrikes are eccentric members of the avian order Passeriformes, the perch-ing birds—what most of us think of as songbirds. Wrens, chickadees, jays, thrushes, warblers, flycatchers, vireos, sparrows, finches, orioles, tanagers—all these and many others are passerines, members of the largest order in the bird world, encompassing about 5,700 of the globe's 9,000 or so species. Many are carnivorous, including the delicate warblers and flycatchers that feed heavily on insects; some, like crows and ravens (the largest of the order), kill small vertebrates when the opportunity arises. But only shrikes have taken preda-tion to such extremes of physical and behavioral adaptation. Their habit of impaling prey on thorns and barbed wire has earned them the collective name "butcher birds," and the northern shrike's scientific name puts a more stylish spin on the same label: *Lanius excubitor,* from the Latin *lanius,* a butcher, and *excubitor,* a guard or sentinel.

There are two species of shrikes in North America. The loggerhead shrike was once common across the heart of the continent, a bird of farm pastures and prairies that has declined precipitously with the disappearance of natural grasslands. The northern shrike is bigger, and remains fairly common in its preferred habitat of open spruce and birch forests, bogs and shrubby tundra, in a belt extending from northern Labrador to Alaska. (In the Old World, the same bird is called the great gray shrike, and for some reason inhabits a much wider range of habitats, from northern Africa across Europe and Asia.)

Although they have evolved a heavy, hooked beak with a shearing notch along the upper mandible, shrikes lack the other defining characteristic of true raptors—powerful, taloned toes. So instead of grasping its prey and killing it with its feet, a shrike bludgeons it to the ground, either hammering the smaller animal with its bill or grabbing it with its weak feet, holding it long enough to deliver the killing blow. While loggerhead shrikes feed heav-ily on insects and frogs, northern shrikes are bold, tenacious hunters of small mammals and birds, occasionally tackling prey like jays that outweigh themselves.

I had seen northern shrikes before, on wind-scoured hillsides in Alaska, but finding one at my bird feeder was a shock. Shrikes are haphazard migrants; although they apparently vacate their Northern breeding grounds in late autumn, in the East they are rare birds almost everywhere south of the Canadian border, and only along the coast can they be considered even marginally regular visitors. But every once in a great while, the shrikes make a major incursion south, and this, as I was to discover, was such a year. Over the next few weeks, dozens would pop up in the counties around my home, more than anyone could remember seeing in decades.

But on this late November day, all I could do was stare in pleasant surprise. The shrike sat motionless for a moment, then bounced up to an adjacent branch, looking carefully around; its tail rose and fell slowly, like the twitching tail of a hunting cougar. Again the shrike flapped up a few feet, and I realized that a knot of house finches was clustered ten feet above it; the shrike was trying to force them into the open. Another upward bound and the finches scattered—some to the neighboring maple, others across the road to a field of standing corn, the shrike in close pursuit. I lost sight of them in the rows of dry, wind-tossed stalks and have no idea whether pursuer or pursued won.

What struck me most about the short-lived drama was not the chase but the finches' immediate, five-alarm response to the shrike's arrival. Not only had the smaller birds never seen a northern shrike before, they had almost certainly never seen a shrike of any sort—and this is a bird, after all, that looks remarkably like a harmless mockingbird, a species common around my property. Yet something about the shrike—its carriage, its demeanor—must have screamed *predator* on a genetic level, and the finches reacted in a heartbeat, listening to a warning buried in their bones.

Because we're a species that likes our climate moderate, we empathize with any bird that keeps a few steps ahead of winter. Migration is innately logical to us—it's what we would do if we were birds. The idea of *not* migrating, if one had the ability, doesn't make much sense to us. And yet for some birds, the worst weather in the world can't pry them loose from their northern homes. To understand migration, therefore, it is also necessary to understand the birds that decline the opportunity.

The only solid object in an otherwise amorphous world of white and gray, a raven sat hunched on a short, twisted spruce tree, neck pulled tight against its shoulders, eyes closed to the pelting snow that lashed in from Hudson Bay in central Canada. The bird had not moved for half an hour, and only the wind tugging at its feathers made it look any more alive than the big, snow-covered rocks that rose in white mounds among the naked branches of the waist-high tundra willows.

Even the nearby bay seemed only half-real in the gathering twilight; the cove fifty yards away was covered with spongy new ice, still supple and plastic near its outer edge, so that the incoming waves rolled through it sluggishly, apathetically, as though through thick oil. With the coming of the subarctic winter, when the sun would barely rise above the horizon, the entire world was sliding into a frozen torpor.

One of the boulders stirred, shedding rounded plates of wet snow like pieces of shell falling away from a hatching egg. The polar bear rose from its bed in the willows, yawning enormously, its black tongue curling like the tail of a cat. Against its buttermilk pelt, the falling snow seemed unnaturally white, as if someone had simply erased part of the animal. The great bear turned and shambled away, and the raven finally roused itself, too, shaking the snow from its feathers and flying off heavily in the same direction, both of them swallowed by the storm.

In summer, the tundra and boreal forests around Cape Churchill, in northern Manitoba, host nearly 170 species of birds, including 26 species of shorebirds alone. But in winter, all but a handful leave—Pacific loons for the western oceans, Hudsonian godwits for Patagonia, gray-cheeked thrushes for Amazonia. Perhaps I have become inured to such migratory miracles, but what struck me as amazing, as I sat through the beginning of a full-fledged Hudson Bay blizzard, is that not all of them get out when they have the chance. About a dozen hardy species stay here right through the dark, abominable winter—gray jays, boreal chickadees, and spruce grouse in the forests of wind-shorn conifers; snowy owls, gyrfalcons, willow ptarmigan, and hoary redpolls out on the open tundra.

And ravens. They are the largest North American members of the crow family, reputed to be among the most intelligent birds in the world—so if a tiny orange-crowned warbler has sense and strength enough to move to Mexico before the snow, why in the name of wind-chill factors don't they?

We'll come back to this question of why, but let's start with how—how do ravens manage to survive in such an overtly hostile environment without migrating? Cape Churchill is at 59 degrees north latitude, more than a thousand miles above Chicago and just a few hundred miles from the Arctic Circle. In midwinter, daylight lasts only a few hours, most of that a sullen twilight rather than bright sun, and even in summer permafrost lies just a few feet below ground. The Cree called this region *Kewantinook,* "the land where the north wind rises," and for good reason. But ravens are found much farther north than this—right across the whole of the Arctic as far as the Queen Elizabeth Islands, which cap the continent, where the sun simply refuses to rise through the long winter months and where temperatures may drop to −70 Fahrenheit. This is exactly the kind of wintry nightmare that all the other migratory birds are hell-bent to escape.

Ravens are corvids, members of the same family that includes crows, jays, and magpies. Corvids are ecological generalists—they live in a variety of habitats and terrains and will feed on almost anything remotely edible. Ravens are primarily scavengers; one author gives the following (yet still incomplete) list of raven foods as "dead elk, deer, whales, seals, fishes, also eats frogs, tadpoles, worms, crabs, shellfishes, which it drops from aloft to break on rocks or ground; eats crayfishes, minnows, eggs and young of seabirds, songbirds, herons; in spring, follows plow for insects; eats berries in fall, also mice, lemmings, and young seals caught through cooperation of pair."

It would be tedious to try to sum up all the different kinds of food eaten by the raven [notes zoologist Bernd Heinrich, who has studied the species in Maine since the mid-1980s]. Judging from only some of the sources, a fairly accurate assessment would probably be the following: all the dead animals it can find, all the live ones it can kill, and fruit and grain if they are available . . . What is of primary interest is not so much what the raven eats, but the different ways it has of getting a meal. Many birds are evolutionarily programmed to feed on one specific resource in one very specific way . . . the raven distinguishes itself by being a jack-of-all trades, but it can become the master of quite a few.

I had a chance to observe the ravens of Churchill over the course of several weeks; I was leading tours to watch the cape's famous polar bears, which congregate here each autumn waiting for Hudson Bay to freeze, and ravens were in constant attendance, both in the small town of Churchill and out on the tundra. Much of our time was spent in a trainlike "bunkhouse" far from town, five long cars perched on giant wheels so we were safely above the reach of the many bears that milled around us each day; at other times, we made excursions in "tundra buggies," boxy, bus-like vehicles built high on dump-truck chassis.

When I first arrived in Churchill, around the middle of October, an unusually warm spell of weather was just ending; the ground was snow-free, and while the freshwater lakes had frozen over, the bay coves were still open. Some of the last migrants were still hanging around—snow buntings, Lapland longspurs, and American tree sparrows, which would soon depart for the northern and central United States, waterfowl like Canada geese, oldsquaws, and red-breasted mergansers that depend on open water, and—much to my surprise—many hundreds of white-rumped sandpipers, a species that usually leaves for its Argentine wintering grounds by early September.

The following week, however, the weather turned meaner; nighttime lows dropped below zero, and while shimmering curtains of auroral light rippled across the sky, the coves and inlets of Hudson Bay began to freeze. One morning, a white line appeared on the horizon: ice was forming out on the bay itself. Snowstorms covered the ground shin-deep, and the waterfowl and the remaining shorebirds left. We'd see a few flocks of willow ptarmigan, white but for their black tails, and an occasional gyrfalcon, snowy owl, or glaucous gull, but most days the only birds we spotted were ravens, which spent endless hours sitting on snow-covered windrows of frozen kelp or perched on the roof of the bunkhouse.

Once, when someone tried to carry a bowl of snacks across the six-foot gap between the kitchen car and the bunkhouse lounge, the ferocious wind scattered potato chips in a plume across the snow, and the ravens feasted. Another day, when an Arctic fox ran past with something in its mouth—it might have been part of an eider duck it had killed—the ravens lifted off one by one to trail it out of sight, probably hoping to pilfer the kill. Yet another time, a fox buried something small in the snow and frozen soil near the bunkhouse, then lifted a leg to urinate on the spot, marking it. No sooner had the fox turned to go, however, than a raven glided languidly to the cache, unearthed the hindquarters of a lemming, and gulped it down.

Polar bears were ever-present, too, usually bedded down in depressions scraped out of the kelp; one morning I counted seventeen around the bunkhouse—a female with two small cubs, subadult males wrestling and sparring on their hind legs, and larger, older males stalking the edge of the ice, testing its strength. The bears of Churchill are forced off the bay in late July, when the pack ice finally melts, and they spend the next four months in what biologists call "walking hibernation," traveling and socializing, but not feeding. With the onset of winter and the freezing of the bay, they would soon be free to once again hunt seals on the pack ice, ending their long fast.

When the bears headed to sea, the small, catlike Arctic foxes would follow them. We often saw the foxes—pelts whiter than the snow, short-legged, with enormously bushy tails—circling bears like planets orbiting the sun. It's an appropriate metaphor; in winter the foxes depend to a large degree on the polar bears, following them far out onto the pack ice and scavenging the remains of the bears' kills, like jackals cleaning up after lions.

According to the Inuit, ravens do the same thing, trailing bears through the perpetual darkness to clean up the scraps left behind. Nor is it necessarily a passive relationship; ravens are believed to lead wolves and grizzly bears to potential prey, and there is anecdotal evidence from Native hunters that they may perform a similar service for polar bears, drawing their attention to seal dens and breathing holes. (Ravens are said to do the same for human hunters.) On the crumpled, frozen surface of the night-black sea, bears, foxes, and birds form a triumvirate of survival.

Even without polar bears, ravens are well suited to life in the Arctic. One nineteenth-century naturalist, Ludwig Kumlien, even saw ravens kill young seals on several occasions, with one bird actually landing in the seal's breathing hole, cutting off its escape, while the other "seemed to strike the seal on the top of the head with its powerful bill, and thus break the tender skull." A wolf-killed caribou, a frozen hare—anything not rugged enough to make it through the winter itself becomes food that helps the raven achieve exactly the same goal.

But while ravens are large, social, and capable of eating a wide variety of food, other birds that tough out the subarctic winter are not. Willow ptarmigan, the ubiquitous grouse of the Arctic that turns white in autumn, spends its time feeding on the small dry buds and twigs of dwarf willows, an abundant resource but surely one of the blandest, most spartan diets in the world. At night, they often burrow into snowdrifts, enjoying the insulation, and by day, the thick mats of feathers that grow from their toes act as snowshoes. Gyrfalcons, in turn, prey on the ptarmigan, while snowy owls are tied to the vagaries of the local rodent cycle.

The smallest resident birds on Cape Churchill are hoary redpolls and boreal chickadees, each about five and a half inches long, each weighing less than two-thirds of an ounce. The chickadees—a duller, browner version of the familiar black-capped chickadee—inhabit the spruce forests, and so reach the northern edge of their range not far from Churchill, where the timber finally gives up its fight with the tundra. The hoary redpoll, on the other hand, approaches its *southern* limit near Churchill; it is a truly Arctic bird, found as far north as the land goes. The color of pale wheat, with a black chin, red cap (or "poll"), and a tiny, pointed bill, the redpoll sometimes irrupts south with its more abundant cousin, the common redpoll, but most stay year-round in the Arctic and subarctic.

In a cold climate, it helps to be physically large, so the amount of skin area (which loses heat) is relatively small compared to one's body mass (which produces heat). That is why so many northern animals are bigger than their more southerly counterparts, and why so many of them have shorter, more compact extremities, which would otherwise also lose heat. In an arctic climate, being small and warm-blooded, especially as small as a chickadee or a redpoll, is a tremendous disadvantage; to maintain a constant body temperature of about 104 degrees Fahrenheit requires a sizzling internal furnace and a nearly constant supply of food to stoke that metabolism. One study of crossbills, which are larger and heavier than redpolls, found that the birds must eat a spruce seed every seven seconds just to stay even with the demands of their bodies. In cold weather, there is time for little else; from summer to winter, a chickadee increases the time it spends feeding twentyfold.

Northern birds may compensate for the cold by growing a thicker coat of feathers in the winter—redpolls double the weight of their plumage—and by roosting in places like the snow caves that form beneath the lowest branches of spruces or inside tree cavities. A chickadee may also, in effect, turn down the thermostat at night, lowering its body temperature as much as 12 degrees, reducing the difference with the air temperature and saving up to 20 percent of the energy it would otherwise use to stay warm.

In winter, polar bears lay on thick rolls of fat that serve as both an emergency food supply and an insulative layer. Arctic birds also accumulate fat, but because they must fly, they cannot add enough to serve as insulation, and most studies have shown that the fat a chickadee or redpoll carries is just enough to see it through the night. Even with a thickened, fluffed coat of feathers and a protected roost, northern birds must shiver continually through the cold weather, the muscular contractions generating enough heat to maintain body temperature. The next time you complain about being chilly, imagine shivering twenty-four hours a day for seven or eight *months* straight.

Which brings us to the original question about ravens and other year-round residents of the Arctic: Why not simply migrate south along with almost everyone else and avoid the cold completely? But remember, cold isn't the problem, the spark for most migrations—food is. If a bird can find enough to eat, it can survive the coldest weather that the planet can dish up—just look at emperor penguins, which actually nest in the Antarctic winter. Most of Hudson Bay's summer residents leave not because it will get cold but because the cold weather cuts off their particular supply of food—by eliminating the insects that flycatchers, warblers, and other songbirds eat, or by sealing off the lakes, ponds, and ocean that waterfowl and shorebirds need. When food is abundant, the hoary redpolls sit tight; only when the seed crop fails do they irrupt south. It is easy to fall into the trap of believing that migration is a requirement of life for birds in the north. But as the ravens, redpolls, and others demonstrate, migration is but one strategy.

If a bird can find enough to eat through the winter—a big "if"—it actually may gain a huge advantage over those that must migrate. It does not face the perils inherent in long-distance travel, nor must it make evolutionary compromises to adapt itself to other, often utterly different ecosystems along the travel route and on the wintering grounds. It can stake out a territory very early in the breeding season, before the migrants have returned, and it can (at least in more temperate latitudes) raise two or even three broods of young during the short summer, compared with the single brood typical of migratory birds. And in fact, while nonmigratory northern birds have an overall higher mortality rate than migrants, those that survive tend to be longer-lived and have higher reproductive success rates, more than balancing the equation.

The key is food. Boreal chickadees have eclectic tastes—not quite so catholic as ravens, but they will take a variety of seeds, including spruce kernels, as well as torpid insects, spiders, and arthropod egg masses that they find by methodically searching through twigs, branches, and bark. Redpolls are seed-eaters and are especially dependent on the seeds of shrubby Arctic birches, borne in catkinlike cones, which they knock from the tree by jumping or flying at them. Then the birds quickly gather the fallen seeds—but although they swallow them, the seeds do not go directly to the stomach. Instead, they are stored in the crop, an expandable pouch in the esophagus that allows some birds to feed quickly on a relatively large quantity of food, then measure it out later to the stomach for digestion, at a steady pace, when the bird is safely hidden from predators. A full crop just before sunset also keeps a redpoll's metabolic fire well fueled through the night.

Other internal adaptations help an Arctic bird make the most of the food it finds. The small intestines of most birds have two dead-end pouches called colic caeca—little more than nubs in pigeons and other seed-eaters, but in birds like ptarmigan that must digest fibrous plant material like willow buds and twigs, the caeca equal or exceed the length of the intestine itself, and they are intricately ridged and folded on the inside, providing excellent conditions for the symbiotic gut bacteria that help the bird break down plant cellulose. In effect, the ptarmigan has two digestive tracts, greatly expanding the body's ability to eke out the most nutrition from the bird's diet; the most fibrous material goes straight through the small and large intestines, with waste being excreted every five or ten minutes, while the more nutritious material is diverted to the caeca, which expel their feces—a larger and softer dropping— only once every day or two. What is most remarkable, the size of the gut changes with the seasons; a ptarmigan's digestive tract may grow by almost half in fall, and shrink again in spring, when more nutritious foods are available.*

Not long after I watched the raven flap off into the storm to follow the polar bear, the blizzard struck with its full power. The winds blew at more than 50 miles per hour, with gusts of 65 or more—approaching hurricane speed—while the blowing snow reduced visibility to only a few yards. The boxy tundra buggy shuddered in the gale as we crept along, the driver and I trying to follow a trail that was fast disappearing beneath the drifts. We moved at less than walking pace, easing down a long point of land, barely able to see the water on either side.

Then, just for a few minutes, the snow eased up, and we could see with the sudden, almost surreal clarity that comes after fog lifts or thick smoke blows away. The bay was a fury of enormous whitecaps combing in, their tops shredded by the wind, and table-sized chunks of ice flung on the rocky beach by the waves. The north wind was only a few degrees off the point, which afforded no protection, but a small flock of ducks were riding the chop anyway, their dark bodies contrasting vividly with the dishpan-gray water. They were common eiders, big, heavy sea ducks that breed here in summer; most were brown females and immatures, but one was an adult male, white and black, with a swatch of pale green on the head. Unlike most eiders, which retreat to either coast in winter, the subspecies that breeds in Hudson Bay stays there year-round, lingering in the narrow crescent of open water, called a shore lead, that remains along the western coast of the bay, diving deep in the inky water for shellfish.

---

* Even with all these advantages, ptarmigan may still cut and run. Rock ptarmigan, which nest north of Churchill, migrate into the area for the winter, while willow ptarmigan, whose populations are cyclic, sometimes irrupt far south of their normal range during peaks of the ten-year cycle. Banding studies in Greenland show that in such years, willow ptarmigan, most of them juveniles, may travel a thousand miles or more, returning north in spring.

Frankly, when I was told this it seemed incredible to me that any waterfowl would be able to survive the dicey weather of Hudson Bay in winter. But then I remembered the spectacled eiders in Alaska, the ones that winter in holes in the Bering Sea pack ice, and I reminded myself that we know very little about even the most common birds. Watching this small flock ignore a blizzard that could easily have killed me—seeing them feed and preen, ducking through the crumbling tops of the largest waves, riding the rest as buoyantly as dry leaves, I could only shake my head, looking out from the protected side of a window on a world I could barely comprehend.

10

Uneasy Neighbors

In winter, the salt marsh hisses. The wind, which is rarely still on the wide, open flats along the Delmarva Peninsula, draws its fingers through the dried tips of the cordgrass, tall and thick enough to hide a man; the sound it makes is sibilant and empty, an endless sigh.

Another sound can be heard over the wind, a shrill gabble that plucks at the ears, falsetto barks and yelps that draw the eye into the distance. At first it seems that the air itself is moving, a rippling pulse of light above the horizon where blue sky and brown grass meet. Moments pass, and the ripple becomes long strands of white that fragment into layers of calcite dots as the melee grows nearer, each flickering like distant stars.

The flock comes closer still—hundreds of white geese with black wingtips, moving against a cold late-afternoon sky of faded denim. They circle, locking their wings and lowering their feet, spiraling down into the marsh; many sideslip, tumbling like falling leaves. The air is full of their clamorous, metallic jabbering. Yet even as the long grass swallows them, more and more pour in from every direction, wheeling by the thousands, then by the tens of thousands—a cyclone of white, a blizzard of wings and sound that lasts for more than an hour, until the sun sinks and the tidal flats tremble under a blanket of murmuring birds. The snow geese have returned to Bombay Hook.

Such sights are common these days along the Delmarva, a bulge of land that hangs like a tear between Chesapeake Bay and Delaware Bay. On the surface, the story of the greater snow goose is an upbeat heartwarmer—a real conservation success story. Reduced by market hunting to a few meager thousand at the turn of the century, then coddled with protection and refuges, the species has surged back, so that now more than half a million descend each autumn from the Arctic to the coastal marshes of the Atlantic seaboard.

But the situation is far more complicated and unsettling—a cautionary tale about how mankind's manipulation of the landscape may tip the natural balance in favor of a single species, pitting its success against the welfare of many others. In truth, the ever-growing multitudes of snow geese threaten to devastate fragile ecosystems thousands of miles apart, forcing game managers to rethink many of their long-held assumptions about how to protect wildlife.

Nor are snow geese the only migratory species dealing with difficulties on their North American wintering grounds. The destruction of crucial habitats, especially wetlands and coastal environments, is squeezing more and more birds into less and less land, spawning periodic epidemics or risking exposure to toxic chemicals that make a mockery of the word *refuge*.

It is easy to forget such nettlesome problems on the tidal flats of Delaware. This was the home of the Kahansink Indians, who in 1679 sold part of the expansive marshes to Peter Bayard, a Dutch settler from New York. He paid for it with one gun, four handfuls of powder, three coats, a kettle, and some liquor, and named the tract Boompies Hoeck, "little-tree point." Whites used the land for many of the same purposes as the Kahansink—trapping muskrats, collecting oysters and crabs, setting traps of woven saplings for fish in the tidal creeks. The cordgrass they scythed and dried as "salt hay," good fodder for their cattle and horses.

The Kahansink and Europeans both hunted the great flocks of waterfowl that came south with the cold fronts of November and December—multitudes of Canada geese, black ducks, pintails, teal, and others. It was for the sake of the ducks and geese that Bombay Hook National Wildlife Refuge was established by the federal government in 1937. Covering nearly 16,000 acres of woodlands, fields, and salt marsh, including Peter Bayard's original purchase, Bombay Hook is one of the stepping-stone refuges that dot the Atlantic flyway, safe havens for migrant birds of all sorts.

In the refuge's early days, snow geese were all but unknown there; relatively few of them were left in existence, and their main wintering ground was far to the south, along the Outer Banks of North Carolina. The wanderers that were seen there during migration always caused a stir among hunters and birdwatchers, and for good reason. Snow geese are lovely birds, not in the gaudy fashion of wood ducks or mallards, all florid greens and purples, but with style and understatement. Only two-thirds the size of a Canada goose, an adult snow goose is pure white, except for inky black wingtips and a corona of faint brick-orange around the face and down the chest, stains acquired while feeding in iron oxide–rich mud. The legs and beak are dull pink, and the neck is ribbed with neat, vertical rows of feathers, like furrows. The bill is short and triangular, marked on each side with black "grin patches," the cutting edges of the beak that look vaguely like lips and are the only graceless note on the whole bird.

Ornithologists recognize two subspecies of snow geese—the lesser snow, which breeds around Hudson Bay, the Northwest Territories, and the Yukon, and the greater snow goose, which nests even farther north, on the islands around Baffin Bay, Foxe Basin, and parts of neighboring Greenland. Both races also have an arresting, dark-colored form, known as the blue goose, in which the body is blue-black and the wings a soft gray; blue geese are very common among lesser snows, which migrate down the center of the continent and winter along the Gulf Coast, but are exceptionally rare among the greater snow geese that winter in the East.

No one knows how many snow geese there were in North America in pre-colonial times. In 1535, Jacques Cartier saw "many thousands of white and grey geese" in the St. Lawrence Valley of Quebec, still an important fall staging area for greater snows. Whatever the original population, it was almost certainly reduced by human avarice. Market hunters, using live decoys, bait, and other techniques now forbidden, took a toll on the snows' wintering grounds, and the geese were also being killed on their nesting territories by Canadian Natives; this was a traditional spring hunt, but the introduction of efficient firearms made the damage worse.

By the time modern conservation laws began to take effect in the first decades of the twentieth century, snow geese were a rarity; the total population of greater snows may have been no more than three or four thousand. But times have changed. At last count there were more than half a million greater snow geese, and several million lesser snows—and both were continuing to increase at about 5 percent a year. Plot that on a chart and you get the sort of swooping, upward curve usually seen with articles on human overpopulation. And like humans, snow geese are consuming the very environment on which they depend.

What do wildlife managers do when an animal is *too* successful, when the creature they were trained to husband becomes a biological wrecking ball, making survival difficult for other, less adaptable species? I was hoping Frank Smith could tell me.

I met Frank in Bombay Hook's small visitors' center, in a classroom-cum-auditorium hung with wildlife posters, stuffed birds, deer skulls, and muskrat pelts. Mounted on one wall was a punt gun, its rusted barrel a metal pipe five or six feet long, the stock a crude, curved piece of wood as thick as a man's thigh—one of the homemade cannons that the old market hunters used to kill waterfowl. This was a small example, but some of the guns were true artillery pieces, mounted with a swivel on the bow of a boat, with a muzzle large enough to swallow a man's fist. Packed with black powder, they would be loaded with lead shot, roofing nails, scrap metal—anything to create a lethal shower of shrapnel. Then in the darkness, the gunners would pole stealthily out into the bay, where waterfowl slept in tightly packed rafts, and pull the trigger. They'd spend hours retrieving hundreds of dead and dying birds from one or two concussive blasts, then head for the city markets to sell their bag; it wasn't sport, it was a living. Punt guns were outlawed by the Migratory Bird Treaty Act of 1918, and the few surviving ones are all museum relics now.

A biologist with the U.S. Fish and Wildlife Service, Frank Smith wore a brown sweater and brown slacks; nearby lay a matching brown coat and cap, both scuffed with wear and emblazoned with the USFWS's emblem. His office was tucked away in a corner of one of the white cinder-block maintenance buildings. We walked past a fellow in coveralls building duck traps, big steel mesh cages to be baited with corn, and into a tiny room, jammed with computers and shelves of scientific journals. Here he sank more comfortably in a creaky chair, leaning back. Frank had an open, blunt face that folded itself into deep creases around his eyes when he laughed, which he did frequently as he talked, dragging his fingers through short, graying hair.

He didn't relax completely, though, until he offered me a tour of the refuge. We slid into a pickup truck and bumped out the dirt road, past stubbled fields and thickets of woods; then the road turned and we were skirting the first of four huge, interconnecting freshwater impoundments, with an eternity of tidal marsh stretching off to our right.

Smith is fiftyish, a native of central Delaware, and he's been a wildlife biologist at Bombay Hook for almost twenty years. Although the refuge plays host to more than 260 species of birds, 33 species of mammals, 27 reptiles and amphibians, and countless fish, invertebrates, and plants, a disproportionate amount of his time is spent dealing with snow geese—something he couldn't have predicted in the early days of his career.

"I was born and raised about fifteen miles west of here, and I never saw a snow goose growing up. I finally saw one snow in a flock of Canadas back in the 1960s, and I thought it was such a big deal," Frank said.

Snow goose numbers have climbed steadily since then. By 1976, more than 9,000 snows were using the refuge each winter; that number rose to more than 20,000 in 1980, 50,000 in 1982, 75,000 in 1987, and almost 90,000 in 1995. A year later, as we met to chat about them, the geese had set yet another record—more than 130,000 snow geese counted on the refuge's land, and about 290,000 in the state as a whole. That is a population explosion of astounding scope.

The same thing has been happening with lesser snows, but to an even more spectacular degree. Since the 1960s, the mid-continent population—those geese that migrate from Hudson Bay down the middle of North America to the Gulf—has risen from about 800,000 birds to between 4.5 and 6 million; a single nesting colony at the mouth of James Bay ballooned from 55,000 pairs to nearly a quarter million in just fifteen years. In other colonies, the nests are packed together at a rate of more than a thousand per square mile—a relentless reproductive engine cranking out gray goslings.

These are salad days for snow geese, and they mostly have us to thank for it. Back in the days of the Kahansink, snow geese were stuck feeding on tough, fibrous marsh vegetation; if the wintering flocks became too numerous, starvation thinned them out and the debilitated survivors raised fewer chicks the next year. That started to change in the 1920s along the Gulf of Mexico, as more and more marshland was converted to rice paddies. Snow geese, it turns out, just love rice, scarfing down the leavings after the paddies are harvested—and they learned to love soybeans, corn, and a rainbow of other crops, too. Around the same time, game laws were toughened and wildlife refuges began popping up across the landscape, usually made all the more alluring by fields planted by managers eager to attract their burgeoning constituency. Instead of being confined to a narrow rim of coastal marsh, the flocks spread out over the whole of the lower Mississippi basin and the southern Plains—elbow room and food for the taking, an avian version of Manifest Destiny. Today lesser snows winter as far north as Iowa and Nebraska. A goose couldn't ask for more.

Along the Atlantic coast, greater snow geese also shifted from marshes to fields for much of their feeding, and they tended to hang farther and farther north each winter, the same shortstopping behavior seen in Canada geese. Instead of laboring all the way to Currituck Sound or Pea Island in North

Carolina, the birds plopped down with an almost audible sigh of contentment on the Delmarva, where the rich, sandy soil grows bumper crops of the very things snows like to eat. They also learned to do very ungooselike things; at the sprawling King Cole cattle-feed lots, not far from Bombay Hook, as many as 48,000 geese have been seen sifting through the manure for what Frank politely calls "recycled" corn.

It was midmorning, and Frank warned me that we would see few snow geese; the massive flocks scatter to the four winds each morning at daybreak, routinely flying to the Eastern Shore of Maryland, across the bay to New Jersey or north into Pennsylvania for the day, then returning—as I had seen them in past visits—in thunderous clouds at sunset. And he was right; we saw at most a few hundred geese, rising and falling among the distant stands of cordgrass in the marsh or loafing in random dozens on mud bars in the freshwater pools.

A blustery wind was churning the open water of the impoundments to whitecaps, and the ducks we saw, mostly mallards, blacks, and pintails, were huddled in the lee of small islands and around the fringe of the pools. My eye was drawn instead to the other side of the road, where the salt marsh stretched like a prairie, soft ochers and buffs stirring in the wind. The tide was nearly full, and the marsh was engorged with gray water, opaque and roiling, flooding the stands of cordgrass. More than two miles of marsh extended beyond us to the Delaware Bay, where two large oceangoing ships were passing; seen across the flat grass, they looked immobile and stranded, like rusted vessels marooned on a dry lake bed.

We stopped, watching a northern harrier coursing just over the tops of the cordgrass, its breast matching perfectly the pale yellow-russet of the vegetation. Known in years past as marsh hawks, harriers are common hunters of the winter shore, searching for rodents or sparrows stranded by the tide. My binoculars compressed the distance beyond the bird, squeezing the miles of cordgrass into tight, foreshortened layers, the creamy tips waving against darker, shadowed beiges and browns. Moving with the hawk, my eye caught another harrier, much farther out; followed it for a time, then hopped to a lone snow goose dropping down, legs dangling and wings flared. Then to a line of ducks, a thin, whirring scrawl against the sky, then to a gull making lazy circles, then to a sparrow popping up and down from the cordgrass like a jack-in-the-box. Only wind and wings moved on the marsh, but it seemed a place humming with life.

Hereabouts, tidal marshes are called "flats," and this one, Leatherberry Flats, is part of one of the finest expanses of salt marsh left along the mid-Atlantic coast. Originally, such wetlands stretched from Florida to southern Maine, but more than four hundred years of development has left them chewed and piecemeal, buried beneath truckloads of fill, bled dry by channels and dikes, relegated to tiny pockets sandwiched between marinas and shore resort communities. Few of the remaining marshes retain as much of their pristine character as those along the Delaware Bay, with some of the best protected by Bombay Hook and Prime Hook National Wildlife Refuges, and neighboring state tracts like Little Creek and Woodland Beach wildlife areas.

"This is as close to wilderness as we have around here," Frank said, looking out on Leatherberry Flats, which runs for miles to the east of the refuge road, its horizon unfettered. "The salt marsh at Bombay Hook is some of the last on the Northeast coast, and some of the best."

The foundation of the tidal marsh is cordgrass, named for its long, tough stems and narrow leaves. Unlike almost all other North American land plants, the two common species of cordgrass can withstand the brutality of a salt bath around their roots. Leatherberry Flats is covered with tall saltwater cordgrass, *Spartina alterniflora,* which may reach heights of eight feet or more. This species thrives with its feet wet, in the lower areas of the marsh, which are inundated twice each day by the high tide. Where the land is just a bit higher and the tide reaches less frequently, it is replaced by two much shorter grasses, salt hay (*S. patens*) and saltgrass (*Distichlis spicata*), both of which are still harvested along the Atlantic coast for livestock food.

Snow geese love *alterniflora*—not the tough, upright stems, which presumably have all the nutritional value of cardboard, but the underground rhizomes, where the dormant plant has stored a summer's worth of chemical energy. Earlier in the day, I had watched a snow goose, still in the gray plumage of adolescence, feeding on cordgrass roots while a white adult stood watch nearby. The goose was dingy, stained with rust on the face and chest; its coverts, the long, tapered feathers that cascade down the back across the folded wings, danced in the wind like nervously tapping fingers.

Planting its wide feet firmly in the mud, the goose probed into the ground with its serrated purplish-pink bill, gripping the unseen root and tugging so violently it almost lost its footing in the wet slurry. Finally, it broke free a pencil-thick section of the rootstock, crisp and white like a water chestnut, and swallowed it. With more wrenching and pulling, the goose worked its way along the length of the foot-long rhizome, coming closer and closer to the upright stem, which finally toppled over. Without raising its head from the mud, the goose began worrying the rhizome of the next cordgrass plant in line.

This, Frank explained, is precisely the problem. Unlike Canada geese, which usually nibble at the tops of plants, snow geese are grubbers, ripping their food up by the roots. When the Atlantic seaboard was wrapped with a cloak of salt marshes, the snow geese could spread out over a wide area. Today, the flocks are concentrated in the few relatively small, protected remnants and are subsidized, in a sense, by the abundance of farm crops in the surrounding countryside. But snow geese still retain their instinctive hunger for *alterniflora*, and as their numbers have snowballed, so too has the damage. The snows are literally eating the heart out of the salt marsh.

We coasted to a stop farther out along the dike road, and Frank pointed through the mud-splashed windshield. "This area's called Money Marsh Flats. I've heard some people, old-timers, say that it got its name because they trapped so many muskrats out of there. But look at it now. No one's making any money in Money Marsh Flats these days."

It was easy to see what he meant. The marsh was simply gone—replaced by a huge hole of open water hundreds of acres in size, like a crater in the sea of cordgrass. This was a spectacular example of what biologists call an "eatout," where snow geese have all but eliminated the vegetation. Eatouts tend to start with a small gap in the grass, like a chink in the tidal flat's armor, where a few geese can muscle down between the dense stalks and begin to yank up roots. With time the eatout grows, attracting more and more geese; wave action may start chewing at the margins as well, further eroding the breach. At first the cordgrass can recover, growing back from root fragments over the summer when the geese are gone, but after taking a beating for a few years it may simply die off, leaving in its place an empty mudflat at low tide and blank water at high.

Back in his office, Frank shuffled through aerial photographs of Bombay Hook, graphic evidence of how snow geese have affected the marshes there. One photograph, the date "5/23/68" scrawled in a corner with a marker, showed Leatherberry Flats from thousands of feet up—the delicate, dendritic tidal sloughs branching off Duck Creek, which meandered in loops and curves through the middle of the marsh. Even from this altitude, and in black-and-white, the cordgrass looked lush and healthy. This was in the pre-goose years, when snows were still an exciting rarity on the Delmarva.

A photograph from March 1992 showed a radically altered marsh. The guts of it were devoured, a two-mile-wide swath of water like a kidney that obliterated several huge oxbows in Duck Creek. Scattered elsewhere across the image were dozens and dozens of smaller holes; the marsh looked like a piece of moth-eaten fabric. In the past seven years alone, researchers at Bombay Hook found, roughly a thousand acres of cordgrass had been transformed into mudflats and open water, with another 600 acres of higher salt-hay marsh degraded to a somewhat lesser degree.

Snow geese seem to be creatures of habit; they tend to return to the same places week after week, season after season, so while the area affected by their grazing was large, it remained relatively static. One explanation may be the difficulty geese have getting through a solid layer of thick cordgrass. During the winter of 1995–96, however, record snowfall flattened the marsh grass and floating ice sheared the stems away. The flocks began feeding in new areas, Frank said, raising the specter of large new eatouts in previously undamaged sections of the refuge.

Although snow geese undoubtedly have an impact on the salt marsh, it's impossible to pin all the blame on them. By most measures, sea levels along the mid-Atlantic coast are rising, both from climate change and because the land itself is slowly sinking; because cordgrass can grow only in a narrow, in-between zone near the crest of the high-tide level, even a modest increase in ocean level can drown it. And in fact, salt marsh is being lost in parts of the East where snow geese never set foot.

It's also wise to remember that what's bad for one set of organisms is often good for another. The loss of cordgrass has been terrible for muskrats, and for the trappers who lost their cash crop; for secretive clapper rails, whose flattened, "thin-as-a-rail" bodies slip easily between the stems; and for black ducks, which feed in the marshes during the winter. But the mudflats created by the eatouts are a boon for migrating shorebirds, which probe the rich, odorous goo for tiny invertebrates.

I once interviewed a refuge manager in New Jersey, who was also struggling to control snow goose damage. As I looked at the pictures of Leatherberry Flats, like a sandwich with the middle eaten away, I recalled what he had said: "Philosophically, you have to remember that habitat isn't created or destroyed, only transformed. As far as black duck habitat is concerned, those eatouts have gone to pot—but I've seen as many as fifteen thousand shorebirds on them. The eatouts are the first place they hit when they arrive." Frank, too, pointed out that in the shallow, freshwater impoundments at Bombay Hook, the geese provide a valuable service by churning up the bottom muck, providing a perfect seed bed for waterfowl foods like nutsedge, millet, and panic grass.

But the damage clearly outweighs the benefits. Refuge managers in several states have tried many ways to limit the havoc caused by hungry snow geese, as have farmers in the surrounding countryside. They tried hazing them with noisy propane cannons, but the geese soon learned that the booming was harmless. They tried buzzing them with planes: the geese simply flushed and returned. At Forsythe National Wildlife Refuge in New Jersey, they experimented with life-sized plywood cutouts of hunters, some painted fluorescent orange. But even a goose can tell that an immobile, two-dimensional human is no threat, and the scarecrows were scrapped. One technique that has shown promise is setting out a grid of wooden stakes, topped with strips of plastic that shimmer and move in the slightest breeze, giving geese the illusion that the whole surface is covered with netting. That's fine for a potato or wheat field, but hardly practical for thousands of acres of soggy marsh.

In truth, people like Frank Smith have relatively few tools to manage snow geese. He's tried planting green crops like clover and wheat to lure them out of the marshes; it works, but only while the food lasts, which isn't long. "See this?" He flicked a color slide of a verdant field of clover across the desk to me. "All it takes is two or three days to look like this." Another slide, this time of a sea of mud, splattered with goose dung; not a speck, not a leaf of green remained. Whatever respite such plantings buy for the marsh is short-lived.

Of course, one way to reduce the damage from geese is to kill some of them, and Bombay Hook, like most national wildlife refuges, is open to limited sport hunting. (This is less paradoxical than it seems, since most of the land for the national refuge system was purchased with money from sales of the federal Migratory Bird Conservation Stamp—the annual "duck stamp" that every waterfowler must buy before going afield. It is, in effect, a federal duck-hunting license, and without it, the refuge system would be very much smaller and very much poorer.)

The snow goose season usually opens in mid-October, with both the number of hunters and the places they can hunt strictly controlled. Frank's annual report for the previous season, which I thumbed through in his office, was instructive. On opening day there were thirty-four parties on hand, totaling eighty-five hunters; they killed fifty-four snows—not a bad showing, given how skittish and wary geese can be.

Two days later, the refuge was again open for hunting, but this time only thirty-three hunters showed up, and the kill dropped to fifteen geese. Two days after that, it was fifteen hunters taking ten geese, then thirteen hunters and seven geese. For the next month and a half, until the season closed, the number of hunters never again rose above ten, and on many days no geese were killed at all. In the end, 152 snow geese were killed—an insignificant fraction of the 89,000 on the refuge that year.

The same story pops up repeatedly, and the results are often even more lop-sided, as at Desoto National Wildlife Refuge in Iowa, where in 1994 hunters shot fewer than 100 of the 800,000 lesser snows present. The fact is, snow geese can be damnably difficult to hunt. Driving into Bombay Hook that morning, I'd seen a pit blind out in the middle of an expansive field, its heav-ily thatched roof barely sticking above the ground, surrounded by a hundred or so white decoys. I mentioned it to Frank, and he chuckled. "A hundred? They need more like a thousand. It takes a lot of decoys to pull in snow geese." Along the Gulf Coast, where hunting for lesser snows is big business, professional guides use disposable diapers, white rags—anything cheap and durable to give the illusion of a massive flock. "Basically, people around here don't want to be bothered with 'em. Too much trouble," Frank said.

Of course, from a refuge manager's perspective, a scared goose is almost as good as a dead one. One reason the kill drops so quickly after opening day is that the geese learn to avoid areas where hunters are active. Yet here again, snow geese have proven frustratingly adaptable. Most of the areas open to hunting at Bombay Hook are in the hard-hit salt marshes, and while the gun-ning pressure keeps the geese out of the flats by day, Frank said, they come right back in after dark, feeding even on moonless nights.

Refuge managers have been working to prevent salt-marsh loss for almost two decades now, but snow goose overpopulation may be having a much more serious effect, one that has been recognized only in the past few years—the destruction of the fragile tundra where the geese nest. In 1997, a coalition of Canadian and American biologists called the Arctic Goose Habitat Working Group sounded alarm bells with a report warning that large portions of the Arctic were threatened with "irreversible ecological degradation" because of snow geese, particularly the mid-continental population of lesser snows.

Freed from the specter of winter starvation, made fat (and thus more fertile) by the largesse of the Grain Belt, the snow geese return to the Arctic in greater and greater numbers each year, and there raise more and more chicks. But because the females have a high degree of faithfulness to the place where they were born, a trait known as philopatry, the breeding colonies have swollen almost beyond belief, distended with expanding generations of geese. At La Pérouse Bay near Churchill, Manitoba, the snow goose colony numbered about 2,000 pairs when researchers began studying it in the late 1960s. Twenty-two years later, the colony exceeded 22,000 pairs, having grown at an annual rate of 8 percent. The story is the same at most other lesser-snow colonies on the western and southern coasts of Hudson and James Bays. At Cape Henrietta Maria, where the two bays meet, just a few thousand pairs nested in 1957, but they had burgeoned to more than 200,000 pairs by 1996, breeding at densities of more than 2,500 nests per square kilometer.

In effect, the geese create a slum. It's bad enough that they grub out the sedges and grasses that are their preferred food, but they also set off a chain reaction that ultimately leads to the utter destruction of their delicate habitat. Scientists call it positive feedback; the rest of us call it a vicious cycle. The intense grubbing reduces the plant cover, which has eked out a tenuous foothold on the thin, largely inorganic soil. Exposed to the sun and wind, the soil dries out, and the salt it contains becomes concentrated near the surface at rates three times that of seawater. Sedges and grasses cannot grow in such hypersaline conditions, so salt-tolerant species like glasswort crowd in. On dune systems, moss takes over, forming mats that blanket the landscape in spring but dry out and blow away in summer, exposing the soil to wind erosion. Geese avoid these invasive plants (the moss is inedible, and glasswort is 45 percent salt by dry weight) and hammer the surviving grass patches even harder. Meanwhile, the hypersaline soil kills grass seedlings and the other plants that grow in it, like willows, and eventually dries into the sort of caked, crusted hardpan usually seen when drought empties a lake. At that point, the geese have long since moved on to new feeding areas, starting the process all over again somewhere else.

The working group's report has an elegant graphic showing the cycle, several curlicues looping back on themselves in a swirl of destruction. But seeing something with your own eyes is worth a thousand positive feedback schematics. La Pérouse Bay lies near the tip of Cape Churchill, so one day during my polar-bear-watching stint I hitched a ride with a helicopter pilot, and we flew out toward the bay. We skimmed along just a hundred feet in the air, passing eight or nine bears, including a huge old male running through the short thickets of dwarf birch and willow. Just as we passed over him, a flock of ptarmigan exploded beneath his nose, a starburst of white radiating out against the red-brown tundra.

At La Pérouse, the pilot circled once around the empty buildings of the research camp, then hovered just a few feet over two room-sized enclosures of chicken wire, which had been built in 1982. I'd seen photographs of these same enclosures in July, when they were lush and green, the kind of flower-spangled carpets that make the Arctic in summer such a visual delight. Even now, at winter's doorstep, they were choked with dead sedges and grasses that poked right through the wire roofs. But outside the fences—and as far as my eyes could see, right to the flat horizon many miles away—was a monotony of cracked, frozen, yellow-brown mud, bereft of almost anything alive.

The Hudson Bay lowlands, where many of the largest snow goose colonies are found, are one of the largest areas of wetlands in the world, and their worth for migratory birds can hardly be overstated. The list of nesting species runs to more than one hundred, many of them waterfowl and shorebirds—green-winged teal and wigeon, shovelers and pintails, whimbrel and yellow rails. The willow thickets harbor Wilson's and Tennessee warblers; the beach ridges and dunes are home to American pipits, horned larks, and Smith's longspur, among many others. The direct loss of nesting cover and food (not only plants but also aquatic invertebrates that once inhabited the hypersaline pools) has knocked the stuffing out of many species. Researchers have seen a drastic decline in the number of formerly abundant birds like semipalmated sandpipers and red-necked phalaropes; yellow rails seem to have disappeared entirely. Elsewhere, white-fronted geese and Richardson's geese (a small subspecies of the Canada goose) have disappeared from areas where they once were common.

"The scale of the problem and associated level of risk to the broader population requires intensive study, including some calculation of the proportion of total range of the species affected by goose damage," researchers write. "It is clear, however, that the interaction is dynamic, and the rapid occupation of new areas by geese increases the threat to other species even as the effects are being calculated." In plain English, there's no time to lose.

A satellite image of La Pérouse Bay in 1993, after twenty years of ferocious assault by snow geese, is telling. It is a false-color photo, in which water is blue, vegetation is greenish-black, and bare soil is red. It looks as though someone has raked the remains of a bonfire across the landscape, leaving a thick bed of glowing embers on the tundra—a firestorm of geese. Biologists say 65 percent of the intertidal vegetation there has been severely damaged or destroyed by geese. And remember—this isn't the largest snow colony, by any means, just the best studied.

"As long as the mid-continent population of lesser snow geese is expanding at the conservative estimate of 5 percent per annum, there is little likelihood that habitat recovery will be possible," the working group concludes. Even if goose grubbing is completely eliminated, grasses take at least 15 years to recover; in some places, where carpets of moss choke out grass seeds on land and sludgy, oxygen-starved salt ponds kill anything that tries to grow in them, the damage may be irreversible.

There are too many snow geese, especially lesser snows; virtually no one argues with that. Nor is anyone arguing with the notion that steps must be taken to reduce the population, and quickly. One biologist has predicted that saving the Arctic breeding ground requires reducing the mid-continental population from nearly 3 million to its 1970 level of about 1 million. But how? The rules surrounding snow goose hunting have already been liberalized—long seasons and daily bag limits of ten geese per hunter, a generosity not seen in American waterfowling since before the Dust Bowl years. Cree and Inuit in the Canadian Arctic, who have been criticized in the past for their traditional spring hunts of Canada geese, are now being encouraged to shoot more snow geese, and to take as many of their eggs as they can eat. Yet even as the number of snow geese has been going up, the number killed by hunters has been going down—from 700,000 in 1970 to 400,000 in 1993.

The Arctic Goose Habitat Working Group and other waterfowl biologists want things loosened up even more, and there is talk of permitting the use of bait, electronic callers, live decoys, and other techniques that have been banned from hunting for generations as unsporting. There are fewer hunters today than in years past; one solution may be to recruit more, perhaps providing government subsidies to offset the cost of decoys and equipment. Unfortunately for those trying to control the snow goose blizzard, I suspect the snows will wise up to whatever new methods the remaining cadre throws at them. There is even guarded discussion of harnessing the profit motive—authorizing commercial harvests of snow geese for sale, something that has been conservation anathema for nearly a century. That makes a lot of people squirm; it was profit that drove the bison to the brink and the passenger pigeon into oblivion. But who knows? If the situation keeps deteriorating, maybe they'll even reauthorize punt guns loaded with roofing nails.

"I get the feeling that it's going to take some kind of natural control—an outbreak of fowl cholera or something like that," Frank Smith said slowly, musing, as we sat in his truck at Bombay Hook beside the flat brown marshes. He'd followed the working group's progress, but seemed pessimistic about its goals. "You look at the geese down in those cattle feedlots, in there among the cows, feeding on recycled corn. It's such a nasty setup—if you ever wanted a disease situation, that would be it. But so far, nothing."

Frank's probably right; in the end, it may be that their own numbers will bring the snow geese back to earth. The question is: How many others will they take down first, buried under their own weight?

I took a solitary walk later that afternoon, out to an observation tower that overlooks the sweep of tidal flats; the wind scored the left side of my face, numbing my cheek. The sky was speckled now with flocks of snows, heading back to the refuge for the night. Some swung low, flapping against the wind, a dance of light and darkness with each wing beat, and despite everything I knew—about the threadbare *Spartina* and the eroding tundra, even the image of geese rooting through manure for digested corn—there was something indescribably wild about skeins of white geese, a visible link to the far Arctic, slanting down into the eternity of marsh.

"They're a neat bird," Frank had said a few hours before. "A lot of the local people don't like them, farmers mostly, but I do. You see a Canada goose, you don't know if it was born on a golf course. But you see a snow goose—well, you *know* where it came from."

Canada geese and golf courses—there's another pairing, like snow geese and manure, that no one would have predicted fifty years ago, when virtually all Canadas nested in the Arctic and their honks presaged the change of the season.

Today, millions of Canada geese still travel along their ancestral routes, but many no longer migrate at all, living year-round in much of the United States and southern Canada. Sleeping on golf course ponds and municipal reservoirs, nesting in parks and back yards, feeding in cornfields and on the neatly landscaped lawns of corporate headquarters, their fast-growing numbers have provoked heated, sometimes vicious debate between homeowners, farmers, town managers, biologists, animal-rights activists, and many others.

Biologists recognize between eleven and thirteen (depending on the source) populations of Canadas, distinguished by their discrete breeding and wintering areas—like the Atlantic population that nests in Newfoundland and Quebec; the southern James Bay population; the Mississippi Valley population in northern Ontario west of James Bay; and the shortgrass prairie population, which breeds on Victoria Island and the coast of the Northwest

Territories and migrates to the plains of Colorado, New Mexico, and surrounding states.

Like most waterfowl, Canada geese hit a nadir in the 1920s and '30s, the result of habitat loss and overhunting; one subspecies, the giant Canada, *Branta canadensis maxima*, which nested in the central United States, was presumed extinct until 1962, when it was rediscovered in Minnesota. Many states established populations of wing-clipped geese of several races, especially the nonmigratory *maxima* geese, in an effort to restore the species on wildlife refuges and in parks—first within its original nesting range, but during the 1960s and '70s the practice became common across the United States.

Originally, my home state of Pennsylvania had no breeding Canada geese. Each fall, migrants from the Atlantic and southern James Bay populations passed through here on their way to the Carolinas and Georgia, but overwintering Canadas were rare, and none stayed to nest. In 1936, a few wing-clipped geese were released at the newly created Pymatuning State Game Refuge, just south of Lake Erie; these began breeding the next year. Over the following three decades, the number of resident Canada geese in Pennsylvania increased slowly, many of them former captives that escaped or were set free by private owners. In 1966, Canadas were captured at Pymatuning and, with some purchased from a commercial propagator, transferred by the state Game Commission to a wildlife-management area in the farmland of Lancaster County. The geese flourished, numbering more than 4,000 by the late 1970s—and they migrated no more than was necessary to stay just ahead of the freezing water. In mild winters, they didn't move south at all. Freed from the rigors of migration, surrounded by a land of plenty, these geese and others reproduced at an amazing rate. By the early 1990s, Pennsylvania had an estimated 200,000 nonmigratory Canada geese.

The same thing was happening all over the Northeast. What had started as small, picturesque flocks in widely scattered locations became larger, messier, more widespread, until by the 1990s it was hard to find a body of water without geese. The resident population in the East was estimated at more than 900,000 geese in 1996 and has been growing at the light-speed rate of 17 percent a year since the late 1980s, while in the Mississippi basin it was more than 1 million. Mankind had transformed much of the American landscape into goose heaven. Take corporate office parks and golf courses, for example; nicely landscaped, with attractive ponds and sweeping lawns, they provide food (Canada geese are grazers and thrive on grass), nesting cover, and security, since hunting is rarely allowed in such places and predators are scarce.

Golfers, in fact, were among the first to complain about too many geese, since fifty geese will produce three and a half *tons* of manure each year, and the birds' greasy, thumb-thick droppings could turn a putting green into a slimy nightmare. Farmers contended that geese caused crop damage, especially in fields of tender winter wheat and lush alfalfa. Municipal water suppliers began to worry about the consequences of thousands of geese roosting in their drinking water supplies—health concerns aside, the nutrient-laden droppings sometimes cause immense algae blooms, robbing the water of oxygen and killing aquatic life—and park managers were caught between complaints about the mess created by geese and protests, if they tried to reduce their numbers, from those who liked watching and feeding the birds. Even bird-watchers grew weary of the legions of geese, referring to them as "sky carp" and "pond starlings."

Cities and towns scrambled for solutions. For a time, the most politically convenient answer was to round up the geese during their summer molt, when they are temporarily flightless, and ship them off to distant states that were trying—against the evidence—to start their own goose populations. By the early 1990s, however, there were only a few states, mostly in the Deep South, still willing to take the refugees, and even they soon realized their mistake, as homegrown goose problems began cropping up there, too.

In recent years, cities like Detroit have taken to rounding up the geese for slaughter, then distributing the meat through food banks to the needy and homeless, a step that has been met by outrage and lawsuits by animal-rights activists. (One group, claiming that geese bound for food banks contained environmental toxins, charged the state of Illinois with "unleashing chemical-biological weapons under the guise of welfare.") Other, less draconian measures have been tried—hazing geese to chase them away from sensitive locations, or shaking their eggs to kill the embryos inside—but these are labor-intensive, have been only marginally successful, and have also incurred the wrath of animal-rights groups. Some municipalities have tried introducing European mute swans, which attack geese and chase them away; the trouble is, the swans become an even bigger environmental problem when they escape and start breeding in the wild, as has happened throughout the East and the Great Lakes.

For a while, wildlife managers thought they had a chemical deterrent—methyl anthranilate, a substance found in Concord grapes that lab experiments indicated was distasteful to birds. But field tests with geese showed they were just as happy to graze grass sprayed with the expensive chemical as that without. Some golf courses now spend thousands of dollars on trained "goose dogs," border collies whose job it is to harass geese on the links until they leave—thus transferring the problem somewhere else.

But wildlife managers were caught in an additional dilemma. While resident numbers were exploding, migrant geese—those birds that still nested in eastern Canada—were dropping, some at alarming rates. The Atlantic population nesting in northern Quebec, which totaled 118,000 pairs in 1988, had fallen 75 percent by 1995, to just 29,000 pairs. In the same period, the southern James Bay population fell, although not as precipitously, from 170,200 to 95,000 pairs. Several years of poor spring weather on the breeding grounds, as well as habitat damage caused by snow geese, and spring shooting and egg collecting by Canadian Natives, were blamed for the declines.

One goose looks pretty much like any other over the barrel of a shotgun. To protect the migrants, the U.S. Fish and Wildlife Service in 1995 took the unprecedented step of closing all fall Canada goose hunting in the East indefinitely, a move that crippled the economies of many small mid-Atlantic communities that depended on revenue from hunters. While the closure had the intended effect—preserving breeding stock until the weather improved, bringing a partial recovery of the Atlantic population by 1998—it also made controlling resident geese all but impossible. States were given the opportunity to hold special early- and late-season hunts, each lasting a couple of weeks, scheduled for before the migrants arrived and after they moved farther south for the winter. But these were almost *pro forma* moves, and few expect them to stem the floodtide of geese.

As with the snow geese overrunning the Arctic, many biologists wonder how long the nonmigratory Canada geese can go on like this, swamping the land like an occupying army. Nature has its own ruthless methods of reining in species that get out of hand, and the seeds of the Canada goose's eventual control may lie dormant within their own bodies, waiting for a critical mass that grows ever nearer. The question is whether, like the snow geese, they will cause environmental havoc before that finally happens.

Of course, for environmental havoc, no organism can quite compare to humans. We've so reduced the safe havens for migratory birds that they are forced to use whatever's left—even if what looks like sanctuary is really a toxic trap.

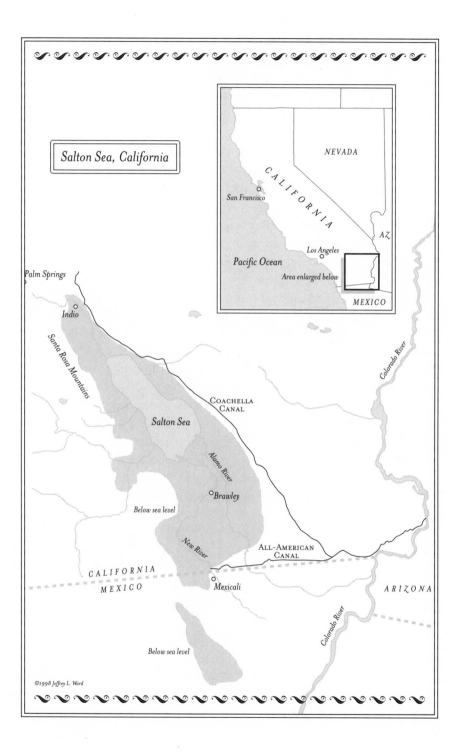

Salton Sea, California

NEVADA

CALIFORNIA

San Francisco

Pacific Ocean

Los Angeles

AZ

Area enlarged below

MEXICO

Palm Springs

Indio

Santa Rosa Mountains

COACHELLA CANAL

Salton Sea

Alamo River

Brawley

Below sea level

New River

ALL-AMERICAN CANAL

Colorado River

CALIFORNIA

MEXICO

Mexicali

ARIZONA

Below sea level

Colorado River

©1998 Jeffrey L. Ward

From the air, the Salton Sea shimmers in the sun, an oasis in the below-sea-level oven of California's Imperial Valley, hard on the border with Mexico. No wonder that waterbirds migrating through this part of the world find it so attractive; no wonder that Salton Sea National Wildlife Refuge boasts more species of birds than any other place in the country outside of Texas.

Salton Sea is an accident; although this rend in the earth known as the Salton Trough (a product of the San Andreas Fault) was in prehistoric times part of the Gulf of California, the modern lake dates only to 1905, when an irrigation canal ruptured, sending a two-year-long flood of Colorado River water into the dry basin. Since then, agricultural runoff has been shunted to the lake, which continues to grow despite evaporation in the ferocious sun, which sends temperatures over 100 degrees for half the year. The Salton Sea is now thirty-five miles long and up to eleven miles wide—the largest body of water west of the Rockies.

The Salton is a striking example of how adaptable birds can be. When the desert basin was replaced by a freshwater lake with trout and insects, multitudes of ducks, geese, shorebirds, gulls, terns, and other waterbirds began using it as a migratory stopover and wintering site. In 1930, Salton Sea National Wildlife Refuge was established on 35,000 acres at the south end of the lake, where the New River empties in from Mexico.

As the lake has expanded, all but 2,000 acres of the refuge have been submerged, but Salton Sea in winter remains one of the best birding sites on the continent. Nearly half a million ducks winter here (up to 190,000 pintails alone), and the highest counts for a dozen or more species on the annual Audubon Christmas Bird Count routinely come from Salton Sea.

But Salton Sea has a problem. It has no drain hole, no outlet—since it is 227 feet below sea level, what flows in can leave only by evaporating, a process that removes water vapor but leaves behind whatever dissolved solids the water was carrying. And Salton Sea water carries plenty. The irrigation water, originally from the Colorado, comes in through rivers like the Alamo and the New, having nourished the fields of the Imperial Valley—and carrying salts leached out of the soil, roughly 4 million tons a year.

As the water evaporates, the salinity of the lake has steadily risen. By the 1920s, the trout that once prospered in the sea died off, and by the 1950s, the state of California was stocking ocean species like croaker and orangemouth corvina; barnacles were accidentally introduced by World War II seaplanes stopping here. But the salt content has continued to climb, so that today, with the salinity 25 percent greater than seawater, the corvina are petering away, and about the only fish that do well in the Salton Sea are tilapia, a basslike African species that must be one of the hardiest fish alive.

If it were only salt, that would be bad enough. But irrigation water, the lifeblood of Western agriculture, carries a much more insidious threat, as was starkly demonstrated farther north, in the Central Valley of California. Up there, the land is underpinned with a natural layer of impermeable clay; unless it is drained off, irrigation water draws salt and other substances from the soil and kills crops. So farmers, with federal cooperation and funding, constructed a drainage system that dumped the wastewater from 600,000 acres into a reservoir abutting the Kesterson National Wildlife Refuge. Everyone thought it was a win-win situation; the farmers got rid of their salty water, while refuge managers could use it—even though everyone knew it contained traces, presumably safe, of agrochemicals and heavy metals—to create wetlands for birds. No one, however, actually tested the water to confirm its safety.

What no one had counted on was selenium, a sulfurlike element that occurs naturally in soil across the West. Vertebrates need selenium in order to survive—but just a little, just the tiniest amount. Even slightly more than that can be hideously toxic, as staffers at Kesterson discovered in 1983. That's the year biologists began finding horrible birth defects among the refuge's nesting waterbirds—missing eyes, missing beaks, deformed limbs, chicks with brains bulging out of their skulls. Half the American coot eggs were affected, three-fifths of the grebes, along with ducks, tri-colored blackbirds, and other species.

Tests confirmed that selenium was the culprit, and that water coming into the refuge contained up to 4,200 parts per billion of the mineral—420 times the dose considered toxic to humans. California declared Kesterson a hazardous waste site, and publicity prompted the Department of the Interior to close it for a time. It also spurred perhaps the oddest, and most tragic, campaign in the history of wildlife protection—frantic biologists burning marsh vegetation to remove enticing habitat, using exploding cracker shells, scarecrows, and other hazing techniques to frighten birds away from the toxic sump that was supposed to be a sanctuary.

Although it was the most dramatic example, Kesterson has not been an isolated tragedy. Similar situations have cropped up across the West, prompting ongoing government monitoring and research into the perils of irrigation drainage. Water is such a scarce and precious commodity in this arid region that birds inevitably congregate wherever it occurs. Increasingly, those are dangerous places, thanks to chemical and biological contamination.

Evaporation ponds, created by farmers trying to deal with selenium poison-ing in their fields, continue to poison birds in the Central Valley. In the win-ter of 1986–87, after several years of drought, the escalating salt content in the Carson Sink at Fallon National Wildlife Refuge in Nevada wiped out 7 million fish and sparked an avian cholera epidemic that killed more than 1,500 waterbirds. At adjacent Stillwater National Wildlife Refuge—once a vast expanse of marsh that has been shriveled by irrigation—white pelicans have picked up so many contaminants that hundreds simply dropped out of the air, dead, in the late 1980s. All this in an area ranked as one of the most important wetlands in the world, recognized in 1988 as one of the first sites in the Western Hemispheric Shorebird Reserve Network—but added to the fed-eral Superfund list two years later for mercury contamination on the lower Carson River.

But in terms of chronic misery, the worst of all may be the Salton Sea, where massive fish kills and the deaths of thousands of birds have become an annual—at times even monthly—event.

Selenium works its way into the lake through Imperial Valley runoff, and once there, the chemical begins to climb the food chain. Like such pesticides as DDT, selenium "bioaccumulates"—that is, with each step up the trophic ladder from plankton to fish to birds, its level increases until it reaches dan-gerous dosages. Tiny copepods and other invertebrates ingest the substance; a small fish feeds upon hundreds of copepods and stores, within its tissues, all the selenium they had consumed. A larger, predatory fish eats innumerable small ones over the years, getting their baggage of selenium, and finally a grebe or cormorant spends its lifetime feeding on the larger fish. By that point, the selenium dose is high enough to interfere with reproduction and other physiological processes, perhaps even to kill.

The irrigation water carries not only salts and selenium but pesticide and herbicide residues flushed off the Imperial Valley's crops. Much of it comes in with the New River, a drainage canal that flows into the Salton Trough from Mexico, about fifty miles away. The New has been called the most contami-nated river in the United States, a vile blend of agrochemicals, raw and par-tially treated sewage, industrial waste, cattle offal, and household trash that is rife with pathogenic bacteria and toxins. The New dumps its sludge in the neighborhood of Salton Sea National Wildlife Refuge, and its mouth is, in a terrible irony, one of the best places on the lake for birdwatching—except that refuge staffers discourage human visitors from going there, citing health considerations.

The Salton Sea is a death trap, not for all the birds that come there, but for many. The tally is appalling. Between 1987 and 1996, more than 184,000 birds died on the Salton Sea, and the actual death toll was much higher, since the annual mortality figures reported by the U.S. Fish and Wildlife Service often include only those individuals collected by wildlife personnel. Every year through that decade, there was at least one, and sometimes as many as three, major die-offs, involving different species and with different causes.

In 1987, 500 ruddy ducks and eared grebes died of avian cholera. In 1989 it was more than 4,500 cattle egrets, killed by salmonellosis. In 1991, 2,000 ring-billed gulls, ducks, geese, and shorebirds—avian cholera again. The next year, 1992, 5,000 dabbling ducks and shorebirds died from cholera, and there was an unexplained die-off of more than 150,000 eared grebes and ruddy ducks. Another 14,000 grebes died in 1994, and 2,000 in 1995; the cause is still unknown, although tests for avian cholera and organophosphate and carbamate pesticides were negative. There is growing suspicion that algae, which plagues the Salton Sea with explosive, smelly blooms, may be producing powerful biotoxins, although other scientists note that selenium poisoning can interfere with the immune system.

In 1996 a typical cholera outbreak killed a thousand ducks, gulls, and shorebirds, but by mid-August of that year, something else was at work. People began finding dead pelicans—beautiful white pelicans, which nest across the Rocky Mountains, and brown pelicans, which are an endangered species in California. Within weeks, biologists patrolling the lake in airboats and walking the shore were finding unprecedented numbers of dead pelicans, as many as five hundred a day. In addition, they were finding sick tilapia with bloated abdomens, some of which were packed off to the federal Northwest Biological Science Center in Seattle.

Scientists there determined that the fish were being eaten alive by acute bacterial infections. "Upon initial examination, gross external signs of bacterial septicemia were evident as the fish showed hemorrhage of the skin and bases of the fins, and extensive ascites [fluid buildup] was evident," a research microbiologist reported. Other specimens collected a few days later "showed the same gross internal and external signs described above. In addition, fish . . . were remarkable in having a hemorrhaged and highly distended abdomen due to extraordinary amounts of fluid in the gut and stomach." Cultures grown from their entrails revealed several types of bacteria, particularly a species called *Vibrio alginolyticus*, which has been implicated in disease outbreaks among saltwater fish elsewhere.

As unhealthy as the Salton Sea has become for birds, it is even worse for fish. Massive kills, on the order of millions of fish, are an annual event on the lake, but they had never before caused the deaths of so many large fish-eating birds like pelicans. What's more, scientists at the National Wildlife Health Center (NWHC) in Wisconsin had shown that the pelicans and other birds were dying from avian botulism, a disease caused by an entirely different bacterium, *Clostridium botulinum*, which produces a toxin that acts like a nerve poison.

Avian botulism (also known as botulism C), flourishes in hot, stagnant, alkaline lakes, erupting into epidemics where waterbirds gather in large numbers, particularly on wintering grounds. It has become especially prevalent in the twentieth century, as wetland destruction has steadily eroded the habitat available to ducks, geese, and shorebirds, forcing them into a relatively few overloaded state and federal refuges, a situation more than one observer has likened to bird ghettos. The disease, which causes paralysis and rapid death, was first reported from Oregon in 1876; for years it was known as "western duck fever," and because it was associated with alkali lakes, early ornithologists believed the dissolved salts in the water were somehow responsible.

Eventually, microbiologists happened on the link between bacteria and botulism—but there really is a connection to alkaline lakes, which are common in the West and are rich in invertebrate life. During droughts or planned drawdowns, when the water level of impoundments is lowered, vast numbers of small organisms die, providing a fertile growth medium for the botulism bacteria, which is always present at background levels. *Clostridium* flourishes, and when dabbling ducks like mallards and pintails root through the muck to feed, they pick up the disease, fanning great die-offs, like the 1997 epidemic at Bear River in Utah that killed 100,000 ducks, or the nearly 40,000 waterfowl, waders, and shorebirds killed by botulism at a polluted reservoir in central Mexico in late 1994.

What was stunning, in the case of the Salton Sea in 1996, was that the deaths weren't among waterfowl grubbing in the mud but among birds that feed on living fish. In fact, researchers found, the live tilapia themselves were carrying hefty loads of the botulism bacteria in their intestines—the first time anyone had recorded such a thing. When the pelicans ate the infected fish— weak, slow, and easy prey—they swallowed a fatal dose of the toxin.

To prevent the spread of the disease, dead birds were quickly incinerated, while more than a thousand sick pelicans were trucked off to rehabilitation centers; injected with an antitoxin, nearly half were saved and later released—at places other than the Salton Sea. By early November, the epidemic appeared to have run its course, leaving more than 14,000 dead birds in its wake, including 8,538 white pelicans, 1,112 brown pelicans, and birds of 64 other species, from loons to egrets. Three yellow-footed gulls, a Gulf of California species that occurs in the United States only around the Salton Sea, also perished. In all, it was the largest die-off of pelicans anywhere in the world, and seemed to doom any hope that the California population of the brown pelican—almost wiped out by DDT in the 1960s—would be removed from the Endangered Species list in the foreseeable future.

Although the botulism outbreak seemed to abate, the bird and fish deaths continued at the Salton Sea, at what the refuge termed "a chronic level." In the first half of 1997 alone, there was another unexplained die-off of a thousand eared grebes, an outbreak of Newcastle disease that claimed 1,600 double-crested cormorants, and hundreds of other birds killed by botulism—in all, more than 5,000 birds of forty-four species. The NWHC concluded that three, possibly four "disease events" were occurring simultaneously.

Why the Salton Sea, with its rotten-egg stench, toxin-rich water, and legions of dead birds, isn't an environmental lightning rod is anyone's guess; too remote and too little known even in the birding community, perhaps. Maybe its genesis, the child of a canal failure, robs it of the concern lavished on natural bodies of water. But birds can afford to make no such distinctions. The waterbirds that migrate through California have few other options; up in the Central Valley only 5 percent of the natural wetlands remain, a situation repeated throughout the West. Places like the Salton Sea are often the only game in town.

One certainty is that, without intervention, the situation at the Salton Sea is going to get worse; the lake will keep growing, the salinity will keep rising, the toxins will keep accumulating, the fish and the birds will keep dying, presumably in greater and greater numbers.* There are several schemes to change that; one would vent the saline water into Laguna Salada, a sometime-lake in a desert basin on the Mexican side of the border that certainly wouldn't

---

* Shortly before this book went to press, Congress approved a feasibility study on ways to "reclaim" the Salton Sea, including $3 million for treatment of agricultural water in the New and Alamo Rivers, and $5 million for wildlife studies at the refuge, which is being renamed for the late U.S. Representative Sonny Bono.

benefit from the toxic gift. Another would bisect the Salton Sea with dikes, creating evaporation ponds to prevent more dissolved solids from reaching the main body. Of course, the evaporation ponds themselves would be attractive to birds, as are those built by farmers in the Central Valley, and no one has said how the birds will be kept away. Perhaps it will mean another round of cracker-shells and hazing, another tragicomic attempt at scaring birds away from a supposed refuge—for their own good.

Part Three

NORTHBOUND

II

The Gulf Express

Aristotle was, shall we say, a creative thinker, even by the freewheeling standards of ancient Greece. In addition to declaring that horsehair turned into living worms, he believed that birds hibernated in holes in the ground each winter.

It's easy to laugh at Aristotle, because we know better. The truth about migration is that birds are conjured from the soft April air of a Gulf Coast sky. The blue is rolled up to make indigo buntings and cerulean warblers, the fog folds in on itself to birth gray catbirds and gnatcatchers, while the orange clouds at dusk give of themselves to create orioles. And the liquid gold of the afternoon sun is measured out, drop by precious drop, to form male prothonotary warblers. Once the sky is full to bursting with these new-made wonders, it lets them fall like snow on the land.

Poetic hogwash, you say? Suit yourself. I've seen it happen.

Slender as a matchstick, Dauphin Island sits in the mouth of Mobile Bay, just one of a string of Gulf barrier islands stretching along the coasts of Florida, Alabama, Mississippi, Louisiana, and Texas. Made of white quartz sand washed down from the Appalachians, the islands are sculpted by wind and tide, which nibble at their eastern ends and build up the western, sending them on a slow-motion migration of their own. Hurricanes shuffle the deck, splitting some islands into pieces, rearranging the contours of others despite the anchoring influence of pine and live-oak forests.

A number of these gems enjoy federal protection, but Dauphin Island isn't one of them. Back around 1971, when the Department of the Interior was assembling Gulf Islands National Seashore, the state of Alabama—not overly fond of Washington government—defiantly opted itself out of the effort to preserve the best offshore and coastline stretches. As a result, the national seashore encompasses Horn, East and West Ship, and Petit Bois Islands in Mississippi, skips entirely over Alabama's 55-mile coast, and resumes again on the Florida side of Perdido Key. Dauphin Island, part of which was divvied into home sites in the 1950s, is accessible by bridge to the mainland a few miles away. Over the years—and despite repeated beatings by hurricanes like Frederic in 1979, Elena in 1985, Opal in 1995, and Georges in 1998—it has grown thick with more than a thousand vacation and retirement homes, T-shirt shacks and hamburger stands.

Fortunately, a little of Dauphin Island's past remains. At the eastern end is the Audubon Bird Sanctuary, a 164-acre remnant of swamp and maritime forest—loose-limbed live oaks draped with Spanish moss, huge magnolias, and tall loblolly pines. The understory is crowded with the fanned leaves of saw palmetto, and thorny catbrier vines twist through it like rolls of concertina wire.

I got there late in the afternoon, pulling into an otherwise empty parking area beneath a grove of pines, and drawing on a windbreaker over my sweater. The wind had been out of the north all day, and it was unseasonably cold for the middle of April in the Deep South. Nor was there much moving in the forest, except the lashing branches high overhead. Where a spring seeped to the surface, creating a damp patch in the dim green shade, I found a few hooded warblers, wearing cowls like yellow-robed monks. A worm-eating warbler (misnamed as any bird could be, since it does not eat worms) was picking methodically through a suspended tangle of dead leaves caught in the lower branches of a bush. With a snap that was audible many yards away, the small, buffy bird cracked the mid-rib of each leaf in turn, prying it back to scan for insects or spiders hidden within—an unusual behavior that allows the warbler to exploit an otherwise untapped food supply.

Except for cardinals and a few titmice, which live in these woods year-round, there were few other birds to be seen. I walked to Gaillard Lake, spring-fed and thus a rarity on a barrier island, hoping to see a few alligators, but the chilly weather had them sulking out of sight. I walked out the boardwalk to the beach, flushing towhees, and swung back through the woods around the swamp, into the open oak and pine woodlands again. It was almost dusk, and I figured that it was past time to get some dinner and find a room for the night. I was disappointed; Dauphin Island is supposed to be one of the best places along the Gulf Coast to watch songbirds in the spring, but there was precious little to see. Maybe in the morning, I told myself.

That's when the sky started giving up its treasures.

I never actually saw the birds come down, but I could hear them, a series of low *whooshes* overhead and around me, like fast pitches that brushed past my ear or the thrumming sound of sticks whirled through the air. An instant later, the lifeless trees were seething with dozens of birds, which cascaded, branch by branch, toward the ground, spilling out into the understory. They started eating without preamble, without stretching or relaxing or preen-ing—feeding with a fervor usually seen only at state fairs during pie-eating contests. Over and over again, small explosions of birds would materialize out of the sky, whirring from on high, beyond the limit of vision and into the trees like bolts, until the woods were stuffed to overflowing with them.

I watched for half an hour, until it was too dark to really see anything but silhouettes and dull shapes. Thousands of songbirds had arrived, dropping straight down from great heights to join the melee, then rolling out in waves through the forest, enveloping me for a few, frantic moments, then passing me by even as the next surge came through. They ignored everything but the imperative to eat, and they certainly ignored me. Binoculars were superfluous; warblers and vireos of two dozen species danced within arm's reach of me on every side. A male prothonotary warbler, as gold as anything from Solomon's mines, dismembered a spider above my shoulder, showering me with a dry rain of tiny, angular legs. A black-throated green warbler buzzed up on trem-bling wings and snagged a mosquito that was on approach to my nose; the warbler hung there a moment, inches from my face, its wings a gray blur, before it veered off to resume the hunt.

What I was experiencing is called a fallout, and it is among the most excit-ing moments in a birder's life. In springtime, millions of songbirds moving north from their wintering grounds in Latin America gather on the Yucatán Peninsula and surrounding areas, then fly across the Gulf of Mexico, making landfall from Texas to the Florida panhandle. The spectacle draws thousands of visitors each April to meccas like Dauphin Island, Grand Isle in Louisiana, or High Island over in Texas, people who eagerly wait for the skies to open up with feathers, with clockwork as regular as any train. No wonder they call it the Gulf Express.

For birders, the diversity is almost as big a draw as the sheer numbers. At least fifty-five species of songbirds make the flight, perhaps more, from tiny ruby-throated hummingbirds to more substantial kingbirds, thrushes, orioles, and cuckoos. They have been drawn from their tropical havens by the subtle shifts of photoperiod and the biological clocks within their own bodies, silently counting off the days until the breeding season. Restless, they take to the wind once more, only this time they are heading north again. Most of them leave Mexico just after sunset and fly through the night and part of the following day. The fastest fliers, like kingbirds, arrive around midday, with the smaller or slower species reaching land in waves over the next several hours.

In good weather they have energy to spare and usually fly past the wide coastal marshes, landing in the larger tracts of hardwood forest twenty-five or thirty miles inland. Only in unsettled weather, with storms and strong northerly winds, do the birds pile up along the coast in great numbers.

In those situations, their frantic feeding behavior is a clue to a much darker side of this phenomenon. Birds know instinctively that the shortest distance between two points is a straight line. Unfortunately, the line they draw between Central America and the great forests of the eastern United States and Canada spans five or six hundred miles of open water, a potential death-trap for creatures that weigh less than two quarters. Making this crossing successfully requires a concurrence of weather, physical conditioning, and simple luck. If anything goes sour, the birds die.

And every year they do die, in anonymous multitudes. The northbound Gulf crossing each spring is one of the great crapshoots in bird migration. In the fall, the birds can pick their moment, departing when the winds shift from the north to give them a boost toward Mexico. But in the spring it is a different story. Most days, the weather pattern is favorable, with clear skies and a fine tailwind from the south. But when the warblers and other migrants lift off from the Yucatán, they have no way of knowing whether a raw cold front is racing down from the Arctic to meet them halfway with towering thunderstorms and icy headwinds that will slow their progress to a crawl, sap their strength, and leave them fighting for their lives. When a storm front hits, they grab for the first land they can find like castaways kissing the sand. Fallout.

So if you're a birder and you want real excitement, you have to hope for a calamity, in the form of bad weather overwashing the Gulf. If you have a conscience, it is sure to bother you, and telling yourself that you have no control over the weather, or that nature's been working this way for thousands of years, isn't going to make you feel any less guilty.

The truth is that, for all its dangers, the trans-Gulf migration usually works very well, even in bad weather, as long as the birds have a place to refuel as soon as they land. In such conditions, the forests that fringe the Gulf are absolutely critical to their survival—which is why so many conservationists are especially worried about this particular link in the migratory chain. Scientists have called the arc of maritime live oak and pine that once rimmed the Gulf from east Texas to west Florida the most important migratory stopover area in North America, but it has been fractured into pathetic slivers—consumed by vacation-home developments, grazing cattle, strip malls, and, most recently, even an explosion of casino construction. Farther inland, the rich, diverse forests of longleaf pine and hardwoods that once supplied the birds with food before the next leg of their journey north are being clear-cut, the trees ground into wood pulp by portable "chipping mills." Elsewhere, the land is replanted with a sterile monoculture of fast-growing junk pine that offers little in the way of sustenance to a weary migrant.

Because most songbirds fly at night and at great altitude, it is difficult to observe their migrations directly. Even so, naturalists had long suspected that passerines were crossing the Gulf of Mexico. In the fall of 1824, an Englishman touring the Mexican coast wrote of his stay near Campeche: "We were visited by great numbers of the smaller kinds of land birds, principally warblers and flycatchers, which reached the ship in an exhausted state, on their migration from the north side of the Gulph of Mexico to the coast of Yucatán. The cabin was never without these pretty creatures, which entered the window in pursuit of flies, that were in great plenty." In spring, people occasionally reported exhausted songbirds landing on fishing boats and freighters in American waters, and of course there were those spectacular fall-outs in bad weather. By the early twentieth century, trans-Gulf migration was an article of faith with most ornithologists.

But not all. In 1945, an ornithologist named George G. Williams, who had worked extensively along the south Texas coast, fired a broadside in the scholarly journal The Auk asserting that trans-Gulf migration was unproven. It was far more likely, Williams argued, that songbirds took a safer, two-pronged route north from Mexico—many following the coast into Texas and others island-hopping from the Yucatán across Cuba to Florida. He dismissed the smattering of eyewitness reports of birds landing on boats, noting that all coincided with storms and north winds. The songbirds, he said, had obviously been blown *south* from their coastal route, out into the Gulf, and were trying to get back ashore.

"Though this theory of trans-Gulf migration is so generally accepted, I can find no modern writer giving actual evidence in support of it," Williams wrote. "Nobody yet has actually disproved the theory that mother snakes swallow their young . . . or that the interior of the moon is made of green cheese. We cannot disprove these theories; all we can do is show how evidence for them is lacking, or how the preponderance of evidence supports another theory."

Williams had a valid point: ornithologists were making a large assumption based on skimpy evidence. But his article—or maybe just that crack about green cheese—obviously got under the skin of an enterprising biologist at Louisiana State University named George H. Lowery, Jr. A year later, in 1946, *The Auk* published a paper by Lowery that rebutted Williams point for point.

"Williams probably succeeded in establishing a reasonable doubt of trans-Gulf migration in the minds of some readers," Lowery wrote. "For this reason I made plans to test the truth of this theory in the only place it could be tested—on the open Gulf, under sunny skies, along the 500-mile seaway from the mouth of Mississippi River to the Yucatán."

Shortly after reading Williams's paper, Lowery signed himself onto a Norwegian freighter bound from Louisiana to Progreso, Mexico. On both the outbound and the homeward legs of the trip, he recorded the species and numbers of birds he saw, in flight and landing on the ship for rest. He also wanted to use a telescope on the deck to watch the face of the moon for nocturnal migrants, but the motion of the ship prevented that.

In all, Lowery recorded twenty-one species, including warblers, thrushes, kingbirds, swallows, sandpipers, orioles, bobolinks, and a peregrine falcon. He also had the crews of Coast Guard cutters counting birds, and he quoted the notes of a friend in the military who kept tabs on birds seen over the Gulf the previous year. Their experiences jibed with his. Once in Progreso, he set up his telescope and watched the disk of the moon, noting the number and direction of night-flying birds. As Lowery's exuberant italics make clear, he felt he had clinched the case: *"All the migrating birds counted at Progreso were flying almost due north, out over the Gulf in the direction of the coast of Louisiana and Mississippi."*

The reason migrants were seen only in bad weather, he contended, was that in good weather they overfly the fishing boats and the coast itself, flying at altitudes beyond the limit of human vision, as was known to be the case over land. "The bulk of spring migration in the Gulf Coast regions is an unseen phenomenon, becoming visible only in bad weather . . . There is little reason to doubt that the *average* bird under *average* conditions can pass from the peninsula of the Yucatán to the coast of Louisiana without tiring unduly and yet continue on for some distance inland."

Williams wasn't convinced, but the rest of the ornithological world was. (Ironically, both Williams and Lowery might have been right. It now appears that quite a few species, such as gray-cheeked thrushes, follow an elliptical route, with many individuals crossing the Gulf in fall but taking a more coastal route north in spring.) Lowery and others continued to study Gulf migration in the decade after World War II, assembling data on storm-grounded migrants, and it was Lowery who, in 1951, first introduced moon-watching networks as a formal means of studying migration. But by the 1960s researchers realized they had a much more powerful tool for following migratory birds—weather radar.

Radar was developed during World War II by the military, using sound waves to create a picture of distant objects. By 1957, eight National Weather Service radar stations were established along the Gulf Coast from Brownsville, Texas, to Miami, Florida, tracking storms instead of enemy planes. But radar, as it turned out, could pinpoint migrants, too, showing a flock of a dozen or so birds as a small dot on the glowing screen and a massive springtime exodus as diffuse green blobs and swirls. Best of all, film records of each day's radar images were permanently archived, creating an ideal data bank covering decades of migration.

Sidney A. Gauthreaux, Jr., at the time also at Louisiana State University, was among the first ornithologists to recognize the potential of radar for studying migration. He confirmed Lowery's assertion that the trans-Gulf route is vital to many species of seemingly fragile birds, and he showed that Lowery was also right when he argued that migrants stay high on their crossing and overfly the coast in good weather.

Now at Clemson University in South Carolina, Gauthreaux has continued to use radar to study bird migration, especially the trans-Gulf flights. In the process, he and others appear to have confirmed what a lot of birdwatchers have been saying for years—that the number of songbirds seems to be declining.

By tediously analyzing decades' worth of radar images, Gauthreaux provided some of the first hard, quantifiable evidence that a decline had indeed taken place. "If one compares the percentage of days with trans-Gulf flights in the spring seasons of 1965–67 with the percentage of days with trans-Gulf flights in the spring seasons of 1987–89, at Lake Charles, LA, the decline over 20 years is almost 50%," Gauthreaux told a gathering of ornithologists in 1989.

That's not to say that most songbirds have become rare. Using radar and visual observation, Gauthreaux has estimated that on peak days migrants will hit the coast of Texas or Louisiana at a rate of 30,000 *per mile, per hour.* If the surge lasts about five hours, which is average, that works out to 150,000 birds per mile, or an incredible 45 million along a 300-mile stretch of coast. The difference between now and thirty years ago isn't so much in the quantity of birds in such a wave, Gauthreaux argues, but the frequency of these events. Today they are the exception, but in the 1960s they occurred almost daily.

The morning after I arrived at Dauphin Island, I drove to an old Indian midden known as the Shell Mounds, where centuries' worth of shucked oyster shells form a series of small, wooded hills, now a park. It was easy to believe Gauthreaux's estimate; the place was crawling with birds deposited by the previous evening's fallout. Male orchard orioles, burnt-orange and black, poked their long bills into roadside flowers, drinking the nectar, and I've never seen so many hummingbirds as were buzzing through the small field at the center of the mounds, chattering in anger like crickets as they contested for the best flowers. Where someone had spread seed beneath a huge live oak, more than fifty blue grosbeaks and indigo buntings fed in a tight knot—all males, all the vivid color of a midday sky, like a living puddle. When a bright red cardinal dropped into the middle of them, the effect was like a Fourth of July decoration.

The same unusually strong cold front that produced this remarkable fallout had spawned a bulging fence of thunderstorms that stretched across the Gulf of Mexico from south Texas to Fort Meyers, Florida—a picket line that, for two days, hammered every migrating songbird unlucky enough to have risked the crossing. About the same time I was enjoying the show at the Shell Mounds, reports were emerging from the Texas coast of thousands of dead migrants washing up on the beaches, in some cases blanketing the sand.

Birders reacted with horror and alarm, but such mass mortality is probably a fairly common occurrence on the Gulf, although it is rarely well documented. One exception was the spring of 1993. On April 8 of that year, a thunderstorm and tornado swept across Grand Isle in central Louisiana, in the coastal marsh country near the mouth of the Mississippi. Two days later, David and Melissa Wiedenfeld of Louisiana State surveyed beaches near the storm path, counting the birds that had washed up. Their findings suggest how terrible a toll weather can exact on trans-Gulf migrants.

The Wiedenfelds found a staggering 7,000 dead birds per mile. By far the most common species were indigo buntings, which made up one in four of the

fatalities. By plotting the area over which dead birds were found, and extrap-
olating from their actual counts, they estimated that more than 4,000 of the
finchlike birds had died, almost all of them males, like the flock I had seen at
the Shell Mounds—probably a result of males and females migrating separately, on slightly different schedules. The next most common victims were
Kentucky warblers, more than 2,500 of them, followed by wood thrushes,
ovenbirds, red-eyed vireos, worm-eating warblers, hooded warblers, white-
eyed vireos, gray catbirds, and orchard orioles. All these species were near the
seasonal peak of their migration when the storm hit.

In all, the pair concluded, at least 40,000 songbirds of forty-five species had
died in the storm—undoubtedly a low estimate, as they admitted, since countless numbers of the dead would have sunk into the ocean without a trace,
while others would have been consumed by gulls, crabs, raccoons, and other
scavengers before the counting started. Forty thousand might seem like an
appalling figure, but it is minor when compared with the number of birds that
make the Gulf crossing every day. According to Gauthreaux's estimate, that's
about what passes over one mile of beach in a single hour on a big flight day.

Among the casualties, the Wiedenfelds estimated that nearly 600 cerulean
warblers died in the storm. This was of particular interest to conservationists,
because this pale blue passerine is ranked among the most seriously threatened songbirds in the country.

You won't find the cerulean warbler on the federal Endangered Species list;
it is not so far gone as to rate that kind of last-gasp governmental life support,
which is usually withheld until it is too late to do much good. But it has an
array of problems conspiring against it, which have sounded alarm bells
among biologists. In 1996, the international umbrella group Partners in
Flight released WatchList, an attempt to pinpoint bird-conservation priorities by systematically ranking the risks each species faces—factors like relatively small population size, tiny wintering areas, long migration routes, or
specialized breeding habitats. When those threats were measured for North
American birds (other than those already listed as threatened or endangered), the cerulean warbler was among the eighteen species—out of more
than 750—considered in gravest jeopardy.

The cerulean warbler takes a hit in almost every facet of its life. Although
it enjoys a wide geographic range in eastern North America, it is fussy about
its habitat choices, preferring to nest in mature stands of floodplain hardwoods that have tall trees and relatively little understory. It is also sensitive to
forest fragmentation, rarely breeding in tracts of woods smaller than five or six

hundred acres, and most common in extremely large woodlands of several thousand acres. It is a long-distance migrant, exposed to all the dangers that entails, crossing the Gulf both in spring and fall on its way to northern and western South America, where it winters from Venezuela to Bolivia. (A very much smaller number of ceruleans island-hop across the Caribbean to the same destination.)

Once in South America, the cerulean warbler is restricted to a thin, snake-like ribbon of misty forest in the eastern foothills of the Andes, within a very narrow band between about two thousand and forty-five hundred feet—the smallest elevational range of any migrant songbird. Nor are they clannish once they arrive. Ceruleans spread out through the forest, joining mixed-species flocks of residents and migrants, but only at a rate of one or (at most) two per flock. This exceptionally low density means that the cerulean war-bler, as a species, needs a lot of elbow room.

Unfortunately, that thin zone of Andean montane forest could hardly be a worse choice in the late twentieth century. The native trees are among the most valuable timber in South America, while the climate—cool and temperate, with abundant rainfall—combined with fertile soil make it ideal for a range of crops, from coffee to rice; it is also perfect for growing coca, the raw ingredient of cocaine. In proportion to its size, researchers say, "the humid forest zone of the montane foothills has been cut more extensively than any other forest type in the Neotropics except the nearly extirpated Atlantic coastal forests of Brazil."

The result has been a disaster for cerulean warblers—the Breeding Bird Survey shows that their numbers fell 49 percent from the 1980s to the early 1990s along nearly three thousand 25-mile routes in the United States and Canada. Of course, the situation is more complex than that; the warbler is actually expanding its range in parts of the Northeast, and in some regions the number of singing males seems to be increasing. But the handful of bright spots are not enough to offset the otherwise dismal picture.

The Wiedenfelds also estimated that the Grand Island storm killed more than 200 Swainson's warblers—named for the same English ornithologist memorialized by Swainson's hawk and Swainson's thrush—another species that the Partners in Flight WatchList ranks at the highest priority for conser-vation. The bird was something of a mystery right from the start, when it was named by Audubon in 1834; it is a secretive ground dweller, a skulker easy to overlook, with a song that resembles those of more common species.

It occurs across much of the Southeast, occupying two very different habi-tats—lowland canebrakes of native bamboo, swamps, and riverbank thickets along the coastal plain and up the Mississippi Valley, and dense stands of rhododendron and mountain laurel several thousand feet up in the Appalachians as far north as West Virginia. It is extremely unusual for one species to occupy two such highly specialized but radically different habitats within a fairly small geographic region, and scientists have tried—and failed—to explain why the Swainson's warbler breaks this otherwise iron-clad rule of biology.

The status of the Swainson's warbler is less clear than that of the troubled cerulean. BBS counts actually show it increasing in some areas, but many ornithologists distrust the statistic, because like so many deep-woods songbirds, this species rarely shows up on roadside BBS censuses. In one year a mere 26 Swainson's warblers were found on just 17—out of 1,832—BBS routes nationwide. With such a small sample, the presence or absence of a half-dozen birds could change the picture dramatically, making any analysis of their population trends little more than guesswork.

Birders and scientists generally agree that the warbler has declined in many areas, especially the Appalachian population, which uses thickets of rhododendron and laurel, as well as mature cove forests, the species-rich hardwood community of the southern mountains. The lowland population seems to be stable or increasing in some areas, declining in others. There, the warbler seeks out bamboo stands—the same habitat favored by the presumably extinct Bachman's warbler. Like Bachman's, the Swainson's warbler winters on Caribbean islands, where much of the native vegetation was destroyed long ago; the fact that it also migrates to the eastern coast of Central America may have spared it its relative's fate.

When I arrived on the Gulf Coast to meet the northbound birds, I was hoping that I might find a Swainson's warbler, which I had never seen. This was partly the incurable itch of a birder to see a new species, a life bird, a thrill that becomes harder with each passing year as the common and easily spotted varieties are tracked down—and partly a longtime fascination with this enigmatic animal. Yet as I worked my way along the Gulf Coast for more than a week, I came up dry each day. Dauphin Island, despite its treasure trove of birds, was no different.

"Well, if you want to see a Swainson's, this is the best place in the world, and the best time to be here. This is the peak of their migration, in the middle of April," Bob Sargent told me, not long after I found him banding a catbird in the shade of a small nylon pavilion. For eleven years, Sargent and his wife, Martha, have run a seasonal banding station at Fort Morgan, on the narrow peninsula that juts out from the east side of Mobile Bay, opposite Dauphin Island. I was joining them for several days to help with the banding chores—a chance, I hoped, to see the details of the migration from the inside, to better understand the complex web of weather and instinct that fuels this phenomenon.

He was right about the Swainson's warbler. An hour later I found my first of many, working its way slowly through the middle of a dense thicket. The bird was already wearing a new, silvery band on its right leg.

"Watch where you put your feet back in here. There's a big old cottonmouth that stays in this thicket, and they're temperamental at this time of year." Bob Sargent led me along a tunnel that wormed its way through a wax myrtle thicket, swinging his flashlight beam slowly up one side of the sandy path and down the other. "On a warm morning like this, the snakes could be curled up right alongside the trail and you'd never see them."

It was five in the morning, still dark, but with a dim violet glow on the eastern horizon. Before the light grew much brighter, and the newly arrived songbirds started moving and feeding, we needed to unroll the twenty-five mist nets that were strung along narrow lanes cut through the forest. A chuckwill's-widow flapped overhead, owl-big and moth-silent, part of the forest's night shift, but already I could hear the chips of songbirds beginning their day's activities.

The nets were stretched between slender metal poles and set in pairs or trios end to end, each net forty feet long and higher than a man's head. I finished opening one net and turned to move on when I heard a rustle behind me; a wood thrush had already been caught, trying to take a shortcut across the narrow path. I quickly untangled it from the net, pulled a cloth mesh bag from the banding apron I was wearing, and slipped the bird inside, feeling the warmth of its feathers against my hand and the rapid-fire beating of its heart.

During the Civil War, Fort Morgan was a strategic location for the Confederates trying to control access to Mobile Bay, one of their major ports. Mines, known as torpedoes, dotted the waters beneath its brick walls. Union gunboats steamed past its defenses anyway in 1864, blockading the city of Mobile. That battle gave us Admiral David Farragut's famous order: "Damn the torpedoes, full speed ahead!" but it ended Fort Morgan's military importance. Today this is a small state park, and while the massive ammunition bunkers still sit in the sun, a gray cliff of stone punctuated by dark, cavelike openings, forest has reclaimed most of the Civil War site. The Sargents' banding station occupies the old stable yards and the yellow-fever quarantine hospital, now nothing more than low foundations and scraps of brick sidewalks in the woods.

"Fort Morgan's a peninsula, not a barrier island, but I think the birds see it as though it was a barrier island," Sargent said as we moved to the last string of nets. "Many of these species have probably used this as a landing point on the flyway for hundreds, maybe thousands of generations.

"When we have a fallout situation we'll have hundreds of just exhausted birds everywhere, very emaciated with no fat reserves at all, birds that used up muscle tissue just to make this journey," Sargent said. "Then this becomes critical, life-saving habitat for them."

Traveling across the Gulf of Mexico puts stresses on a songbird that are difficult to comprehend. The key to surviving the trip is preparation—laying on sufficient stores of fat before leaving the Yucatán. A ruby-throated hummingbird, gobbling insects along with nectar, can add 10 percent of its weight per day. A typical warbler weighing about fifteen grams, roughly half an ounce, can add six or seven grams of fat for fuel, a process that may take days or weeks, depending on the richness of its feeding grounds.

Each gram of fat is good for about 125 miles of flight, ornithologist Paul Kerlinger has calculated, and a Kentucky warbler flying 600 miles across the Gulf will use four or five grams on its eighteen-hour nonstop flight, arriving up to 35 percent lighter than when it took off. Other studies have shown that migrating songbirds lose up to 4 percent of their body weight an hour during these marathon flights. "Tailwinds reduce energy consumption by nearly half, to two or three grams. Imagine how much fat would be required if the warbler experienced a headwind," Kerlinger writes.

Even with ample fat reserves, a migrant's body needs rest after an extended flight, a chance to metabolize the poisonous lactic acid that builds up in muscle tissue (the same toxin that produces the "burn" in a runner's legs). Such birds may need only a day's rest and a quick feed before heading farther inland; from this point on, they will make shorter flights at night, burning at most a gram or two of fat at a time. Truly emaciated birds, those whose bodies ran out of fat on the crossing and began to catabolize muscle tissue in a desperate bid to stay aloft, face a much longer period of recuperation. Some will not be in shape to breed when they finally reach the nesting grounds, and many, especially the older individuals who have made this grueling trip three or four times before, won't make it at all. They will linger on this spit of land, showing up in the nets day after day, unable to muster the energy for one more trek north.

"If you care greatly for these birds, as we do, there's a special sadness in that. Even though you know it's natural, even though you know all these birds eventually reach the end of their lives, that makes it no easier to accept," Bob Sargent said, opening the last net. We were ready for the day.

About a dozen people work at the Fort Morgan station, all volunteers—the Sargents' daughter Donna and her husband, Duane, who runs the satellite station across the road, and a changing cast of friends who take a week or two of vacation each spring and autumn to help out with the operation. Most of them have been doing so for years, all living amiably in an old house rented to them by the park.

After we finished opening the nets, we started around the meandering trails again, checking for birds. The first few rounds were the most productive—each net was hung with little bundles, like fruit that occasionally fluttered. There was an understated excitement about this process, never knowing what the next net held. In this one, two veeries, a catbird, a Kentucky warbler, and a hooded warbler. In the next, a yellow-billed cuckoo, a rose-breasted grosbeak, and another Kentucky. In the next, a pair of summer tanagers, the male rosy red, the female greenish—twin apples, one ripe and the other not. And so around the circuit, until we had dozens of birds, each in its own mesh pouch, dangling by its drawstrings.

The process for removing a netted bird was relatively simple, and although most of my experience involved larger species like hawks and owls, the steps were the same for a diminutive songbird. First, figure out from which side of the net the bird had entered, so you removed it the same way—"finding the pocket," as banders say. Gently take the legs in your fingers and disentangle the toes, then the tail, the wings, and finally the head. From what looks like a hopeless mess, an experienced bander can extract a bird in seconds, like slipping a butterfly from its chrysalis.

Most of our captures that morning were white-eyed vireos, chickadee-sized birds with yellow spectacles and pale irises that give them a slightly addled, manic look. White-eyes bite hard, and since their stout bills have hooked tips for tearing apart caterpillars, they get your attention in a hurry. Bob referred to them, only half-jokingly, as "hateful little boogers," but their nips were more of an annoyance than anything else. The birds to watch out for, I discovered, were the powerful seed-crackers, like cardinals or grosbeaks, which can chop divots out of a bander's fingers. The best solution was distraction—offering the bird a small cotton roll of the sort used by dentists, which gave it something squishy and fingerlike to chomp on while it was being handled.

The Sargents have been banding passerines for years, although they are best known for their work with hummingbirds; they (along with Geoffrey Hill at Auburn University) proposed that western rufous hummingbirds are rapidly evolving a new migration route that takes them to the Southeast instead of Mexico in winter. Near the banding tent, they had set up a hummingbird trap—a large wire-mesh cage on a stand, with a nectar feeder hanging within it and a trapdoor on the side. When a hummer entered to feed, someone would reach up, give the long trigger line a tug, and drop the door.

Bob Sargent's a big guy, six three and 240 pounds, with a shaved head and a benevolent but very impressive demeanor. He has massive hands, yet he handles the hummers, no bigger than a peanut, with incredible dexterity. One habit, however, startled me the first time I watched him remove a bird from the trap. To free his hands so he could reset the door, Bob casually transferred the hummer to his mouth, gently clamping the bird between his lips. For a moment, he looked like a giant out of a children's fable, about to eat the tiny hero. How the hummingbirds perceive all this is anyone's guess, but they seem rather nonchalant about the whole affair. Once the crew is done banding and color-marking them (with a dab of colorful typing correction paste on the forehead), they are offered sugar water in a tube; most eagerly sip and return to the feeder in the trap not long afterward. Many times we would see a hummer zoom into the cage, then recognize the spot of purple, orange, or white on its head and let it feed in peace.

A lot of banding stations are closed to the public, but Bob and Martha do everything they can to encourage visitors to stop, to watch, and to ask questions. Groups wandered constantly down the net lanes, which are also great trails for birding; they were simply warned not to touch the birds, as someone would be along in a few minutes to collect them. Not everyone listens; one afternoon a man came chugging up to the tent, red-faced, to tell us that a bird was caught in a distant net. Yes, Bob said amiably, a crew was making the rounds and would be there directly. No, no, the man insisted, the bird was horribly tangled and needed immediate help. He grew increasingly agitated and abusive, until with a sigh Bob stood up and followed him into the woods.

When we reached the net, the man's wife was jerking repeatedly on the wing of an ovenbird that was barely visible inside a cocoon of black mesh.

"Get your hands off that bird!" Bob roared, all semblance of amiability gone, descending on them like a thunderstorm. "You were specifically told *not to touch the birds!"*

The woman tried to protest, but it was like arguing with a boulder, and she and her husband finally walked off, turning now and again to sputter about how rude and cruel the banders were. Sargent ignored them, trying to undo the damage they had done. By pulling at the bird's wing, the woman had tugged it through several layers of netting, which its struggles had wrapped tight. Try as he might, Bob couldn't find the pocket where it had entered, and the ovenbird's movements were becoming feebler. Finally Bob began popping the threads, ripping a hole in the expensive net until the warbler was free. Its wing hung at an odd angle, and at first Bob was afraid it had been sprained, but when he released it, unbanded, it flew well enough.

Such unpleasant encounters are notable only for their rarity. The banding station is a minor tourist attraction, with crowds of people each day during the spring and fall banding seasons: buses of school kids, organized birdwatching tours, elderly snowbirds who retired to nearby Gulf Shores, vacationers with their families passing through. Today was typical. At the folding table Bill, a longtime volunteer, was banding a worm-eating warbler, his hands moving with practiced speed, using calipers and metric rulers to take a series of measurements, which his wife, Susie, typed directly into a laptop computer. The Fort Morgan banding station will band 2,000 birds in a typical two-week season in April, all their vital statistics stored just a keystroke away. Bob, meanwhile, had a crowd of people circling him as he reached into a bag and pulled out a white-eyed vireo. True to form, the vireo latched on to his hand, catching the soft webbing between the fingers and drawing a pinprick of blood.

"This hateful little booger is a white-eyed vireo," he said with a smile, so they'd know it was a joke. "He's a survivor. The majority of the birds we study here are in sharp decline, but this species lives in second-growth, so it's doing better than a lot of these birds. But not forever. Not for long." A few of the people in the crowd were birders, but most weren't; this might be their first encounter with the marvel of migration. Visitors to Fort Morgan get a strong dose of conservation, with emphasis on the importance to songbirds of the fast-disappearing coastal forests. But mostly they are just dazzled by the birds, which few of them have ever seen close-up before. One after another, Bob produced jewels from the white mesh bags: A male blue-winged warbler, yellow with a black streak through its eye, like smeared mascara. A scarlet tanager, blood red with black wings. Bob released the bird, which disappeared in a puff, and the people shook their heads slightly and sighed, as if waking from a dream. By then he was holding a wood thrush, spotted breast and chestnut back.

"These birds have this blind faith, if you will, in what their ancestors have done for them in the past. Every year they pack up everything they own, as far as their species goes—the very essence of what they are, wrapped up in this migration northward each spring," he told them. "And in this migration they have all the genetic information that makes up that species; there are no remnants left anywhere for these birds to fall back on."

He was on a roll now; everyone was transfixed by the picture he was painting. "They're trusting they can make that migration like their ancestors did. And they do. They come north, they raise their babies, they head back south carrying all that genetic information as if it was luggage. It's a remarkable trust. It is a ticket that was bought for them by their ancestors, and it's a ticket that was bought by many of these birds dying off, so what you're left with are the ones with the right genetic information. This whole matter of migration is a refining process."

Bob drew a six-year-old girl from the circle and showed her how to hold a wood thrush for release, her stubby fingers gently caging its body until he told her to open them. Her sister, even younger, stepped up, and Bob opened her right hand flat, palm up. Onto this he carefully placed a male hummingbird. "Now, just hold your hand real still, don't move, just stay real, real still, that's right, just don't move . . ." Bob spoke quietly, in a singsong voice almost like a chant. The bird lay there for a second, three seconds, almost ten seconds as we all held our breath, then burst into buzzing wings and was gone. The girl's smile was as sudden and brilliant as lightning.

The PR is primarily for the birds' sake, of course, but a few years ago it helped save Fort Morgan itself from the kind of sneak attack that has ruined so much of the Gulf Coast for birds. As Bob Sargent tells it, one day in 1994 he received a phone call from someone he knew, a state official whose name would be best left out of this story. A group of private investors had drawn up an elaborate plan to develop Fort Morgan State Park—a large restaurant, a six-story hotel, a marina, parking lots, twelve "period-style" homes to be leased to private individuals—all of it sitting smack on this last scrap of coastal forest where the Sargents band each year. Even worse, the investors had quietly managed to do much of the initial groundwork without making the plan public. Now, the informant said, it was on the fast track to approval.

He could give Bob ten minutes to look at the plans himself, but no more than that.

By luck, this all transpired the same weekend as the annual meeting of the Alabama Ornithological Society. Bob got an aerial photograph of the park and, on a clear plastic overlay, superimposed the outlines of the proposed development over the live oaks and pines. When he stood up at the meeting with his news and his map, the result was mayhem, which quickly rippled out into the general public, many of whom had visited the banding station over the years.

"Within that first week, there were something like 4,000 personal messages to the governor of Alabama, Jim Folsom, Jr. That certainly got his attention early on. But over the ensuing month, he received some 12,000 additional messages, that's my understanding—and this did not include messages that went to individual state legislators or district commissioners, or any of the other elected officials," Bob recalled. "The newspapers got wind of it, and it became highly publicized. It became a genuine stink, and it was genuinely effective."

When at last the plan came before the Alabama Historical Commission—which has jurisdiction over the property and was reportedly angered because it had been left out of the loop—the board dismissed the proposal almost out of hand. Bob was relieved, but he knew it had been a close call. "We got blind lucky, to be perfectly honest with you," Bob admitted. Without that anonymous whistle-blower, the project might have gained unstoppable momentum before the public became aware of it.

The population in Baldwin County, Alabama, which includes Fort Morgan, grew 20 percent in just five years following the 1990 census, rapidly swallowing any land, like Bon Secour National Wildlife Refuge down the road, that isn't publicly owned. Driving between the fort and the booming town of Gulf Shores, twenty miles away, I started counting PROPERTY FOR SALE signs, but I lost count after four dozen or so. Knowing that sooner or later a monstrous hurricane is going to level this stretch of coastal development, like Frederic did in 1979, only adds to the absurd tragedy of it all. And as if vacationers and retirees hadn't chewed up enough of the Gulf already, now this ancient landscape must contend with gambling, in all its garish, tasteless excess. Along Interstate 10, which parallels the coast, you're hardly ever out of sight of a billboard proclaiming this or that casino, with more popping up like mushrooms after a rain.

Increasingly, the birds must rely on the few bits of protected, public land, like the barrier chain of Gulf Islands National Seashore. But they may be a poor substitute for the rapidly disappearing hardwood forests on the mainland. Researchers found that most songbirds stopping on Horn and East Ship Islands left quickly, and those that stayed gained little, if any, weight. Songbirds stopping in live-oak and hackberry forests, on the other hand, stayed an average of two days, gaining 3 to 5 percent of their weight in fat per day. The difference appeared to be the habitat. The hardwood forests had an abundance of high-quality food, such as caterpillars and other soft-bodied insects, that the dry, pine-covered islands simply don't provide. Lovely and undisturbed though they are, the islands offer little more than a landing place for weary migrants. It also suggests that the industrialized pine plantations that have replaced enormous areas of wild hardwoods in the South, especially Alabama, are similarly unsuitable for migratory songbirds.

"You can't grow 150-year-old live oaks in five years, if you have a change of heart," Bob told me one day, as we sat under the shade of those old trees. "If you lose these special habitats, there will be bad storm years in which you will lose great numbers of the breeding population—birds that would have been salvaged had they been able to stop here and rest on the way inland. And at the rate these neotropical birds are declining, that's a luxury we simply don't have."

Three hundred and fifty miles west of Fort Morgan and Dauphin Island, six or seven hours by interstate, is the low, wet country where Texas and Louisiana meet, a watery near-wilderness of freshwater and tidal marsh that stretches like a prairie. About the only dry land along this part of the Gulf are cheniers—old beach ridges, narrow as snakes, but some of them twenty or thirty miles long, made of sediment carried down the Mississippi River. Just a couple of feet above sea level, cheniers provide a toehold for live oak, hackberry, prickly-ash, locust, mulberry, and other trees that could not otherwise survive in this marshy environment; in fact, the Cajun word, pronounced *shen-year*, means "place of oaks." Not surprisingly, they are supremely attractive to trans-Gulf migrants.

Like most coastal forests, the cheniers have had a rough time of it. The woods were cut for housing and farms, and many cheniers were cleared for cattle; even today, the cows are ferried across from the mainland to graze for the summer, eating away at both the understory plants and seedlings that would someday replace the forest. Once entirely forested, the cheniers now have less than 5 percent of their natural, undisturbed woodland still intact.

Perhaps the most famous chenier of all—at least in birding circles—is High Island in east Texas, along the coast between Beaumont and Galveston. It isn't really an island, but with the emerald marsh all around it—beach and ocean a mile to the south and the Intracoastal Waterway to the north—this long, wooded hummock certainly looks like one.

It looks like one to the migrants, too. Let's say you're a bird, a little bitty Tennessee warbler that weighs as much as four pennies. You left Celestún, Yucatán, at dark the previous day, and ever since you've been flapping without rest, twenty beats a second for eighteen hours. At first you had a tailwind from the south, but in the middle of the night you encountered a squall line, with battering rain that forced you lower and lower toward the deadly sea. You pushed through that with scant feet to spare, salt spray from the white-caps stinging your eyes, but when the rain ceased the wind did not, hammering you in the face, cold and gusty as you stubbornly regained altitude. The little deposits of fat beneath your skin—pale, yellowish mounds under your wings and in the hollow of your neck, which you accumulated by eating uncounted small tropical insects—are nearly gone. Soon your body will start catabolizing muscle tissue, a desperate act of self-cannibalism with only one purpose, to get you across the 600 miles of fatal water.

Finally, from your vantage point at 4,000 feet, you can see a dark rim on the horizon, like a lid on the ocean. Land. Are you overjoyed, relieved, delighted? If you are, the celebration is premature; over the next half hour, as you draw closer and closer, you see nothing but flat, empty marsh beyond the breakers. That's fine if you're a grebe or a gallinule, but you're a forest bird and you need woods—trees for cover and tree-dwelling bugs for food.

That's where the cheniers like High Island come in, and why this place is so wildly popular with birders. Its two small sanctuaries, which total less than 200 acres in size, hold more birds in spring than almost anywhere else on the continent.

Most of High Island is taken up by the village of the same name, a fairly typical, out-of-the-way Gulf Coast hamlet with a gas station and a small restaurant that depends heavily on folks driving down Highway 124 to the beach. But for a few months each spring, High Island turns its attention almost entirely to birds and birders. In that respect, this tiny community has been on the cutting edge of a rapidly growing national trend—catering to the tens of millions of people who enjoy watching migratory birds.

Because I hadn't made reservations months in advance, there was no chance of finding a motel room in or anywhere near High Island, so I stayed near Port Arthur, almost an hour to the north. I left for High Island well before daylight and drove through a fearsome thunderstorm down the long, straight two-lane that lances through the marshes. With the slashing rain, visibility was poor; at one point, a flash of lightning illuminated the road a second before I would have hit a large alligator crossing the macadam. As it was, I swerved enough to miss all but the tip of his tail, while still avoiding the deep water-filled ditches on either side.

The rain ended at dawn, as I pulled into the empty parking lot at the Louis B. Smith Sanctuary, having driven past the turn once. Easy to miss, it's just a woodlot on a shady residential street, with small homes on either side. Yet this and a 122-acre tract a mile or so away, both owned by the Houston Audubon Society, are among the most popular birding sites in North America. By 8 A.M. the place was crawling with people, and more than seventy cars were crammed in the small lot and lined both sides of the street. At the entrance, volunteers collected a nominal admission fee and handed out maps; on a blackboard someone had been keeping track of the species seen there so far that spring, and the total was approaching three hundred. Nearby, on a small set of wooden bleachers overlooking a room-sized pond, five or six mostly older men sat with their cameras trained on low branches, hands poised on the focus rings of big-barreled lenses as wide as dinner plates and as expensive as my car. When a brilliant yellow prothonotary warbler hopped into position, flashes fired and motordrives whirred.

At most birding sites—national wildlife refuges, national parks, and the like—birders are usually outnumbered by casual visitors, well-meaning folks who are likely to ask in a loud voice (scaring everything within a hundred yards) what you're watching. High Island is different. It is internationally famous in birding circles but still unknown to the general public, so that almost all the visitors are committed to their hobby—people with thousand-dollar binoculars and field guides worn in holsterlike pouches riding low on their hips. Many of these folks take their hobby very seriously—not in a joyless way, but with an intensity and drive that even I, a lifelong birder, sometimes find a bit off-putting. I encountered lots of older couples, retirees by the looks of them, and somewhat fewer younger couples; many women in small groups and duos; and a roughly equal number of men who seemed to be working the woods alone.

There were two groups of people on organized bird tours, a dozen or two in each, both squads led by celebrities of the birding world; these men moved along with their hovering entourages, all of whom walked when they walked, stopped when they stopped, looked where they looked. The most intense of all, however, were the three Englishmen I met, lone wolves all, in their genuine green Wellies and waxed coats. Brits are the most knife-edged of all the world's birders; I've know English "twitchers" (as they are called) who, right off the plane, were more adept at identifying warbler flight notes or subtleties in confusing fall shorebirds than most Americans who have been birding here their entire lives. Serious British birders bring a rigor and a discipline to the pastime that Yanks rarely achieve, but one wishes they'd lighten up a bit. These fellows didn't birdwatch, they *prowled*, always on the hunt for a rarity that no one else had seen. Watching them, I got the impression that the lesser birds—meaning, in this case, the hundreds of glorious parula and Tennessee warblers, orchard orioles, wood thrushes, and summer tanagers that blew in with the storm—were just in the way.

These few wet blankets aside, there was an air of subdued festivity among those of us slogging the muddy trails. It was an all-right morning by High Island standards, nothing for the record books—which meant there were more songbirds than most of us had ever seen in one place in our lives. Conversations were conducted in whispers; I have never, church included, seen so many people be so quiet and act so considerately. As people passed on the trail, they'd give a little preview of what lay ahead: "Watch on your left, there's a Louisiana waterthrush by that puddle." "Have you seen a cerulean yet today? Well, there's a male down by that honeysuckle clump." "Any sign of a prothonotary? It'd be a lifer for me." This last was from a woman in her fifties, and when I was able to turn, point, and say, "Right there," just as the warbler hopped into view, I felt like a genie bestowing a wish.

High Island is one stop along the Great Texas Coastal Birding Trail, the brainchild of the Texas Parks and Wildlife Department. Taking a cue from states and provinces that create self-guiding auto tours around a historic or natural theme, Texas mapped out a 600-mile route along the Gulf, identifying 300 prime birding sites along the way, each no more than a half hour's drive from the next. The route and the sites are marked with snazzy road signs (a black skimmer silhouetted against the sun), and the state has published a comprehensive trail guide. Now neighboring states are planning to do the same, eventually forming an interlocking circuit that could keep a birder busy for months.

Birds are big business—bigger than anyone ever imagined. Every five years, the U.S. Fish and Wildlife Service surveys wildlife-related recreation. For decades, this meant hunting and fishing, but during the 1970s and '80s, so-called nonconsumptive uses like birding and nature photography exploded in popularity. In 1996, the agency reported, nearly 63 million Americans watched, photographed, or observed wildlife of some kind, in the course of which they spent $31 billion on everything from sunflower seed to air travel, binoculars and cameras to restaurant meals. A different assessment, the 1994–95 National Survey on Recreation and the Environment by the University of Georgia, found that 27 million Americans considered themselves active birders. That's a 155 percent increase since 1982–83, making birding one of the fastest-growing recreational activities in the country, well ahead of golf, downhill skiing, and hiking. (The difference between 63 million and 27 million may be a matter of degree; many people who love feeding the birds each winter don't identify themselves as birders, even though, in a very real sense, they are.)

Of course, migratory birds also support a multimillion-dollar business catering to hunters, especially waterfowlers. Roughly 3 million people hunt ducks, geese, doves, woodcock, and other migratory game birds, spending $1.3 billion in the process—and federal excise taxes on their guns and ammunition, combined with state and federal licenses like the Migratory Bird Hunting and Conservation Stamp, provide hundreds of millions of dollars more for land acquisition.

The impact of birds on a local community can be profound. In 1992, one study found, High Island attracted 6,000 birders from the United States and abroad, who spent $2.5 million there—not bad for a town with little in the way of tourist amenities. A few years later, the figure had climbed to $6 million. The lower Rio Grande Valley of south Texas, which boasts two world-class national wildlife refuges (Santa Ana and Laguna Atascosa), attracts more than $90 million in birding-related tourist money each year, researchers Paul Kerlinger, R. H. Payne, and Ted Eubanks found. Texas, with more than 600 species of birds, is the nation's most popular birding destination; by one reckoning, a single rare species, the yellow-green vireo, brings in about $100,000 a year in local spending.

Nor is the phenomenon restricted to Texas. Birding tourists spend about $4 million to visit Malheur National Wildlife Refuge (NWR) in Oregon's high desert, one of the most remote stretches of real estate in the country, and $1 million to see Quivira NWR in Kansas. "Avitourists," as they are sometimes called, drop $10 million a year in Cape May, New Jersey, and the 100,000 birders drawn to Chincoteague NWR in Virginia spend a similar amount.

Overall, birders stopping at national wildlife refuges spend between $25 and $160 per visit. And as Kerlinger, Payne, and Eubanks point out, birding income is crucial to local communities like Chincoteague because "much of the spending comes during the 'shoulder,' or non-peak, seasons, when beachgoers are absent. At this time restaurants, motels, and other businesses would be close to empty without the birders visiting the refuge."

No other aspect of commercial birding has grown as spectacularly as local birding festivals. In 1985 there were just five annual festivals in the United States devoted to birds, but by 1997 that figure had mushroomed to more than sixty, in every corner of the country and almost every month of the year. (Another fifteen were scheduled for Canada.) Snowbound in February? Watch whooping cranes in Texas or bald eagles in Washington State. Preholiday blahs in November? The Rio Grande Birding Festival, which promises 400 species of birds and field trips in Mexico, might be appealing; in 1996 it attracted 1,800 people and generated $1.6 million for the host community.

Some festivals focus on individual species—like the Wings Over the Platte festival each March in Nebraska, timed to coincide with the arrival of half a million sandhill cranes, or the Kirtland's Warbler Festival in tiny Mio, Michigan, the center of this endangered songbird's breeding range. Most, though, take place during spring and fall migration, hoping to capitalize on the large numbers of birds passing through their areas, or—like the Southwest Wings Birding Festival each August in the Arizona mountains—on a unique roster of breeding species.

For years, conservationists have fought a generally losing battle between the environment and economics, arguing that the natural world shouldn't have to justify its existence. That is true, but only recently has anyone examined wildlife in objective, purely economic terms. Not surprisingly, it turns out that nature pays its freight in ways we can scarcely imagine. Take insect control; scientists working on the cheniers of Louisiana, studying competition between migrants for food, built wire cages to keep songbirds away from portions of hackberry trees. When they compared the number of bugs in these exclosures to the number in the surrounding forest, they found a substantial difference. The birds were making big inroads on the local insect population.

That may be even more true on the summer breeding grounds, where adults of dozens of species scour the woods for bugs to feed their chicks. One study indicated that hungry songbirds permit faster tree growth and delay major pest outbreaks, "services that may be worth as much as $5,000 per year for each square mile of forest land." In Missouri, scientists found that white oaks covered with netting to keep birds away suffered twice the insect damage of unprotected trees.

So tell me, what is a blackburnian warbler worth, orange and ebony like a jungle tiger? A pair of scissor-tailed flycatchers? A flock of orioles? I suppose that depends on how you measure such things. We can weigh the tangibles and the intangibles, retail sales and spiritual renewal, tourist dollars and check marks on a life list. But in the end such measures are pointless; we should probably just stand aside and watch with quiet humility as another generation of travelers flies north, compelled by a priceless bravery buried deep in their genes.

12

Heartland

Wind is all there is, and all there ever was, on the open prairie. At least that's how it feels on a ragged March day, chilly despite the clear blue sky. There is no place to hide from the wind, no coulee or hollow where it will not reach you, no place you can go where it will not moan across the openings of your ears or make your eyes sting and water.

Beyond me, almost at eye level, hung a red-tailed hawk, pale and washed-out as the sandy soil, held in place by the wind like a child's kite. It didn't see me, hidden as I was in a clump of stunted junipers that grew on the knoll of this high prairie hill. The hawk was looking down, watching the earth for ground squirrels or sharp-tailed grouse, letting the wind nudge it slowly in my direction.

I was sitting in the middle of the Nebraska Sandhills, the largest dune field in the Western Hemisphere—more than 19,000 square miles that run up into South Dakota and west to the Wyoming line, empty land where there is little to slow the wind. Not that you see sand, of course; these dunes were sculpted during the last ice age, when the wind—always the wind—carried the sand and dust down here from the shores of great glacial rivers to the northwest. The dust kept blowing, to make the fertile loess soils of the eastern Plains; the sand stayed here, to make dunes. When the climate warmed, grass, the other constant of the prairie, bound the dunes into place as rounded hills, some of them many hundreds of feet high. Off and on, over the centuries, droughts freed the dunes from their grassy prisons, and the sand marched again, only to be confined by roots when the rains returned.

The hawk and I looked down on countless lakes and marshes, long fingers of cobalt blue amid the yellow grass. Beneath the sand lie beds of hard, mortarlike sediment that trap water, so that just under these ancient dunes is a mid-continental lake containing billions of gallons; wherever there is a hollow or a valley, the water shimmers on the surface. Even though it was only the third week of March, already the wetlands were stirring with newly returned birds—red-winged blackbirds bullying each other with loud, croaking calls and flashing red epaulets, ghostly gray harriers hunting for meadow mice, flocks of ducks and geese, coots and rare trumpeter swans.

The heartland of North America is one of its greatest migratory corridors, especially in the spring. Warmed by the first thaws, dotted with wide rivers and abundant wetlands, full of food for hungry travelers, it is a safe and beckoning highway for open-country species. Waterfowl roll across the Plains and up the Mississippi and the Missouri basins, heading for the glacier-dug pothole lakes of Minnesota, the Dakotas, and central Canada; shorebirds leapfrog over the prairies on their way to the Arctic. Grassland birds return to the hayfields and pastures, and to the few remaining examples of wild, natural prairie—bobolinks and upland sandpipers from Argentina, dickcissels from Venezuela, lark buntings and clay-colored sparrows from Mexico, LeConte's sparrows from Gulf Coast fields.

The hawk drew nearer, still unaware of my presence. But a sound high overhead made me twist and arch my neck to look back, and the redtail saw the movement, tilted its wings into the wind, and hurtled away. I wasn't watching the hawk anymore. The thin, almost triumphant calls came from a chevron of birds passing a thousand feet over my head—long-legged, long-necked, with wings as wide as eagles', arrowing toward the north. The cranes were back.

A sandhill crane is a big bird, with a heavy, powerfully built body, long legs, a muscular neck, and a lance-shaped bill. The cranes are about four feet tall, but look bigger, an impression that is strengthened by their six-foot-wide wings. Their feathers start out dull gray, but become stained with rusty iron oxide, especially on the wings and back, a condition the cranes (for reasons unknown) intentionally enhance by smearing mud on themselves—the only animal besides humans known, in a way, to paint itself. The crown of the head is bare and bright scarlet, and when folded at rest, the inner wing feathers curve up and out like the bustle on a nineteenth-century gown.

There are about three-quarters of a million sandhill cranes in North America, making this the most abundant of an otherwise globally threatened family. Most nest in Canada, from western Quebec to British Columbia, and from the Great Lakes states north to the Arctic islands and Alaska; smaller numbers also breed in the Rockies, the Great Basin, the Cascades, and the Southeast. In winter, most migrate to a fairly small region along the western Gulf Coast through Texas, southern New Mexico, and Arizona, and down into Chihuahua and Durango in Mexico.

From this common refuge, the northbound cranes converge, like the pinched waist of an hourglass, on the central Plains, before fanning out again across Canada and Alaska. Along the way they stop at traditional staging grounds to rest and feed, building fat reserves that will tide them over migration and the first hectic weeks of courtship on the still-frozen tundra and bogs of the North.

No staging ground is more important for the cranes than the Platte River of central Nebraska, fifty miles or so south of the Sandhills. From late February until early April, about half a million of them, the world's largest assembly of cranes, gather on an 80-mile-long stretch of the river known as the Big Bend Reach, around Kearney and Grand Island, where the Platte sags down like a weighted belt.

The Platte River is likely to fool—and disappoint—an Easterner seeing it for the first time. Zooming down Interstate 80, you see a bridge and a sign, PLATTE RIVER, but the cottonwood-lined channel is barely wider than a good-sized creek in the Appalachians or the Ozarks, and a good deal shallower. This is the fabled Platte, the highway of the West along which the Oregon Trail was blazed? Then, just as disillusionment sets in, the car flashes over another concrete bridge with a shallow stream, and past another sign: PLATTE RIVER. And another. And another, all within a mile or so of each other, until you finally realize the truth. The Platte is braided across the land like loosely knitted cords, each channel finding its own random path across the flat plains, parts of the greater whole.

Many places in North America claim to be a crossroads, but the Big Bend of the Platte has a better right to the title than almost anywhere else. "Here the east-west human transportation corridor—the Pawnee buffalo trail, the Sioux Holy Road, the pioneer trails, the transcontinental railroad, the interstate highway—crosses the ancient north-south migration route of the continent's birds and animals," wrote University of Nebraska professor Allan Jenkins. As he points out, the Big Bend Reach sits on one of the most telling divisions in North America, the twenty-inch rainfall line that separates lush tallgrass prairies to the east from the more arid shortgrass regions to the West. North and South, East and West, wet and dry, human history and natural history, all come together in this place.

In 1864 an emigrant named Jesse Shoemaker claimed land on a large island in the Platte, south of what is now the city of Grand Island. He grazed cattle there, the forests that once covered it having been cleared for railroad timbers. Today, Shoemaker Island is mostly hayfield and croplands, some wet meadows of sedges and native tallgrasses like big bluestem, and the buildings of a half-dozen farms scattered along its tapered length. Just downstream is Mormon Island, equally big; together, they run for sixteen miles, with the Platte's channels, edged with old, raw-boned cottonwoods, girdling them on either side.

The evening after I left the Sandhills, I drove slowly down the single dirt road that runs the length of Shoemaker Island. It was a clear day, the sky a pale blue that faded to cream near the horizon. Western meadowlarks sat on fenceposts, yellow breasts toward the lowering sun, and poured out intricate melodies as flocks of juncos flitted through the roadside brush, mixed with Harris's sparrows with bills as pink as bubblegum against their black faces.

Everywhere I turned, there were cranes. This isn't just a figure of speech but the literal truth; for hours, wherever I'd gone along the Platte between Grand Island and Kearney, there had been flocks of cranes in view, ragged lines and skeins of them in the air, and groups of dozens or hundreds scattered around the open fields. But now, with sunset coming on, the cranes that had flown far off into the countryside to feed in harvested grain fields were pouring back into Shoemaker and Mormon Islands by the thousands, from every direction, preparing to spend the night on wide, shallow stretches of the river.

I followed the road around a right-angle turn, and on my left, the fencerow fell away to reveal an old wheat field maybe a half mile long. Into this single field were crowded tens of thousands of cranes, standing in gray ranks like weathered corn. They were extraordinarily skittish birds, and those within a hundred yards of the car jumped into flight as I came into view, then glided to a landing closer to the main flock. Hundreds more were landing every minute, planing down at a shallow angle, bugling and calling. When an especially large flock would begin its approach, the clamor was almost deafening, as the incoming birds sideslipped and tumbled like falling leaves, spilling air from their wings, then straightening out an instant before impact and thumping down, one after another.

A sandhill crane is a living, flying expression of the concept "curve"—the swell of the wings' trailing edges, the undulation of the neck and the belly, the humped shoulders, the reach of the long, fingerlike primaries that push against the weight of the air. Only the bill and the legs are straight, and even these are mitigated—the bill by its exquisite taper, the legs by their bulging knees and curving, partly folded feet.

Many of the cranes on the ground were dancing in couples, strengthening old pair bonds in advance of the breeding season. One of the endearing things about cranes, besides their somewhat human size and lifetime monogamy, is their exuberant courtship display, a ritualized performance of leaps and bows, flapping wings and pumping heads. Dancing is especially common on the Platte in March, when the hormones are running at flood. Daybreak is the time of heaviest dancing, but as dusk approached, many pairs were springing into the air, first one partner, then the other, a movement multiplied hundreds of times so that the vast flock seethed.

I wanted to know how many sandhills were in this one flock, but counting them individually was nearly impossible; for one thing, we were at the same level, and the farther tiers merged into a single, leaden mass. The best I could do was count parts of the fringe, guess what percentage of the whole flock that portion comprised, and multiply. I tried several times, from several vantage points, and the estimated totals ranged from 22,000 to 31,000 cranes. However accurate my guesswork was, those figures aren't improbable; up to 70,000 have been counted here using aerial surveys, and up to 500,000 cranes pass through the Big Bend at this time of year.

By luck, I'd hit the peak of the season. For weeks the numbers of cranes in the area had been building, as more and more took advantage of mild, southwind days to move from the wintering grounds to Nebraska. On average, the cranes spend about six weeks along the Platte, feeding ravenously on waste grain left in the fields the previous fall; they must gain about a pound, mostly fat, before they leave. This doesn't sound like much, but the cranes, for all their size, weigh only seven or eight pounds, so that's the equivalent of a person gaining twenty or twenty-five pounds. By the third week of March, the cranes—now sleek and refueled—grow increasingly restless, and when a period of strong southerly breezes sets in again, most of them race north, emptying the fields and sandbars almost overnight.

After days of listening to cranes, making notes, drawing analogies and comparisons, I finally decided in frustration that their call is next to impossible to describe. It isn't a bark, or a yodel, or any of the usual descriptions, and the sound of a single crane at close range is utterly different from that of a group in the distance. At times, I was reminded of the clamor of gulls at the shore, but that was a pale approximation. The closest I could come for a far-off flock was fingernails drawn along the teeth of several combs, but with a rich, melodic sound, like delicate bamboo chimes struck with small mallets—a cumbersome analogy—and even that's wide of the mark.

Aldo Leopold, in his classic book A Sand County Almanac, described crane music by turns as "a tinkling of little bells," "the baying of some sweet-throated hound" and "the clamor of the responding pack," "a far clear blast of hunting horns," and finally, "a pandemonium of trumpets, rattles, croaks, and cries that almost shakes the bog with its nearness." Roger Tory Peterson's famous field guide suggests a "shrill rolling garoooo-a-a-a; repeated." A different guide book offers this: "Long, hollow, bugling call given on ground and in flight; heard for miles." One of the definitive scientific works on sandhill cranes notes that the call has a "mean fundamental frequency [of] approximately 0.56 KHz for males and 0.93 KHz for females; can be heard up to 4 km away." But the same authors, having catalogued the various ways crane calls can sound, admit that "these adjectives do not fully convey the volume or quality of the sound produced by a mature Sandhill Crane." I suppose I'm not the only one who has thrown up his hands in defeat.

Sandhill cranes are one of the loudest birds in the world, and their secret is coiled like a snake inside their chests. Rather than simply connecting the lungs to the outside world, an adult crane's trachea, or windpipe, loops along the breastbone, forming a tube which, if stretched out, is almost as long as the bird itself. Like someone speaking through a metal pipe, this coil deepens and enriches the crane's voice—and the bony rings of the trachea, which fuse to the breastbone, make the whole thing vibrate during vocalization, amplifying the resonant calls and adding harmonies to them. The result is pure magic, guaranteed to raise gooseflesh on someone hearing it for the first time. (This same trick of tracheal elongation is also found in some other cranes, trumpeter swans, guinea fowl, and curassows, the last a neotropical family of turkeylike birds.)

The sun had set, and the cranes were becoming restive. Now the incoming flocks were gliding to a landing beyond the trees, out in the river itself, where they could sleep safe from coyotes and other predators. Just as I was wondering how much longer the birds of this great flock would linger in the field, dancing and feeding in the twilight, a small crop-duster plane appeared behind me, barely clearing the tops of the trees on the far side of the island.

The pilot tipped into a sharp bank, gunned the throttle, and, to my utter disbelief, strafed the crane flock.

The result was mayhem. Every crane leaped into the air, and the noise was an assault of panicked sound that drowned out even the plane, which shot over the nearer trees and was gone. I clapped my hands over my ears for an instant to shut out the blare. The cranes rose up a couple of hundred feet in twin eddies, as though they were leaves caught in the crop duster's slipstream, then spilled over the treeline to the river beyond. Within a few minutes, the sky was empty, but the murmur of the multitudes on the distant Platte continued until it was too dark to see.

Security is paramount to a sandhill crane. Because they do not swim, they are unable to use deep water, as do ducks and geese; nor can they stay on land, where mammalian predators would be likely to find them. The Platte, with its shallow water and ever-shifting sandbars, offers the perfect compromise. The cranes roost in places where the water is fast-moving and only a foot or so deep, or on sandbars that have been cleaned of their vegetation by floods. Researchers (aided, at times, by Nebraska Air National Guard pilots flying jets equipped with infrared detectors) have found that the cranes prefer areas where the channel is at least 150 yards wide, and they avoid places where the river is narrow, choked with trees, or near a bridge or a road.

In the days when the Pawnee lived along the river they called Kisparuksti, the Platte was a much different watercourse. Snowmelt and spring runoff up in the headwaters, in the Colorado and Wyoming Rockies, once sent scouring floods down the Platte every spring, ripping away cottonwoods and willows, rearranging the sandy channels almost from hour to hour and replenishing them with more sediment; in those days, the Platte eroded its banks at an annual rate of nearly five acres per mile. The landscape was almost devoid of trees; pioneers along the Oregon Trail often compared the Platte's shores to the seacoast. The floodwater flowed into the spongelike marshes and wet meadows, then soaked down to recharge the subterranean aquifer, which kept the meadows moist even after the floods subsided.

But as early as the 1860s, irrigation ditches started bleeding the river, and between 1909 and 1940 a rash of dams plugged its two tributaries, the South Platte coming out of Colorado and the North Platte from Wyoming. Dammed and diverted, today's river is a dim reflection of its old self; nearly three-quarters of the Platte's water now goes for irrigation or municipal use, siphoned off long before it reaches the Big Bend.

Water is one of the touchiest subjects in the West, the difference between barren land and rich, irrigated fields, between profit and loss. Water law is an odd and arcane legal zone based on the idea that a patchwork of individuals and entities can control rights to specific allotments in a river or stream down to the last drop. The Platte was spoken for quickly, and those who got there first, putting the water to "beneficial use"—which has traditionally meant things like irrigation, industry, hydropower, and drinking supplies—have the highest priority on water rights. That's known as the prior appropriation doctrine, or as it's more commonly put, "First in time is first in right." Those with more recent claims must wait for senior holders to slake their thirst, as it were, and in times of drought, the most junior rights holders can be left dry.

Wildlife never enjoyed legal standing when it came to water rights. Because plants and animals weren't held to be "beneficial," it wasn't legally possible to make a claim based solely on ecological reasons. That has begun to change; Nebraska now officially considers recreation and fish and wildlife as beneficial uses, and its lawsuit with Wyoming over planned diversions of the North Platte, which has reached the U.S. Supreme Court, argued in part that environmental considerations must be taken into account. But under the doctrine of prior appropriation, nature is a Johnny-come-lately for its own water, subordinate to all senior rights holders. In dry years, wildlife may still go begging.

The demand for Platte basin water hasn't slackened. In recent years, fourteen new water projects have been proposed for the river, which together would have taken 80 percent of the Platte's remaining flow. Some are still being actively pursued, while others, like Denver's Two Forks Dam project on the South Platte, have been stopped for now. Two Forks, which would have flooded scenic Cheesman Canyon and a world-class trout fishery, as well as reducing downstream flows on which the cranes and other animals depend, has been thumping around like a ghost in the attic for more than thirty years. In the 1980s the U.S. Army Corps of Engineers rated it "the most environmentally destructive" alternative for Denver's water needs, but granted it a permit anyway; after the U.S. Environmental Protection Agency vetoed the dam in 1990, its proponents filed suit in federal district court, trying to have the decision overturned. They lost, but no one will be surprised if Two Forks rears its head again as Denver's thirsty population continues to grow.

The amount of sand sweeping down the Platte, crucial to maintaining sandbars, has dropped by two-thirds in this century; instead, it piles up behind reservoirs. The dribble of water one sees beneath the interstate bridges is a sham, a sadly weakened river which, if it no longer rampages each year with loss of life and property, is no longer able to cleanse itself, either. Where once the braids of water and sand spread over a mile-wide channel, empty of trees, they are now all but pinched off, like sclerotic arteries, their margins supporting thick forests of cottonwoods—great for bald eagles and Bell's vireos, but rough for the cranes. The loss of huge sandbars has been crippling as well for piping plovers and least terns, both of which nest on the open beaches, and both of which are now covered by the federal Endangered Species Act (ESA).

In part to avoid lengthy and costly ESA reviews for each new or reauthorized water project, the governors of Nebraska, Wyoming, and Colorado, with the Department of the Interior, agreed in 1994 to a basinwide recovery program that, for the first time, tries to balance the needs of wildlife against the needs of humans. The first step, expected to take up to thirteen years, calls for increasing the water flows at Grand Island some 130,000 acre-feet annually by juggling stored water in upstream reservoirs and holding it for low-flow periods. The plan also calls for holding up to 200,000 acre-feet more specifically to aid endangered species. While not everyone believes the agreement goes far enough, proponents say it is the first step toward actually restoring habitat for cranes and other wildlife along the Platte.

Change is certainly needed. Where once the sandhills could spread out along more than 200 miles of river, today almost the entire mid-continental population must shoehorn into four relatively small staging areas comprising less than 80 miles of streambed. The biggest section is found along the roughly 40 miles of channel between Grand Island and Kearney, followed by the Overton–Elm Creek stretch just upriver. Another staging ground lies above the town of North Platte, where the North Platte River comes in, along with a tiny spot above Lake McConaughy near Lewellen, also on the North Platte.

Even within the surviving staging zones, however, good roost sites are hard to find. The cranes must jam together where conditions are right, raising the possibility of diseases like avian cholera or botulism. What's more, farming, flood-control structures, and gravel mining pits have destroyed all but a quarter of the wet meadows and native prairies along the river—habitats crucial for feeding.

Gravel and corn have economic worth; cranes, until recently, did not. But that has changed, along with the boom in birdwatching and nature appreciation, and today sandhill cranes mean a lot of money for towns on the Big

Bend Reach. The Wings Over the Platte festival, begun in 1989, is a sellout most years, featuring seminars, crane safaris, and historical tours. The local visitors' bureaus produce "crane tour maps" detailing backroad routes to the best viewing areas, and the state has created elaborate, wheelchair-accessible viewing platforms at some of the river crossings. By some estimates, 100,000 people a year come to see the cranes, which is great for local coffers but can be a problem for birds as shy and sensitive to disruption as sandhills. Most visitors, however, seem to take to heart the rules spelled out in "Crane-watching Etiquette" brochures that are available everywhere, urging birders to stay in their cars and not to approach the birds.

One day I drove south from the Platte to the comfortable town of Hastings. The Hastings Museum is an eclectic, enjoyable mix of old and new: a state-of-the-art IMAX theater and a group of horse-drawn hearses, for example, and an expansive display of antique firearms not far from what surely must be the most complete collection of kitschy salt-and-pepper shakers in any public museum. On the second floor of the museum is a diorama in which ten mounted whooping cranes are frozen in a tableau with a few sandhills, against a background painted to look like the muskeg of northern Canada. It is an image from North America's primeval days, when the whooping crane nested from Illinois and Iowa northwest across the Dakotas and into northern Alberta; in winter they could be found from the Atlantic seaboard and Gulf Coast to the central Mexican highlands. They were never common; the pre-settlement population may have been no higher than 1,400. But in spring, along with the multitudes of sandhill cranes, flocks of the much bigger, more dramatic whoopers funneled north through the Big Bend Reach of the Platte.

The Hastings diorama is the largest exhibit in the world of whooping cranes, and at one time there were nearly as many mounted cranes in this glass-fronted box as live whoopers in the wild. The birds were shot for the museum in the early 1900s, at a time when the species was already acknowledged to be a hairbreadth from extinction; by 1926 the entire world population was thought to be just a dozen pairs, and by 1941 it was just fifteen cranes.

For long, dangerous decades, that number didn't change much. By the 1950s there were still only about twenty wild whoopers, plus a handful in captivity. All the wild birds wintered at Aransas National Wildlife Refuge, on the marshy Texas coast—all the species' eggs in one basket, if you'll pardon the cliché, vulnerable to a disease outbreak, pollution from oil and gas drilling, or a hurricane like the one that wiped out a tiny, nonmigratory flock of whooping cranes in Louisiana in 1940.

Even worse, no one knew where the great cranes went to nest. They were seen each spring along a narrow, hundred-mile-wide corridor running almost due north from Aransas, up through Nebraska and the Dakotas. Although the birds once bred in the northern prairies, no nests had been found in the United States since 1895. Presumably the survivors were breeding somewhere in Canada.

Finally, in 1954, a Canadian biologist helping to investigate a fire in Wood Buffalo National Park in northern Alberta spotted a pair of adults with a rusty orange chick, or colt. In all, the whoopers were making a 2,600-mile trip each year, fraught with dangers for the inexperienced youngsters—among them predators, disappearing wetlands, poachers, electrical wires with which they collided, and fences that tangled their long legs. Few of the chicks made it to maturity; in fact, biologists discovered that, like sandhill cranes, the whooper pairs were laying two eggs but raising only a single chick to fledging, then losing most of them on migration.

In the nearly fifty years since that discovery, the whooping crane is still endangered, still in a very precarious position. Over the years, hundreds of eggs were filched from wild nests, with no harm to the population, since only one from each nest lived anyway. Those raised in captivity bred, but only with human assistance in the form of artificial insemination; because their wing tendons were cut to prevent them from flying away, the males were unable to balance on the females' backs for mating. Recognizing the dangers of having only one wild flock, which undertakes a long migration each year, scientists have tried to establish other populations. Whooping crane eggs were placed in the nests of sandhill cranes in Idaho, but there were many problems; most of the chicks were killed by predators, and the surviving females scattered—natural dispersal among cranes, but behavior that ruined chances for breeding. Eventually the project was scrapped, and attention shifted to Florida, where attempts are being made to establish a nonmigratory flock, which now numbers more than seventy birds. (Because cranes, like waterfowl, must be taught to migrate, the Florida whoopers stay where they are put, much as the original population along the Gulf Coast once did.)

In that same period, the wild whooping cranes that breed in Alberta have done remarkably well; by 1996, more than 200 cranes were migrating to Aransas for the winter, including 47 mated pairs. Those whoopers begin to grow restless in March, especially when the wind is from the south. They make more and more flights, circling high as if in anticipation, until, one day, they are gone.

On their way north, some of the cranes stop on the Platte. For several days before I arrived on the Big Bend, there were reports of an adult in the company of a large flight of sandhill cranes, sending billows of excitement among the hordes of crane-watchers. At the Crane Meadows Nature Center, a recently converted gas station along I-80, someone had posted glossy photographs of the whooper on the bulletin board—a starkly white bird half again taller than the gray sandhills, its red crown and mustache electric in the late-day sun. In another picture, the crane was flying, its seven-foot wings tipped with black and glowing with backlight.

That crane left the same day I arrived; people had seen it circle higher and higher with a group of sandhills, then set off for the north. But there was always the chance of spotting another. One would think that a five-foot-tall white bird would stand out in a crowd, but a whooper can be tough to find, an ivory needle in the haystack of half a million sandhills. Still, I tried. I'd never seen a whooping crane, although doing so is as easy as booking a plane to Texas. Somehow, though, paying for a guided boat tour at Aransas, being led to the cranes as though by hand, held little appeal for me. But to see one on migration, to chance upon it in the wide plains of the Platte, to witness a small part of a 2,000-mile trek reenacted each year since the Pleistocene— *that* possibility got my juices flowing.

So for a week I methodically scanned every flock of sandhill cranes I saw, on the off chance that a whooper was hiding somewhere in the back, screened by lesser beings. I heard that a rusty immature had been sighted up near Kearney, at the National Audubon Society's Rowe Sanctuary, and that an adult had been seen one evening on Shoemaker Island. Wherever they were, I wasn't.

My last day in Nebraska, a warm front moved into the state. The temperature rose into the low eighties, and a powerful wind from the south blew grit and dust in my eyes all day. It also blew the cranes north—scraggly lines and overlapping Vs of them by the thousands, dropping their clarion calls behind them as they went. That evening, the flocks feeding along Shoemaker Island Road were much diminished, although there were still seven or eight thousand sandhills in one field of corn stubble.

Several times I thought I caught glimpses of white from the far edge, nearly a third of a mile away. Unfortunately, the field in that area dipped slightly and the cranes in the foreground blocked my view. Then they rose into sudden flight, a ripple of movement and sound that swept the field in a wave of rushing wings and loud yelps as they circled and came back. A squadron of fifty flew in, and my heart skipped a beat, for there was a white shape in the middle of them. But when the crane landed, I could see that I had been fooled; its body was white, but its head, neck, and bustle were gray, with a peppering of gray feathers on its wings. It was a partially albino sandhill, not a whooping crane.

Someday—if the Platte hasn't been sucked dry, if the dun legions of sandhills still take their annual rest in its shallow waters, if this undeserving world is still blessed by whooping cranes—someday I'll come back and try again.

It is a prairie's gentle deceit that you think you see everything there is in a single, sweeping glance, when in fact you see very little at all, even if you spend a lifetime looking.

That quick glance doesn't show you the sharp-tailed grouse hen, sitting on her dozen eggs in a hollow beneath the stems of last year's blue grama and buffalo grass. You don't see the elk bedded down in the small aspen grove, the Plains garter snake sunning itself beside an old badger hole, the way the sunlight creates a silver nimbus around the fluffy seedheads of the pasqueflowers. You don't see the history of the land—the flaked spearpoints of Knife River flint, lost by Assiniboine hunters and buried deep in the loamy soil, or the bits of bison horn, still black and shiny like coal, that a ground squirrel unearths as it digs its burrow. You might not even recognize the humped and rolling landscape of northwestern North Dakota for what it is—the handiwork of a glacier a hundred centuries gone.

The French trappers and traders called this area *Coteau du Missouri*, the hills of the Missouri, but that river just to the south had no role in their formation. About 12,000 years ago, the retreating glacier left a belt of stagnant ice, twenty-five miles wide, a hundred yards thick, stretching from southern Saskatchewan down into South Dakota. The glacier surged south once again, burying the old ice beneath a thick, insulating layer of glacial "till," the jumbled mix of stone, sand, and soil it bulldozed across the land, then vanished for good with the moderating climate.

Protected beneath its blanket of earth, the stagnant ice survived the warmth. Forests of spruce and tamarack grew on the layer of till (just as "drunken forests" grow today on the unstable surfaces of old glacial ice in Alaska), but wherever the covering of soil was thin, the ice below started to melt. The sunken pockets became lakes and ponds, the bottoms of which

accumulated thick mats of organic matter—trees and shrubs from the surrounding forests, the remains of sphagnum and other aquatic plants. That
insulated the ice below the lakes even more, at the same time that the ice
elsewhere was disappearing at an increasing rate.

Several thousand years after the ice was first buried, it finally melted away.
The lakes drained, of course, and the layer of glacial till collapsed to the
bedrock below—leaving the thick bottom sediments of the former lakes to
protrude from their surroundings as low hills. Between them, on the sunken
till, new lakes formed. The climate dried; the forests were replaced by grassland. Here, then, is the ultimate deceit of the coteau prairies: land and water
have traded places. What now are hills were once lakes, what are now lakes
were once hills.

From the southern Canadian plains, through northern Montana, the
Dakotas, western Minnesota, and northern Iowa, the relicts of the glacier
sparkle in the sun like millions of scattered coins—the prairie potholes, as
these lakes and marshes are known. Some are barely as big as a garage; others
cover hundreds of acres. Some are permanent even in drought years, others
disappear with the heat of August. In all, some 25 million potholes originally
covered this 300,000-square-mile area, sometimes at a density of more than
eighty per square mile. They are vital to migratory birds; this part of the world
is known as North America's Duck Factory, breeding ground for at least half
its waterfowl, as well as grebes, rails, coots, shorebirds, gulls, terns, and many
other wetland species. The surrounding grassland, where the native plants
haven't been plowed under for corn and wheat, are home to pipits and
longspurs, meadowlarks and sparrows, prairie-chickens and shrikes and many
other birds.

Lostwood National Wildlife Refuge is in the middle of the Missouri Coteau,
and its nearly 30,000 acres are some of the most pristine mixed-grass prairie
left in the United States. For hours, I had been wandering its backcountry in
a happy trance, hiking the low hills and skirting dozens of pothole lakes,
without a single hint of the twentieth century visible in any direction. It was
the last week of May, a time of fevered activity; the place positively hopped
with bird life, most of the species having arrived only in the previous few
weeks.

Male bobolinks performed their courtship flights, soaring against the seam-
less blue sky, then hovering on vibrating wings, their metallic songs cascading
out: *bobolink-bobolink-bobolink-bobolink,* like someone performing riffs on a
banjo. Bobolinks are among my favorite birds—snazzy dressers, with their

Prairie pothole region
of United States and Canada

Hudson Bay

Alberta

Saskatchewan

Manitoba

Lake
Winnipeg

Ontario

Lostwood NWR

Coteau
du Missouri

Lake Superior

Montana

North Dakota

Minnesota

Idaho

South Dakota

Wisconsin

Wyoming

Iowa

©1998 Jeffrey L. Ward

Nebraska

Illinois

sable plumage setting off an ocher triangle behind the head, and white rump and wing patches: a blackbird in evening dress.

From somewhere high above the bobolinks, I could hear—just barely—the lisping, descending song of a Sprague's pipit, a slender, sparrow-sized bird of the northern Plains once known as the Missouri skylark. Male pipits perform a remarkable display flight, climbing more than 500 feet into the air, then slowly spiraling to earth, then repeating the exercise again and again, all the while drizzling their delicious music for the benefit of unattached females hidden in the knee-high grass below.

"In the spring mornings the rider on the plains will hear bird songs unknown in the East," wrote Theodore Roosevelt of his ranching experiences in the 1880s in western North Dakota. "The Missouri skylark sings while soaring above the great plateaus so high in the air that it is impossible to see the bird; and this habit of singing while soaring it shares with some sparrow-like birds that are often found in company with it. The white-shouldered lark bunting, in its livery of black, has rich, full notes, and as it sings on the wing it reminds one of the bobolink; and the sweet-voiced lark-finch [the lark sparrow] also utters its song on the air." Roosevelt, a keen naturalist, was right. Song flights and other aerial displays are common among grassland birds, for unlike forest species, they have no convenient perches from which to sing. Instead, they make the sky itself their billboard.

Just as they come in many sizes, prairie potholes also come in a variety of chemical compositions, depending on their water source, the underlying soil, and other factors. Some are spring-fed and fresh, others are more transient, filled only with snowmelt and rainwater. Some of them are quite alkaline, collecting salts and other minerals leached out of the soil, their borders white with dried crust. These are surprisingly rich in insect life and therefore very attractive to birds. Piping plovers nest on the refuge, and part of the alkali flats along Upper Lostwood Lake are fenced off to protect the small, endangered shorebirds; earlier in the day, I'd watched through binoculars as refuge staff combed the lakeshore for plover nests, with their almost invisibly camouflaged eggs; they moved like people threading their way through a mine field.

The saline lakes also attract large numbers of avocets, which are so delicate that they seem more like paintings than living birds. An avocet is a collection of tapers; its legs are slender and blue-gray, its wings and tail at rest come to a single, fine point, and its bill is hypodermic, exaggerated and long, with an odd upward turn three-quarters of the way to the tip. In summer, avocets develop a smooth, rusty tone on the head and neck, with black-and-white patches on the wing. A single avocet is breathtaking, a flock of them beautiful beyond belief.

Avocets breed in loose colonies across much of the West, wherever there are shallow, alkaline pools. The nest is a simple, grassy bowl with a normal clutch of four eggs; while one parent is incubating, the other stands guard or feeds. The peculiar bill shape of the avocet is an adaptation for feeding in the shallows of the saline pools; the head is lowered until the bill tip is parallel with, but just beneath, the surface of the water, then scythed from side to side. Inside the bill, fine lamellae—comblike fibers—filter out the minute crustaceans and other invertebrates that thrive in the salty water.

As the avocets feed, they are often accompanied by Wilson's phalaropes, another abundant nesting species on Lostwood's alkali lakes. Little bigger than thrushes, phalaropes have long, thin bills and the ducklike habit of swimming, rare in shorebirds. Unlike the avocets, which migrate to the Gulf Coast and Mexico, the phalaropes travel all the way to the altiplano, that high desert of southern South America, where saline pools are also common, and to the pampas. In North Dakota, phalaropes sometimes follow the avocets, pecking up the bits of food the larger birds have missed, although they have their own unique method of feeding—spinning in the water like a top, which creates a vortex and concentrates food for easy capture.

Phalaropes are best-known, however, for their reversed sexual roles, in which the female is more colorful, initiates courtship, and leaves incubation and chick-rearing to the male; a female may also mate with, and lay eggs for, several successive males through the course of the summer. In most shorebirds, though, the sexes are identical, and it wasn't until scientists started capturing, sexing, and color-marking individuals that they realized reversed sexual roles are the rule in several other species, including spotted sandpipers. In fact, it is hard to find a breeding arrangement that *doesn't* show up in shorebirds, from simple monogamy to variations on both polyandry (a female mating with more than one male) and polygyny (a male mating with more than one female). The male may take full responsibility for the chicks, the female may, or it may be shared. In some species, like the sanderling, any or all of these patterns may show up.

Biologists have struggled to find explanations for the diversity of shorebird breeding arrangements. One key seems to be migration distance; long-distance travelers tend to be those species that practice polyandry or reversed roles, and among which the females leave the breeding grounds early—in the case of Wilson's phalaropes, by mid-June. Female Wilson's phalaropes gather at sites like the Great Salt Lake by the hundreds of thousands to loaf and feed on brine shrimp, preparing for their journey back to the altiplano. Perhaps the physical strain of producing eggs, coupled with the demands of flying thousands of miles, require that these females pare their investment in the breeding process back to the absolute minimum.

The prairies are vital, not only to the shorebirds like avocets and phalaropes that nest in them, but to those passing through on their way farther north. Even species that travel different routes in the fall converge on the heartland in springtime.

In fall, most adult American golden-plovers, like those I had seen at Izembek in Alaska, leave the western Arctic and fly east. They first stop around Hudson and James Bays, then move on to the Canadian Maritimes and New England. Having fattened on invertebrates and the autumn cornucopia of tundra berries—crowberries, blueberries, cloudberries, and others—the plovers leave the Northeast coast and fly southeast, across the western Atlantic to South America; there is evidence that some may fly nonstop all the way to the upper Amazon in Brazil. Finally, after a trip of 8,000 miles, they reach their wintering grounds in the southern cone of South America.

In spring, however, they do not take the same route back north. Instead, many of the plovers head up the center of South America, then through Central America, up eastern Mexico, and into the United States. By April, they are pushing in a wide front across the Plains states, a movement that reaches its peak the first week of May.

The golden-plover is famous for its elliptical migrations, but a number of other shorebirds, including the Hudsonian godwit and the white-rumped sandpiper, do essentially the same thing. Others make a narrower ellipse, crossing the Caribbean and the Gulf on the northward trip. Once in America, they join still others that wintered along the southern coast, and the whole menagerie rolls north over the prairie states.

There are probably several reasons for such loop migrations. In fall, the prevailing winds work in favor of strong birds that can tolerate a western Atlantic crossing. Leaving land in the wake of a cold front, they have northwesterly tailwinds, which carry them south over Bermuda—and once they pass through the turbulence of the frontal zone, the birds find themselves in the northeasterly trade winds, which nudge them back toward landfall in Venezuela.

In spring, both the northwesterlies and the trade winds would work against them; it's easier to cut across the Caribbean islands or the Gulf of Mexico, or to dispense with a water crossing entirely. The land route has other advantages. In spring, with seasonal rains and snowmelt, even rather arid parts of the Plains have an abundance of water, and the early spring warmth means plenty of insects and other protein-rich arthropods for food.

One of the shorebirds that makes an elliptical migration is the Eskimo curlew, which I had sought (without much hope) in Argentina and whose near-ghostly trail I have followed, almost accidentally, over the years—to the northeast Canadian coast, where they once fed by the millions each fall before crossing the Atlantic to South America; to Galveston Island in Texas, where a few of the northbound survivors are seen once or twice every decade; and to Mormon Island along the Platte, where one was spotted in the spring of 1987.

Exactly a century before that, a twelve-year-old Omaha boy named Lawrence Bruner witnessed the great flights of curlews passing through

Nebraska—coming, as he recalled as a college professor years later, with the blooming of the willows and leaving before the wild plum flowers had fallen, roughly the end of April and the first week of May. The curlews were especially fond of feeding in burned-off prairie, a habit on which others had also commented.

Myron Swenk, a turn-of-the-century ornithologist who left the best account of the species in Nebraska, wrote: "These flocks reminded the settlers of the flights of passenger pigeons and the curlews were given the name of 'prairie pigeons.' They contained thousands of individuals and would often form dense masses of birds extending for a quarter to a half mile in length and a hundred yards or more in width. When the flock would alight the birds would cover forty or fifty acres of ground."

It is believed that curlews fed mostly on grasshopper egg masses buried deep in the soil, although Bruner and other witnesses said they also spent much time feeding in freshly plowed cornfields. Some of these observations of "prairie pigeons" were made incidental to blasting them out of the sky; curlews, golden-plovers, godwits, and other shorebirds were treated much as passenger pigeons back East had been, with no-holds-barred slaughter. There are a number of accounts of gunners filling wagon beds with heaps of dead birds—then dumping the birds out to rot and filling the wagons all over again because the shooting was too good to stop.

The Eskimo curlew never recovered from such excesses, and other shorebirds that migrate up the center of the continent barely escaped a similar fate. On a single spring day in 1821, John James Audubon reported, 48,000 golden-plovers were shot near New Orleans—remarkable only because someone interested in birds was there to record it. Like that of all shorebirds, the plover's recovery is hampered by a low reproductive rate; shorebirds generally lay only one clutch of four eggs each year, and their long migrations exact a harsh toll on youngsters, so that populations take a very long time to grow. Buff-breasted sandpipers probably numbered in the millions in the days of the buffalo, but by the 1920s they were nearly extinct, and in the seven decades since then, their total population has risen to only about 15,000.

Just as avocets and phalaropes seek out alkaline pools for breeding, waterfowl have their own preferences for nest sites, based on the salinity of a pothole. Gadwalls like the more alkaline lakes, while blue-winged teal and most other puddle ducks like the freshwater pools. It was easy to see why. I squished through the mud of one small pond, as mallards complained querulously and two male shovelers pursued a female overhead, wings whistling as they circled the pothole. The coffee-colored water was writhing with movement—mosquito larvae near the top, their tails poking up through the surface film to breathe; diving beetles corkscrewing down with silvery bubbles of air clasped to their bellies, like scuba tanks; and animated, dustlike bits of life that I couldn't identify until I caught one in a drop of water on the tip of my finger, reversed my binoculars, and looked through the "wrong" end. Like a powerful magnifying glass, they showed that I held a tiny crustacean, a copepod or some closely related creature, with a tear-drop shell and flailing legs.

The potholes are food factories for waterfowl. While seeds, grains, and green plants are favorites of adults, ducklings need protein, and the pothole lakes provide it in spades. Mosquitoes and their larvae are especially abundant. "Beside many of the reedy ponds and great sloughs out on the prairie, they are a perfect scourge," Roosevelt wrote. "At sunset I have seen the mosquitoes rise up from the land like a dense cloud, to make the hot, stifling night one long torture." The mosquitoes bother the ducks, too; I watched one drake pintail flicking his head repeatedly, as though having a seizure, trying to discourage the orbiting multitude.

The mosquitoes also made it hard for me to remain still and watch the birds, although I finally found the perfect spot—a low knoll overlooking a small marshy pond, where the breeze kept the bugs at bay and I was hidden from view by tall grass. Two pairs of ruddy ducks were courting, and a more ridiculous sight would be hard to find anywhere. Ruddies belong to a group known as stifftails, and they seem comically misproportioned—the short tail poking straight up in the air, the head overly large. The drakes were dark chestnut, with white cheek patches on their black faces and bills such a bright pale blue they might have been plastic; with their swollen, fluffed necks, the males looked like weight-lifters, with small tufts of feathers above each eye. The hens were drab brown and acted as though they'd seen it a hundred times: Thanks, no, I'm really not interested—can't you guys take a hint? But the males kept it coming, pumping their heads rhythmically, splashing water with their garish bills. When they finally rushed their intended mates, the females dove and disappeared for good.

Yellow-headed blackbirds called from a patch of cattails, the prettiest birds on the potholes, the males looking as if they'd been dipped, headfirst, into yellow paint. Calling, they lifted their heads skyward, wings partly fanned, every feather on their bodies standing out in exertion: *kee-yuk kee-yuk kee-yuk*, then a sound like a gate squeak and a room buzzer. American coots dabbled around the edges of the pond, ubiquitous on potholes, the root of many an insult: You old coot. Crazy as a coot. Coots are rails, with feet that are lobed—each toe equipped with flaps along the edges, not webbed like a duck's. They are a dull, dark gray, the size of chickens—black footballs at a distance, with their white beaks shining like headlamps. One took off, running along the surface as though its life depended on it, wings flailing, legs churning, froth everywhere, a fat man in a panic. Then—a miracle—it was airborne, fluttering madly, as if it knew it was out of its element and had a long way to fall.

This had been a year of record-breaking snow cover and heavy spring runoff, with devastating flooding across the northern Plains; even in this, the last week of May, there were still dirty snowbanks along some of the roadsides. Every hollow and low spot in the land was full of water, and duck populations—which rise and fall with the droughts and rains—were at a thirty-year high. But that isn't always the case on the prairies, and drought is just as essential to the health of a pothole as is a wet year. In a dry season, air can get at the matted, mucky sediment that makes up the pond bottom, allowing the dead stems and other organic matter to rot—a process that would, in a permanently filled wetland, use up much of the dissolved oxygen and reduce the amount of invertebrate life it could support.

A dry year is fine, from the pothole's perspective, but not the kind of prolonged, relentless drought that squeezed the northern Plains in the Dust Bowl of the 1930s. It happened again, to a lesser extent, in the 1980s and early '90s. Potholes withered and vanished, and farmers plowed them right under, permanently destroying them. It's an old story; farmers and road-builders have been draining, ditching, and plowing the prairie potholes since the land was taken from the Indians. North Dakota alone has destroyed more than half its wetlands, with losses in certain parts of the state approaching 90 percent and up to 20,000 acres more destroyed each year. Throughout the prairie pothole region, nearly three-quarters of the original wetlands have been converted to other purposes, mostly agriculture.

Prairie duck populations hit rock bottom in the 1980s, the lowest level since annual surveys had begun thirty years earlier. Hunting usually just skims the excess from a population, but during such a drought there was a real risk of harming the breeding stock, and bag limits and season lengths were cut back drastically; some waterfowlers, contending that the regulations didn't go far enough, started a "voluntary restraint" program, asking their fellow hunters to kill few, or no, ducks. But it didn't help, not that much. In the summer of 1990, aerial surveys of waterfowl breeding grounds in the United

96   SCOTT WEIDENSAUL

States and Canada showed a paltry 25.1 million ducks. Only rain would matter.

And then the rains came. As the years from 1985 to 1993 had been abnormally dry, beginning in 1994 the seasons were unusually wet. Heavy snows melted, soaking into the earth and recharging springs, raising water tables that had sunk too far down to help the shallow potholes. Rain and snowmelt fed streams, flooded river bottoms, filled dips and sloughs that had been plowed and planted for years. But ducks need more than just water; they need upland habitat nearby where they can hide their nests, areas that are big enough to confuse predators like foxes and skunks. Farmers tended to plow right to the edge of each pothole, leaving tiny islands of nesting habitat—an easy target for carnivores, whose populations had risen as fur prices fell and trapping became less common.

To combat this habitat loss, the U.S. Congress authorized the Conservation Reserve Program (CRP) in 1985, encouraging farmers to take marginal, highly erodible land out of production by leasing it to the U.S. Department of Agriculture. Over the next ten years, more than 36 million acres—an area the size of Iowa—were enrolled in the CRP and planted with grass or other cover, 8 million in the prairie pothole region. Tougher wetlands-protection measures in the Farm Bill and Clean Water Act helped, too. And at the same time that the CRP was being started, wildlife agencies and private organizations in the United States and Canada (later joined by Mexico) were launching the North American Waterfowl Management Plan (NAWMP), which set a goal of restoring the continent's duck and goose populations to 1970s levels. The plan named the prairie pothole region as the highest priority, and by 1994, when the drought finally broke, the partnership had protected more than 1.4 million acres of wetlands and upland habitat, with another 480,000 acres restored or enhanced.

With water and habitat, the ducks came roaring back. Unlike large, long-lived birds like cranes, which raise only a single chick each year, ducks lay an average of nine to eleven eggs, giving them an unmatched ability to bounce back from low populations. By the spring of 1997, after just four wet springs, surveys of key breeding areas led biologists to predict a continental population of 42.6 million ducks, the highest level since 1955, with most species above the goals set by the NAWMP. Mallards were at 9.9 million, shovelers at 4.1 million, gadwalls at 3.9 million. Even pintails, which had been declining for years, reversed that trend, rising 30 percent in just a year to 3.9 million birds—still below goal, but a big improvement. (The news was not uniformly positive; scaup, for example, have remained more than 25 percent below long-term averages.) After the breeding season, biologists predicted, more than 90 million ducks would head south—the largest number in modern memory.

It would be wrong to feel complacent about the long-term stability of waterfowl. This tide of rising populations will last only as long as the wet years do—and drought is a certainty on the Plains. Despite the gains made by the NAWMP and other initiatives, most of the region's surviving wetlands are unprotected and badly degraded. Lostwood, with 70 percent of its land still virgin prairie, is an exception; just outside its boundaries, in the fields along Highway 8, the giant tractors till right up to the edges of the potholes, creating swirling, abstract designs in the dark earth as they plow on the contour. The fencerows and small, fallow nooks and crannies, which once sheltered nesting ducks, have vanished, to make it easier to use the outsized machinery of modern agriculture. Elsewhere, the ponds and marshes sit in heavily grazed pasture where cover is sparse and nests are trampled, or in hay fields that will be cut while the hens are still incubating.

When the rain and winter snows forsake the Plains again, the ducks will tumble. How far will depend on how well their remaining habitat is safeguarded. Not even the CRP land is inviolate; although Congress reauthorized the program in 1997 for another ten years, should it be taken off the books, millions of acres of prime habitat would probably vanish again beneath the plow.

The Conservation Reserve Program was reauthorized for the same reason that most of the national wildlife refuges concentrate their energies on ducks—because waterfowl are big, visible, and have powerful constituencies fighting for them, well-funded private organizations like Ducks Unlimited. State, federal, and provincial government agencies recognize the political clout of sportsmen, who are the source of most of their funding. Not surprisingly, most of the conservation attention in the prairies has gone toward game species like ducks and geese. Yet the group most in need of help in the pothole region—in fact, the cadre of birds facing the worst population decline in all of North America—don't quack. They are the grassland songbirds, and their crisis has been as nearly invisible as it is frightening.

Taken as a whole, grassland birds have declined faster, for a longer period, and over a wider area than any other group of species—they have fared worse than neotropical forest songbirds, worse than marshland birds, and certainly worse than waterfowl. The United States has lost 50 percent of its wetlands since colonial days, but the trend for native grassland is nearly apocalyptic.

Iowa, once blanketed in tallgrass, has only one-tenth of one percent left today. Minnesota has lost 99 percent of its prairie, South Dakota more than 75 percent. Most was converted to cropland, or to hay fields planted with nonnative grasses like smooth brome.

The Patuxent Wildlife Research Center, a federal facility, has documented the declines over the past thirty years, using Breeding Bird Survey data. Take the dickcissel, a lovely finch with a canary-yellow breast and a black goatee:

> Populations have generally declined since the mid-1960s ... This species is well adapted to residing in agricultural landscapes, inhabiting hayfields, pastures, weedy fallow fields, and the weedy margins of ditches and roadsides. However, the conversion of these habitats into cultivated fields and the more frequent mowing of hayfields contributed to the declines in some areas ... Changing land use practices as well as persecution may also be affecting Dickcissels during the winter months in northern South America.

Lark bunting, black with white epaulets, a wanderer across the Plains: "Despite their nomadic movements, population declines predominate throughout their range."

Grasshopper sparrow, a secretive bird that sings like an insect: "Population declines prevail throughout most of this range, although some local increases are evident in the Pacific coastal states ... The survey-wide indices indicate a fairly consistent decline since the 1960s."

Henslow's sparrow, a Northeastern and Midwestern species: "The survey-wide indices exhibit a constant decline since the mid-1960s."

Bobolink: "BBS data indicate that Bobolink populations have generally declined throughout their breeding range ... Areas with increasing populations tend to be small and very locally distributed."

Eastern meadowlark: "Some of the most consistent declines of any grassland bird on the BBS ... The long-term trends are almost entirely in a negative direction, with only a few states and strata [geographic regions] showing non-significant increases."

Western meadowlark: "Increased only in the Chihuahuan Desert stratum but declined in 10 states/provinces, 11 physiographic strata, the Eastern and Western BBS regions, U.S., Canada and survey-wide." In other words, it's bad news just about any way you slice it.

The rates of decrease are sobering. Mountain plovers have shown an annual decline of around 3 percent, which might not sound like much, but it adds up—in the past twenty-five years plover populations have dropped a whopping 63 percent.

Yet, despite the dire circumstances, most of the attention on declining birds has been focused on colorful, neotropical forest songbirds like warblers, thrushes, and orioles. There may be a number of reasons for this. Forest song-birds are well known even by those who are not birders; everyone is familiar with a Baltimore oriole, but few have heard of LeConte's sparrow. Forest birds tend to be colorful and dramatically plumaged, ready-made poster children for the cause of conservation. If you're trying to catch the attention of the public or policymakers, better to use a scarlet tanager or a painted bunting, rather than the relatively drab, confusingly similar grassland sparrows and finches—the sort even birders lump as "LBJs," little brown jobs.

Also, some of the first clear evidence for widespread songbird declines came from the radar images of trans-Gulf migrants, and concern immediately centered on the destruction of tropical wintering habitat, a hot environmental topic in its own right. Funding for research flowed primarily to those studying neotropical forest migrants. Yet many of the grassland species most in jeopardy are temperate migrants, like the mountain plover, an earth-brown shorebird that nests in the shortgrass prairies of the western Plains and winters in the Southwest and northern Mexico.

The relative lack of attention to grassland birds may also reflect an unconscious regional prejudice. Most of the declining songbirds breed in the eastern hardwood forests—the same ecosystem with which a majority of birders and conservationists have firsthand experience, since it surrounds the continent's major population centers. Prairies and desert grasslands, on the other hand, have fewer residents, fewer birders, fewer researchers—and thus fewer advocates clamoring for action, although some farsighted scientists have been raising the alarm for years.

In parts of the East, the slump in grassland birds may be only a return to normal. From the Appalachians north and east, the land was originally cloaked in thick, nearly unbroken forest, and grassland species like bobolinks and vesper sparrows must have been restricted to old burns, river meadows, and coastal marshes. With settlement and the clearing of the forest for farmland, species like the upland sandpiper became abundant where they were once all but unknown. Now, that trend has reversed itself; many old farms have passed out of cultivation, growing up with brush and trees, while development has swallowed others.

In the East and the Plains, even where farmland remains, much of it is inhospitable to birds, who cannot nest in soybeans or corn. After native prairie, hay fields are the most favored nesting sites, but these pose their own hazards. Taking in a hay crop was once long, hot work for the entire family, and each field was generally cut only once, in midsummer. Today, a single man on modern machinery can take several cuttings from a field during the course of the growing season, the first often coming in late spring or early summer, just as nesting season peaks.

Ironically, strip mining is providing some of the best grassland habitat in the East. Reclaimed mines, seeded with grass, may furnish hundreds of acres in a single sweep, perfect cover for northern harriers, short-eared owls, upland sandpipers, vesper and Henslow's sparrows, and other beleaguered species. The old strippings constitute an unlikely but vital refugium for prairie species.

But if grassland bird declines in the Northeast represent a return to an earlier order, they highlight how seriously out of balance the prairies have become. It isn't just agriculture; even on virgin prairie, the age-old cycles have been broken.

One of the elemental forces on the wild prairie was fire, sparked by lightning or set by Indians, sweeping through the dry grass, turning old thatch to nourishing ash, and leaving the ground open to the sun. Fire is what kept the forests at bay in the tallgrass prairie and weeded out the woody shrubs in the mixed- and shortgrass communities. But today range fires are stamped out, and cattle sometimes consume the fuel before it even has a chance to ignite, allowing trees and brush to take over.

That's what happened at Lostwood. After settlers moved into the Missouri Coteau around 1900, they broke the soil for crops and worked to kill the dangerous fires; a lookout tower, built in the 1930s, still stands on the highest hill, a sentinel through the decades. Even after the farms failed in the Dust Bowl years of the 1930s and were bought by the government for a refuge, the fires were suppressed—tipping the ancient balance between grass and wood. Snowberry, a low, tangled shrub that bears pink blossoms in midsummer, spread like the fire that had once reined it in; aspens expanded from tiny copses to large woodlands, spreading by root clones and wind-borne seeds. Eventually, snowberry covered as much as three-quarters of Lostwood, and the remaining native prairie grasses were being choked out by alien species like Kentucky bluegrass, brought in by farmers.

Only fire could restore the balance, and now it is used, ever so cautiously, a tool that must be wielded with care. Sitting on one of Lostwood's hills, a long-ago glacial lake in this topsy-turvy land, I could see smoke on the horizon many miles away—just a smudge at first, then a great, billowing cloud like an erupting volcano, thick and turgid, leaning with the wind that smeared its top into haze. After a time I could smell the scent of burning prairie grass, sweet in my nose but sharp in the back of my throat.

It was a prescribed burn, one set by wildlife managers to rejuvenate the prairie, like one whose aftermath I had examined earlier in the day—charred aspens and scorched ground, the crumbling remains of carbonized snowberry stems. And among them, a haze of green grass shoots surging through the blackened earth, ready to renew their ancient covenant with the sun. Above me, on exultant wings, a bobolink was singing.

13

Hopscotch

It's midwinter, and the muddy bottom of the cold, dark Delaware Bay is littered with half-buried forms, only their spiky tails protruding from the silt. They are horseshoe crabs, whose lineage extends back some 300 million years, and as the winter nor'easters churn the surface waters of the bay to froth, they pass the months torpid and almost immobile beneath small mounds of sediment.

To us, the deep bay is a frigid, forbidding place, where the warmth of spring, even simple daylight, are alien concepts. But the changing seasons do make themselves felt here. By March and April, the crabs stir from their lethargy and begin to crawl toward the shallows, moving up from the deep waters and in from the continental shelf beyond the bay mouth. They are armored, ancient-looking creatures whose forbidding appearance—a domed, shovel-faced shell like a horse's hoof, a triangular abdomen with spiny serrations, and the long, pointed tail known as a telson—belies their utterly benign character.

At the same time the crabs begin their migration, another journey is about to unfold nearly 7,000 miles to the south, along the coast of Patagonia and Tierra del Fuego. Immense flocks of red knots, which arrived on the mudflats in November after flying from the Arctic, grow restless as the long, antipodal days erode. Plump sandpipers, they are members of the subspecies *Calidris canutus rufa,* which nest in the barren archipelagos of central Canada, as much as a thousand miles above the Arctic Circle. (Another subspecies, *C. c. islandica,* nests even farther north in the Canadian High Arctic but migrates east to Europe.)

The knots pass the winter on the empty, windswept beaches and hardpan tidal flats, feeding on tiny shellfish. When they arrive, they wear the somber plumage of the nonbreeding season, all dull grays and browns, but as the months pass they molt in a fresh set—new wing and tail feathers to replace the worn and tattered quills that carried them from the Arctic, and new body feathers with the bright chestnut hue of the breeding season.

In March, as the horseshoe crabs are stirring, the knots begin to leapfrog up the Argentine coast, congregating first at the Valdés Peninsula and Golfo San Matías, where they feed on mussels, then moving some weeks later to the extreme southern shore of Brazil, around Lagoa do Peixe near the Uruguay border. Others may move straight across the heart of South America, over the pampas and the Amazon, to Venezuela. Those in Brazil lay on fat at a dizzying pace, chowing down on clams and snails, sometimes more than doubling their weight in a month.

And from Brazil? No more preliminaries, no more leapfrogging. The evidence—from banding, color-marking, tagging—indicates that the knots perform a curving, 7,000-mile flight along the eastward bulge of South America, across the western Atlantic, and finally to the East Coast of the United States, flying between five and eight days without rest or food.

These twin migrations—one local and slow, the other global and swift—collide in late May on the sandy beaches of the mid-Atlantic coast, especially on the shores of the Delaware Bay. It is then, around the highest tides of the month, that the massed horseshoe crabs emerge from the water in solid waves of gray, armored rank upon rank. The larger females, some two feet in length and weighing three or four pounds, drag smaller males that have clamped hold of their abdominal spines; sometimes still other males find a toehold on the same female, or grasp the first male, so that a single hen crab may be pulling a line of three or four hopeful suitors, all jockeying for a chance to fertilize her eggs. The female crabs clamber for position at the high-water mark, digging pits in the sand and laying their eggs—billions of greenish globes the size of tapioca beads. Latecomers plow up the nests of those that have finished spawning, until in places the beach has an olive tinge, more eggs than sand. Crabs push and shove, bullying each other out of the way, waving their telsons in the air where they have been flipped upside down; the beach looks as though it is blanketed by animated cobblestones, paving and repaving a seaside road.

Into this melee come the knots, exhausted and starving from their marathon flight. With them are semipalmated sandpipers, ruddy turnstones,

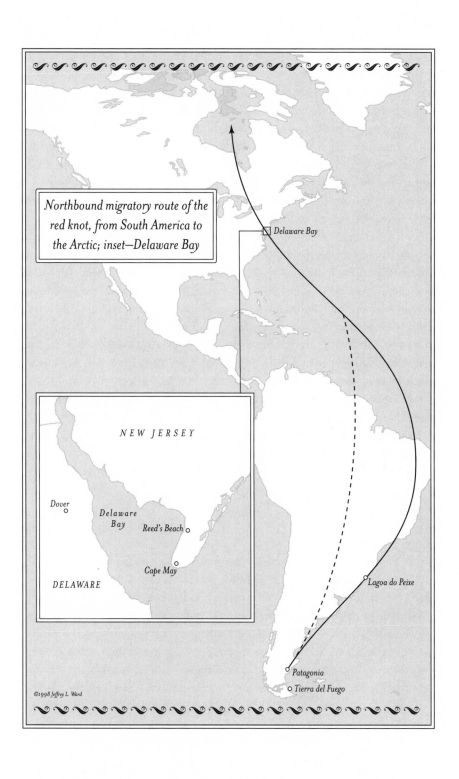

Northbound migratory route of the red knot, from South America to the Arctic; inset—Delaware Bay

Delaware Bay

NEW JERSEY

Dover

Delaware
Bay   Reed's Beach

Cape May

DELAWARE

Lagoa do Peixe

Patagonia

Tierra del Fuego

©1998 Jeffrey L. Ward

sanderlings, and other species, perhaps a million birds in all, most of them also having just completed long, overwater flights from South America. Nothing is more critical to them than a good meal—and they find it, for the crab eggs are a feast beyond measure, enough and more for the weary birds. As in Brazil, they rapidly gain back their lost weight, padding themselves with fat, because the bay is not their destination, only another pit stop on an annual circuit. Within two weeks, they must be airborne again, making another 2,400-mile leap to the Arctic breeding grounds. The fat they accumulate here must be enough not only for the flight but also to carry them through the beginning of the breeding season, when they have more on their minds than food.

For this remarkable system to work, there must be a profligacy of crabs, an orgy of Roman proportions, and more eggs than stars in the sky. It is not enough that there merely be some horseshoe crabs on the beach, or even a lot—there must be so many that the bay shore *seethes* with them like an invading legion, multitudes squandering their biological capital with utter, reckless, bacchanalian abandon, churning the carefully buried clutches of eggs to the surface, where the birds can find them. And even then, with the quarreling, frantic, probing, gobbling crush of a million birds that carpet the beach so that at times one cannot see the sand beneath them—even then, the shorebirds only skim the cream, the tiniest part of the tiniest part of this incredible bounty. In a few weeks, the tides will again rise high enough to reach the surviving eggs and liberate the minute larvae inside, flushing a new generation of crabs into the bay.

But by then the shorebirds will be gone, living on the wind once more.

No other group of birds illustrates the dependence of migrants upon small, widely spaced islands of habitat and resources as vividly as do shorebirds. The Charadriiformes, the order to which plovers, sandpipers, avocets, oystercatchers, and phalaropes belong, is perhaps the most highly migratory order of birds in the world, exceeded only by the pelagic tubenoses. Of the forty-two species of sandpipers that breed in the Western Hemisphere, for example, all but ten are long-distance migrants, and twenty-five of them travel from the high latitudes of North America to the cone of South America or to the South Pacific islands.

Their habitat requirements for feeding and resting are precise; some prefer freshwater marshes, others tidal estuaries or sandy beaches, wooded riversides or rocky coastlines. Baird's sandpiper likes the drier, upland margins of fresh-water wetlands, while Wilson's plover is a bare-beach specialist, avoiding veg-etated areas. Sanderlings are famous for chasing waves right at the surf edge, picking up tiny crustaceans tumbled by the rushing water. Marbled godwits are drawn to muddy pools with grassy borders, rock sandpipers to bouldery shorelines draped with seaweed—and so it goes for each species, which has evolved to fit a particular, sometimes very narrow set of circumstances.

Nor is habitat the only benchmark; the amount and quality of available food is even more critical, since many of these birds make a rapid series of nonstop flights, each spanning thousands of miles, and must be able to refuel quickly and efficiently. And *that* introduces the matter of timing to the equa-tion, because the populations of invertebrates (the food of choice for most shorebirds) rise and fall dramatically with the seasons. A shorebird arriving at the wrong time may find an empty larder and bleak prospects.

In practical terms, that means most of the world's surface is useless to a shorebird—too wet, too dry, too forested, too mountainous, too farmed, too urban, too this, too that. Much of the wetland habitat on which many species depend has been lost. So the relatively few places that still suit the birds' needs are important beyond measure, and the very best of them—like the Delaware Bay, the Copper River Delta in Alaska, the Bay of Fundy in eastern Canada, or the Coppename estuary in Suriname—may serve as gathering places for hundreds of thousands, even millions, of migrants. You don't have to be a conservation biologist to understand the need to protect such places.

How birds use these disparate links in the migratory chain is known as stopover ecology, a maddeningly complicated subject, the study of which is relatively new. It affects not only shorebirds but almost every migratory species—songbirds that have just crossed the Gulf or which must transverse Southwestern deserts, raptors crossing the fragmented forests of Veracruz, seabirds migrating between rich oceanic upwelling zones, and sandhill cranes looking for safe roost sites on the Platte.

The game of global hopscotch highlights, more starkly than any other aspect of the birds' lives, both the raw wonder of their migration and the way humanity's mounting pressures affect the animals themselves. And nowhere is *that* duality, equal parts natural spectacle and ulcer-spawning worry, as clearly revealed as along the Delaware Bay.

I had an epiphany about stopover ecology one May about ten years ago, when I'd gone to New Jersey to feel sorry for myself. My marriage had col-lapsed a few months before, but I was coming out of the shock and figured the change of scene would do me good. I'd birded along the Delaware Bay for years and had encountered impressive groups of shorebirds and horseshoe crabs on occasion, but my timing was always off, always a couple of weeks too early or too late for the big show. Now I had all the time in the world. I checked the calendar, saw that the new moon was a few days later, and packed the car.

There are two high tides every day, roughly twelve and a half hours apart, the handiwork of the moon's gravitational pull. But twice each month, when the moon and sun line up and exert their gravity in concert, the ocean rises even higher than usual—the neap tide, when the moon is new, and the spring tide, when it is full. The crabs, which had been dribbling onto the shore for weeks, would make a mass spawning around the neap tide, laying their eggs where the sea wouldn't reach for another two weeks. The vanguard of the shorebirds had already arrived, and their numbers were building steadily each day. The big flights would be airborne, somewhere between Brazil and the beach, a trip of between 130 and 190 hours. This time, I'd be there to see them when they arrived.

Reed's Beach is a tiny bayside cottage community about ten miles north of Cape May, mostly still shuttered tight in the middle of the week before Memorial Day. The tide was slack, just starting to come in, and the beach itself, which curves north and west for a mile beyond the village, was largely deserted, littered with hundreds of upended horseshoe crabs unable to right themselves, dead or slowly dying in the sun. A few small clusters of birds were milling around, hanging out with their own—twenty or thirty red knots here, a handful of ruddy turnstones over there, little cliques of sanderlings and semipalms keeping to themselves with a clannishness typical of shorebirds. But at nine o'clock in the morning there was nothing to indicate that this beach was the stage for a world-class wildlife marvel.

Six hours later, as the high tide was just starting to ebb, everything had changed. I drove to the end of the road and parked near a rock jetty, where I sat and watched the waves roll in sluggishly through the endless cordon of horseshoe crabs mobilized at the waterline. Down the crescent of beach, the air rippled with birds, and the ground between the dunes and the ocean had a patchy, parti-colored look—great swaths of brown and russet where thousands of knots were scrambling, smaller pockets of harlequin-patterned turnstones, dunlin with their black bellies, and clots of white laughing gulls standing like football players in huddle, heads down to feed, their crossed, ebony wingtips sticking up like so many hundreds of pairs of scissors as they dug through the sand for eggs.

The noise hit me, a cacophony of coarse gull yelps and musical shorebird whistles, as did a fish-roe smell on the warm afternoon air. I looked down and saw that the sand was blanketed with green-gray eggs that had been floated free by the tides and deposited here by the currents in thick windrows, the source of the odor. More shorebirds were flying in from the salt marshes behind the beach, mostly knots with their robin-red breasts and lithe, powerful wings, and turnstones with mahogany shoulders and black-tipped tails. Wave upon wave, they landed among the flocks already feeding, each one somehow finding a bird-sized opening in a crowd that looked as tightly packed as the bristles on a brush. It was easy to believe what I'd read, that 80 percent of the hemisphere's red knots (including every single *rufa* in the world) and 30 percent of all the sanderlings pass through here in a few weeks.

The sight, the sound, the smell: taken together they were nearly overwhelming. Then panic swept the beach, a phenomenon birders call a "dread"— maybe sparked by the predatory silhouette of a distant falcon, maybe just an explosion of jitters in these naturally edgy creatures. Whatever its cause, the birds roared into flight with a hurricane's voice, wheeling and eddying like water in a boulder-strewn river. It was as though the beach had exploded into its constituent atoms. Yet even in their alarm, there was the precision of a close-order drill—no collisions, no confusion, thousands of birds moving like one encompassing, amoebic organism as they whirled back to resume their meal, the false alarm forgotten.

The foundation of this biological pyramid, *Limulus polyphemus*, is not a crab at all. Already old when the dinosaurs were young, horseshoe crabs have scarcely changed in 300 million years; a fossil from the Permian period would be instantly recognizable by anyone who's strolled along an Atlantic beach. They are not crustaceans, like true crabs, but members of the Chelicerata, the same branch of arthropods to which spiders and scorpions belong. The relationship goes no further, though; horseshoe crabs lack venom, fangs, stingers, and all other offensive weapons. The five pairs of large legs, which are hidden from view by the carapace, are tipped with small claws, which can, at worst, give you a weak pinch if you're curious enough to stick your fingertip in them.

The Atlantic horseshoe crab is found from Mexico to New England, but is most abundant in and around the Delaware Bay, which has nearly 150 miles of beach, protected from the worst excesses of ocean surf and perfect for spawning, as well as tidal estuaries where the minute, planktonic larvae can feed and grow. A bottom-dweller, the adult crab rummages through the sediment for shellfish and worms, cutting a meandering furrow as it goes. A few precocious crabs reach the spawning beaches in April, and a handful show up right through the summer and fall, but by far the greatest breeding activity comes in late May and early June, the same time that the shorebirds peak. Marine biologists censusing horseshoe crabs have found as many as a quarter million on a single beach on a single night. (*Limulus* prefers to mate under cover of darkness, and the daylight high tide attracts only a portion of the crabs waiting offshore.) The females dig a series of nests, each a hand-span deep and containing about 3,600 eggs; in repeated spawning bouts over the course of several weeks, each one will lay as many as 80,000 eggs.

Biologists Gonzalo Castro and J. P. Myers, who knew from earlier studies that each shorebird eats about 8,300 eggs per day, calculated that the flocks passing through the Delaware Bay in spring required nearly 1.2 million pounds of crab eggs to keep them fat and healthy. That seems like a lot, but other researchers found that the top two inches of sand on Reed's Beach contained more than 100,000 eggs per square meter—almost a pound, so many that the birds have almost no impact on the crab population. As many as a million eggs per square meter lie buried at least six inches deep, below the level the shorebirds can reach, and those nearer the top are replenished constantly by spawning crabs, some of which continue to breed for weeks after the shorebirds depart for the Arctic.

This sort of organic extravagance is the overriding theme at all the major shorebird stopover sites that have been identified and studied. The upper Bay of Fundy, between New Brunswick and Nova Scotia, teems with a tiny amphiopod named *Corophium volutator*, a.k.a. mud shrimp, a minute, slender crustacean that builds tubular shelters in the silty bottom of this, its only home. The bay's fifty-foot tides expose them to as many as a million hungry semipalmated sandpipers, perhaps 95 percent of the hemisphere's total population, along with at least thirty-three other species of shorebirds, whose arrival in late spring and midsummer coincides neatly with the mud shrimp's twice-annual spawning periods.

In the western reaches of Prince William Sound, Alaska, the attraction is herring roe, which in May accumulates knee-deep in tidal pools, providing food for tens of thousands of black turnstones and surfbirds—a phenomenon that was unknown until frantic biologists (including my old friend Stanley E. Senner) stumbled across it while assessing the damage from the *Exxon Valdez* oil spill. A hundred miles to the east, where the Copper and Bering Rivers empty into the sound, the mudflats hold uncountable numbers of small shellfish and crustaceans. Each spring, as many as 20 million shorebirds pass through the Copper River delta, making it perhaps the most important stopover site in the world for this family of birds. Back in the late 1970s, while he was doing his master's research at the University of Alaska, Senner and his colleagues studied the importance of the delta for two common migrants, dunlin and western sandpipers, concluding that it provided critical fat reserves that allowed the birds to continue their migration into northwestern Alaska and supported them well into the breeding season.

Here's another similarity between Delaware Bay and the Copper River Delta—both sit beside shipping channels alive with oil tankers. The worst possible nightmare seemed to have come true in March 1989, when the *Exxon Valdez* dumped its 11 million gallons of oil into Prince William Sound, just weeks before the start of the spring migration. Fortunately, winds and currents carried the crude away from the Copper River system, although beach oiling was heavy in some parts of the sound where the herring spawn.

On the Delaware Bay, tanker traffic is also constant, supplying the refineries near Wilmington and Philadelphia; the bay is the largest oil transfer point in the East, with more than 12 billion barrels moving through it in the course of a year, virtually all arriving by water. As I sat at Reed's Beach, I could just make out long, low-slung ships on the horizon, and shivered to think what a mistimed spill would do to this spectacle. Other potential threats include pollution from the heavy industrial development that lines the upper bay and agricultural runoff laced with pesticides from its more rural watershed; both PCBs and DDT are still found in the bay and may work their way up the food chain from the horseshoe crabs through their eggs to the birds.

The birds have only about two weeks, on average, to double their weight before they must be on the wing again, and a finite number of daylight high tides during which they can feed most efficiently. The fact that Memorial Day weekend falls smack in the middle of the stopover period guarantees a collision between their needs and the recreational desires of thousands of pasty, sun-starved humans. The state of New Jersey has built special viewing platforms at discreet distances from the best beaches and erected flashy signs that plead with visitors to steer clear of the birds, but not everyone gets the message.

As I drove out the narrow road to Reed's Beach the next morning, I passed an old rattletrap pickup truck driving out, its bed piled high with horseshoe crabs. I'd heard that some fishermen used them to bait their eel pots, and I remember thinking that at least it's lucky there is no great commercial demand for horseshoe crabs. Which proved two things: that I am a lousy historian, and that I should avoid a career in prophecy.

How do you preserve a group of animals dependent, not on just one or two places, but on dozens of them, scattered across two continents? Where a single break in the circuit can cause the whole system to collapse? That's the dilemma facing shorebird conservationists.

"The crux of the conservation problem is that a majority of the shorebird species that migrate between North and South America tend to concentrate in enormous numbers at a relatively small number of migration staging areas. In some cases, perhaps more than half the population may gather at single sites, attracted by predictably abundant but ephemeral food resources," Brian A. Harrington warned in 1986. "The conservation priority is clear— major staging sites in North and South America must be protected."

Harrington, a shorebird biologist with the Manomet Center for Conservation Sciences (formerly the Manomet Bird Observatory) in Massachusetts, has been studying red knots for more than two decades and was among the first to document the importance of the Delaware Bay for this and other species in the early 1980s. Harrington, J. P. Myers, and a few other prescient scientists organized the first international efforts for shorebird conservation in this hemisphere, starting with the International Shorebird Surveys (ISS) in 1974. Almost a decade later, biologists involved with the Pan-American Shorebird Program traveled to South America to band, tag, and color-mark plovers and sandpipers, which were later spotted in migration in the United States and Canada, confirming links between specific sites.

With data from the ongoing ISS and other research programs, biologists proposed a series of reserves up and down the hemisphere, protecting the most important of the stopover and wintering sites, ranking them based on the number of birds they attracted. If a marsh or estuary supported at least 20,000 birds or 5 percent of a population, it would be considered a regionally important reserve. Those with at least 100,000 individuals, or 15 percent of a population, were held to be of international significance. And those places drawing at least half a million shorebirds, or 30 percent of a species' population, were judged of hemispheric importance. The first such reserve, not surprisingly, was the Delaware Bay, so designated by the governors of New Jersey and Delaware in May 1986.

And thus was born the Western Hemisphere Shorebird Reserve Network (WHSRN), which by 1998 included thirty-four sites in seven countries, overseen by a loose partnership of more than 120 organizations that cooperate to

monitor shorebirds and their habitat. The network was modeled on the Ramsar Convention on Wetlands, an accord adopted in 1971 in Ramsar, Iran, and
the only international treaty focusing on a particular ecosystem. Ninety-two
countries have joined the Ramsar system, pledging to designate at least one
site for inclusion on a list of wetlands of global significance and working to
protect that site's ecological character. Obviously there is some overlap
between Ramsar and the WHSRN, and many sites designated under one are
recognized under the other.

WHSRN status, of itself, provides no legal protection to a site—it represents no binding treaties, carries no legislative restrictions, and while many of
the locations are otherwise safeguarded as national parks or wildlife refuges,
others are privately held, or cover a patchwork of public and private owner-
ship. The network's strength comes from drawing attention to the value of a
particular stopover or wintering site, bringing expert assistance to bear on
management problems, focusing research, fostering contacts between scientists and land managers in distant countries, helping to shape conservation
plans to defend the place, and applying public pressure when development
proposals would threaten the integrity of a site. WHSRN's weight is moral
and scientific rather than legal, but that can be potent.

It has to be, because there are plenty of threats. Cheyenne Bottoms in central Kansas is both a WHSRN hemispheric site and a Ramsar wetland, and
part of it has been protected as a state wildlife management area since 1957;
as many as half of all the shorebirds that migrate up the Great Plains in spring
stop here, up to 600,000 at a time, including a large percentage of the world's
long-billed dowitchers. Yet irrigation withdrawals from the Arkansas River
and deep-well pumping that has depleted the underground aquifer have left
the reserve chronically short of water, with invasive cattails crowding into
the fertile mudflats where birds feed on bloodworms. In 1989, the year before
it was declared a WHSRN site, Cheyenne Bottoms went completely dry, and
drought plagued it through the early 1990s, until wetter weather and a court
order mandating greater water flow for the reserve came to the rescue.

Now Cheyenne Bottoms is threatened by plans for a nearly half-million-
square-foot pork-processing plant near the refuge, which will draw millions
of gallons of water a day from the failing aquifer and create waste-water lagoons
both at the plant and at nearby hog farms that could rupture and flood the site.

Across the network, threats run the gamut from illegal hunting and pesticide contamination to large-scale industrial development. At the Tierra del
Fuego reserve in southern Argentina, the most important wintering site for

Hudsonian godwits, offshore oil drilling and tanker loading is a major worry. Gray's Harbor in Washington, which attracts a third of a million shorebirds in spring, is being squeezed by urban expansion and alien plant species like cordgrass and giant reed, which are crowding intertidal areas. Great Salt Lake, which, in addition to its importance to a million shorebirds, attracts more than 3 million waterfowl and 500 wintering bald eagles, faces the prospect of a new freeway slicing through the fertile marshes of its eastern shore. Marismas Nacionales, an expanse of lagoons, marshes, and mangroves on the Mexican coast, is being chewed apart for shrimp farms, while rice farming threatens the Bigi Pan hemispheric reserve on the Suriname coast, where as many as a million semipalmated sandpipers stop in migration. Some of the threats are almost surreal, like proposals to stretch enormous barriers across the Bay of Fundy and exploit the tidal surge for hydroelectric power, drastically altering the ecology of the bay in the process.

The official WHSRN site profile for the Delaware Bay Hemispheric Reserve lists oil traffic, water pollution, and human disturbance as the primary threats. It says little about the horseshoe crabs, except to speculate about possible toxins in the crab eggs and their effects on shorebirds. But by the mid-1990s, scientists who study the interactions between birds and crabs were noticing a disturbing trend. There seemed to be fewer and fewer crabs on the spawning beaches each spring.

It is impossible to actually count how many crabs live in Delaware Bay, and estimates of the total population range from 1 million to as high as 4.1 million. Those figures are extrapolated, in part, from counts of breeding crabs in spring. Starting in 1990, a group of about a hundred volunteers, including oceanographer and horseshoe crab expert Carl N. Shuster, Jr., have surveyed the main spawning areas on preselected dates in May or June. The first two years, they recorded peak counts of more than 1.2 million crabs, most of them on the New Jersey side of the bay, which has traditionally offered better breeding conditions. But after 1991, the surveys showed a steep decrease—to less than 400,000 crabs in 1992 and 1993 (very few of them in New Jersey) and barely 100,000 in 1995.

Did the crab population crash? The answer isn't as simple as the numbers would make it seem. The survey, while a fine effort, is a fairly crude snapshot of just one or two days during a breeding season that spans months; in 1995, for instance, the peak spawn occurred in early May, prior to the census, so most of the crabs were missed by counters that year. (The count technique has since been amended to keep tabs on the most important beaches over longer periods.) Bad weather and rough surf can keep the crabs off the shore even during what ought to be the peak high tide, something that happened during the unseasonably cold spring of 1994.

The beach count wasn't the only evidence of serious trouble, however. Deepwater trawls by the Delaware Division of Fish and Wildlife, which collect a random sample of crabs throughout the year, showed a sharp drop in numbers, especially among adult females. Clam-dredging surveys on the New Jersey side echoed that finding. And the same researchers who sampled crab-egg densities in 1990 along New Jersey's shoreline repeated the exercise in 1995 and 1996 and found a staggering 90 percent reduction in the quantity of eggs available to feeding shorebirds in the upper few inches of sand.

But there was one trend involving horseshoe crabs that was heading *up* at a steep angle—the number caught by commercial fishermen. Horseshoe crabs have no food value themselves, but they make a cheap, convenient bait for other species. Watermen had always used a few to bait their eel pots, but as the Asian market for American eels increased in recent years, so did the demand for crabs, especially prespawning females stuffed with eggs; the crabs also became popular bait for shellfishermen targeting whelk, also known as conch. Crabs were often hand-picked from the beaches, allowing harvesters to select the large, egg-bound females, although trawlers worked the deeper water in late summer and fall, taking both sexes and in far greater numbers.

Significant crab fishing in New Jersey waters began about 1990 and grew rapidly, with the catch climbing to 100,000 animals in 1993 and more than 600,000 in 1996; that year, an estimated 900,000 crabs were caught in the Delaware Bay as a whole—perhaps as much as a third of the total population. What's more, most of the decline was showing up on the New Jersey side of the bay, where the heaviest fishing occurred—and where most of the shore-birds traditionally gathered. (Unlike New Jersey, Delaware had earlier banned commercial collecting on all state and federal beaches, leaving just a few miles of private beach open to the harvest, but like New Jersey, it allowed unrestricted dredging in its waters.)

For all their fecundity, horseshoe crabs are at unusually high risk from over-harvesting. They are slow-moving and must expose themselves to land-based predators like humans in order to mate. Waves and their jostling neighbors flip many while they are spawning, and one crab in ten dies of exposure on the beach. According to some studies, an egg has only one chance in 130,000 of making it to maturity, a glacially slow process that takes a horseshoe crab nine or ten years. By selectively removing adult females before they had a chance to breed, scientists felt, fishermen were risking a collapse of the crab population from which it would take years, perhaps decades, to recover.

Everyone was jittery as the 1997 spawning period approached, and with it the spring shorebird migration. It soon became clear that a crisis was at hand. Shorebird numbers were far below average on most of the normally crowded beaches, and almost all the red knots, along with most of the ruddy turnstones and sanderlings, deserted New Jersey entirely. Some of the birds shifted to the Delaware shore, where collecting pressure was less intense, and where crab numbers—though much lower than had been previously found across the bay in New Jersey—remained relatively stable.

Some biologists worried that the birds, forced into smaller foraging areas with less to eat, might not be getting enough food to allow them to finish their migration and breed successfully; others, including Brian Harrington, noted that shorebirds captured on the bay were actually heavier than average that spring, and warned about jumping to conclusions. Almost everyone agreed, however, that the decline in horseshoe crabs was real, as was a long-term decline in both overall shorebird populations using the bay and populations of several species in particular, including red knots, sanderlings, and semipalmated sandpipers.

Pressure had been building on New Jersey for several years to toughen its crab-fishing regulations. Crabbers were already restricted to collecting on beaches only at night, to avoid disturbing the birds, and were allowed to pick crabs only two days a week. But unlike Delaware, which barred collecting on public beaches and which levied a hefty license fee, New Jersey granted fishermen much greater access to spawning areas and charged only $2 for a commercial crab license in 1995.

As word of the worsening situation spread through conservation and birding circles, New Jersey governor Christine Todd Whitman was urged to take immediate action. She did on May 30, 1997, imposing a sixty-day moratorium on all horseshoe crab fishing in state waters, following that up two months later with permanent regulations designed to drastically curtail crab fishing, including a complete prohibition on dredging or trawling (which accounted for 60 percent of the harvest), and restricting hand-pickers to just two days a week in backbay areas only.

The conservation community, although a shade disappointed that the regulations didn't go farther, seemed impressed by Whitman's fast response. While she may have been swayed by concern for the delicate ecological web, cynics wondered if economics might have played a crucial role, too. The crab harvest, while an $850,000 industry, couldn't hold a candle to the importance of birding in South Jersey, where the pastime is by some estimates a $10 million-per-year powerhouse.

Surprisingly, human health was also a factor. Horseshoe crab blood contains a substance known as Limulus amebocyte lysate (LAL), a clotting agent that reacts to the presence of deadly gram-negative bacteria. By federal law, a wide variety of injectable drugs, vaccines, and medical devices must be tested with LAL for bacterial contamination—and the only way to get the stuff is by capturing live crabs, carefully bleeding a few ounces of their blood for processing, and returning them, alive, to the surf. The Delaware Bay, with the largest population of crabs in the Atlantic, was the best source of LAL, but their decline forced some of the biggest collectors to cut way back on production.

Not everyone was pleased with Whitman's actions. To New Jersey watermen, among whom annual income is generally low and to whom crab fishing was a small but important part of the yearly round of employment, the moratorium and new rules seemed like an overreaction. They took their case to the New Jersey Marine Fisheries Council, which overturned the governor's proposed regulations—only to reverse itself a few weeks later after legal action by conservationists. In the end, the Atlantic States Marine Fisheries Commission, a federal body, announced in November 1997 that New Jersey, Delaware, Maryland, and Virginia had agreed to develop a joint management plan for horseshoe crabs. But in the meantime, unlimited dredging still continued on the Delaware side of the bay.

There are no guarantees that a plan balancing the needs of crabs, birds, and humans will come out of this process, or that one will come quickly enough to prevent a wholesale collapse of the horseshoe crab population in the bay. Everyone hopes that calamity will be avoided. But if such a disaster does come to pass, it will be the second time, not the first.

Horseshoe crabs were only marginally important to the Lenape Indians and other tribes that originally inhabited the land surrounding Delaware Bay. There isn't much meat on the animals, except for some clumps of muscle in the abdomen, but dead crabs buried in maize fields fertilized the growing corn, a technique the Lenape passed on to Europeans before they themselves were forced off the bay.

By the middle of the nineteenth century, the "king crab," as it was widely known, was the foundation of an enormous fertilizer industry, with an estimated 4 million harvested in 1870 alone, a fairly typical figure for the day. Old photos of the fishery have a bizarre quality to them; one from 1908 shows a series of slapdash wooden pens at Cape May holding as many as 18,000 crabs, while another from the 1930s shows thousands of them, dead and stacked as neatly as dinner plates, forming right-angled walls on Bowers Beach in Delaware. Over the years, tens of millions of crabs were dried and ground up for fertilizer or as protein supplements for cattle and poultry feed.

The king crab industry finally collapsed in the 1950s from a lack of crabs. Carl Shuster, Jr., the horseshoe crab expert who helps with the spring beach counts, guesses that by then there may have been only a few tens of thousands of crabs left in the entire bay. When I interviewed Dr. Shuster a few years ago for a newspaper article about the crabs, he recalled that when he started his research in Cape May in the 1940s, he never saw the mass spawning so famous today. In fact, he told me, he'd walk the beaches and see only a relative handful of crabs.

This is one of the strangest riddles about the crabs, and perhaps the most important. Conservationists today fear that a horseshoe decline will have devastating effects on migrant shorebirds. But during the first half of the twentieth century, the crab population was, at most, a fifth of what it is today. How did the birds survive that debacle? Could their dependence on crab eggs be a recent development? Are they more flexible in their dietary requirements than we have always believed? Or is there another, less obvious explanation?

Except for two casual and incomplete references, early naturalists on the Delaware Bay didn't comment on a feeding frenzy of shorebirds, a truly remarkable omission. Even though the Delaware and New Jersey bay coasts were remote backwaters until well into the twentieth century, it's hard to believe these astute observers overlooked something of this magnitude. In 1934, a note in the old National Audubon Society journal *Bird Lore* mentioned thousands of shorebirds feeding on "king crab spawn" along the bay, and three years later ornithologist Witmer Stone, in his two-volume work, *Bird Studies at Old Cape May*, passed on a secondhand report about the same thing.

It is at least theoretically possible that the phenomenon is of recent origin, dating back only to the past few decades as crab numbers recovered from overfishing. I don't buy that, and neither does anyone else I've spoken to. If the fertilizer crews were plucking 4 million king crabs a year off the beaches, year after year in the late 1800s—a figure three or four times the total estimated crab population in the 1990s—the spawning orgy must have been considerably bigger and more impressive, the load of eggs on the beach even more staggering. Hungry birds, quick to exploit any advantage, wouldn't have passed up a ready food supply like that.

So if the crabs were there, where were the birds? The answer may lie in a little-appreciated chapter in the history of American wildlife, the market hunting of shorebirds in the late nineteenth and early twentieth centuries. Shorebirds? Usually talk of old-time market hunting brings waterfowl to mind—punt gunners blasting canvasbacks on the Chesapeake Bay, for instance. But in those days, almost anything with feathers was considered table fare, including songbirds; the menu of an 1886 game dinner in Chicago lists "Jacksnipe, Blackbirds, Reed Birds, Partridges, Pheasants, Quails, Butterballs, Ducks, English Snipe, Rice Birds, Red-wing Starlings, Marsh Birds, Plover" along with buffalo loin and ragout of bear. Robins were popular ("Every gunner brings them home by bagsful, and the markets are supplied with them at a very cheap rate," Audubon observed), as were "reed birds" and "rice birds," both colloquial names for the bobolink.

Shorebirds were a backbone of the market hunting industry, however. They flew in vast flocks, responded to wooden decoys and to whistled imitations of their calls, and many had the habit of circling back in response to distress cries from their own species, so that wounded birds felled by the first volley would pull their comrades back again and again to the slaughter. And slaughter it was, especially for incredibly abundant species like American golden-plovers and Eskimo curlews. There was no closed season, and attempts by states like Rhode Island to curb the worst excesses, like spring shooting, were ignored or repealed. Massachusetts banned "fire-lighting" shorebirds, but the practice remained common on Long Island until past the turn of the century; roosting flocks of red knots, known as "robin snipe," were blinded with torches, grabbed by hand, and their necks wrung.

It was an interstate trade of substantial proportions, with railroad shipments from the Great Plains and along the East Coast supplying the cities with shorebirds at a dime apiece. In 1890, an avid sport hunter named George Mackay, who loved to shoot shorebirds but who was growing worried about the commercial kill, noted that just two game dealers in Boston received more than 23,000 shorebirds that year from Nebraska and Texas, including 10,000 upland sandpipers and 5,000 Eskimo curlews. Taxidermists set up shop along the Atlantic coast as well in those days, buying shorebirds to stuff for the millinery trade; Stone wrote of two at Long Beach, one at Beach Haven, and another at Barnegat—all in New Jersey—paying ten cents each for "sand snipe," as least and semipalmated sandpipers were known.

In 1916, the United States reached a historic treaty with Canada on bird protection, which was implemented two years later when Congress passed the first federal migratory bird laws. That finally closed the season on all shorebirds except woodcock, common snipe, yellowlegs, black-bellied plover, and golden-plover; the plovers were added to the protected list in 1926, and yellowlegs the year after. In fact, the move to safeguard wild birds in general stemmed in no small part from the dismal state of shorebird populations after a century of relentless market gunning. By World War I, the Eskimo curlew, American golden-plover, long-billed curlew, upland sandpiper, buff-breasted sandpiper, and a dozen or more other species seemed headed for extinction.

While all but (perhaps) the Eskimo curlew escaped oblivion by a whisker, it has been an arduous and uncertain road to recovery, stymied by the slow reproductive rate and hazardous migrations characteristic of the group. At the time of his first visit to Cape May in 1890, Stone wrote: "The birds were far less numerous than at present, and infinitely more wary," a condition he ascribed to overhunting. Even in the 1930s, when Stone wrote his books, shorebird numbers continued to fall, and he specifically mentioned that knots were hard to find, leading him, a bit wistfully, to wonder what the shorebird flocks must have been like in the days of early ornithologists like Alexander Wilson.

The low point for horseshoe crabs, therefore, coincided with rock bottom in shorebird populations. There were few eggs to eat, and few birds to eat them. The speculation among some shorebird biologists is that the relatively small flocks of surviving birds would have dispersed more widely across Delaware Bay, switching to alternate sources of food in the absence of crab eggs, spending more time in tidal marshes and less on the beaches. Once the crab fishery shut down in the 1950s and the spawning orgy began to recover its former heft, the slowly increasing knots, turnstones, and other birds would have returned to the beaches in growing numbers.

If this is true, it suggests that the shorebirds are more adaptable in their feeding requirements than conservationists have thought. But while this notion is comforting, there are two wild cards in this assumption. There is no way of knowing what effect being forced to switch to other foods might have on the knots and their kin. The Delaware Bay is the final stop before they reach their nesting grounds, and the fat stored up there is essential not only for migration but to carry the birds through the first weeks of breeding. Alternate food supplies along the bay might keep the birds alive, might even provide them with enough energy to reach the nesting grounds, but not enough to successfully reproduce.

Nor is the Delaware Bay the same place it was fifty or sixty years ago. Coastal development has hit this region hard; rising ocean levels have degraded tidal marshes, pollution has ruined once-fertile shellfish beds, and beaches that used to be isolated and deserted now are rimmed by vacation homes and invaded by off-road vehicles, joggers, swimmers, kids with Frisbees, unleashed dogs, and jet-propelled watercraft. A lot of experts suspect there isn't as much bird food, and they know there isn't as much peace and quiet as there was the last time the horseshoe crabs crashed. The options for the knots and other shorebirds are a lot less attractive this time around. The birds have many more mouths to feed and far fewer ways to fill them.

The largest group of red knots that Witmer Stone ever saw on the New Jersey coast was a flock of 150 in 1898, a blip by modern standards. What would that old gentleman have said if he could see tens of thousands of knots on those same beaches a century later? Maybe he'd congratulate us. Maybe not. I wonder what he would think about the recurrence of their perilous state, so long after he and other pioneering conservationists thought they had safeguarded the migrants against the threat of guns and greed.

"While we realize that we, at best, can see but a mere fragment of the hordes of these beautiful birds that thronged our meadows and beaches in the time of Wilson, we should be grateful for what protection has done for them," Stone wrote sixty years ago. Even then, he saw new dangers on the horizon; "the craze for draining the marshes which seems to have obsessed those in authority and bids fair to destroy all the wildlife on our coasts that we have taken such pains to preserve."

And then, in closing, Stone makes a statement as true today as it was then, the reason why conservationists can never let down their guard: "As one menace is disposed of, another seems inevitably to develop."

14

Catching the Wave

Springtime nights are cold in the Chihuahuan Desert, cold enough to make you glad for a fleece jacket and warm gloves, and gladder still for the first rays of light coming over the high limestone cliffs of the Sierra del Carmen, just across the Rio Grande in Mexico. The rising sun sets everything in this stark place in dramatic contrast—casting long, skeletal shadows from the scraggly limbs and scarlet flowers of the ocotillo; illuminating the flat pads of prickly pear with their shields of long, crisscrossed spines; even gilding the wizened creosote bushes with their miserly leaves redolent of toxic resin.

Most of Big Bend National Park, down in the southwest heel of Texas, looks like this: majestic but seared, withered, rock-paved, brown. Yet here and there, spaced barren miles apart, are patches of green, magnets for a starved eye that leaps to them hungrily.

If you think that an oasis should have palm trees, then Dugout Wells will be a disappointment. There is one palm, a short, unremarkable thing, but most of the half acre or so is taken up by a dozen big cottonwoods and a thicket of huisache, mesquite, and other thorny, nearly impenetrable shrubs. Dugout is a natural spring, fed by water that seeps through a buried gravel bed from the high Chisos Mountains to the west. It got its name when early settlers dug a crude pit house for shelter and sank shallow wells in the damp soil around 1900; later, they installed a windmill to make the seasonal water permanent. The settlers (and the tiny schoolhouse that gave Dugout Wells the nickname "cultural center of the Chisos") are long gone, but the water remains, creating a bright green dome in the otherwise brown expanse of desert.

The sun was rising, and as I hiked through a low canyon toward the oasis, desert birds were coming to life. A cactus wren chattered from the top of a head-high prickly-pear stand, in the middle of which was the wren's untidy nest. A flock of scaled quail skittered crazily away, bursting into flight like a string of firecrackers going off, *pop, pop, pop, pop*. A black-throated sparrow sang from a creosote bush, its trill as thin as the morning's warmth.

The wren, quail, and sparrow live in the Chihuahuan Desert year-round, adapted to the daytime heat and nighttime chill, to the lack of water and the paucity of vegetation. But to migratory birds, especially small passerines, the Chihuahuan Desert is a formidable barrier, one that can be as challenging as the western Atlantic or the Gulf of Mexico. Songbirds that migrate up from Central America and Mexico on their way to breeding territories in the western United States and Canada must transverse enormous spans of desert—the Chihuahuan in Texas and New Mexico, the Sonoran and Mojave farther west in Arizona and California, and the Great Basin across parts of Nevada, Utah, Oregon, and Idaho.

Just as an overwater migrant must find land, so these desert travelers—many of which are adapted to high, cool mountain forests—must find a suitable haven when daylight comes. Dugout Wells, shining and green in the morning sun, fits the bill perfectly.

The transition from desert to oasis was sharp; I stepped out of the creosote bush flats, across a dusty, one-lane track that encircled Dugout Wells and into the shade of the thicket. Following a trail that looped through the brush, I found myself surrounded by furtive movements and quiet call notes. A yellow-breasted chat sat at the top of a small acacia, whose tiny, globular flowers were the same color as the bird's underside. Bell's vireos, small, gray, and hyperactive, darted through the foliage, as did yellow warblers and white-crowned sparrows.

As I rounded a bend in the path, coming to the largest cottonwood in the grove, there was a startling explosion of wings and alarm cries as dozens of birds dove for cover. I froze, and almost immediately they began to filter back, drawn by an imperative that overcame their natural caution—thirst. Near the base of the tree, a patch of ground the size of a card table was damp, with a spot of open water near its center. That the puddle was only the size of my palm and scarcely half an inch deep made not the slightest difference; to birds that had just flown for hour after exhausting hour from Mexico, this was the most important twelve square inches in the whole Chihuahuan Desert.

Within moments, the birds returned in force. I counted more than a dozen species at or around the water hole, including such migrants as Bell's and

Cassin's vireos, hermit thrushes, and yellow-rumped, orange-crowned, and yellow warblers. A male Wilson's warbler plopped himself in the middle of the puddle, tail high, head low, and beak open in a futile attempt to claim the whole thing, but a much larger thrush simply shouldered him out of the way and sank down, soaking its belly feathers. A pair of mockingbirds, permanent residents of the oasis, tried the same thing with a bit more success, but they were bailing the ocean with a teacup; as soon as they turned to chase one interloper away, several more poured in behind their backs, stealing quick sips of water.

Like shorebirds hopscotching from one stopover site to another, north-bound songbirds must find appropriate habitat at every stage of their journey. That is relatively easy in the East, where forests are widespread, but in the Southwest wooded habitat is a much scarcer commodity. Migrants tend to gather around desert springs, along river corridors, or in high-altitude forests like those in the Chisos Mountains.

The morning before, up in the Chisos around 7,000 feet, I'd been hiking through forests of oak and piñon pine at daybreak, when I found myself in the middle of several large flocks of migrating passerines, dominated by Townsend's warblers and Hutton's vireos—the former golden and black, the latter small and greenish, almost dead ringers for ruby-crowned kinglets. Hermit thrushes fed on the red berries of Texas madrone trees, while the warblers and vireos searched methodically among the emerging leaves and flowers of the oaks, pulling out small caterpillars hidden inside. As I labored up steep switchbacks on the trail, one of these mixed flocks would overtake me, immersing me in sound and movement for a few minutes, then quickly passing on. A few minutes later, another flock would come in on their heels—and another, and another, until the sun was high and the birds had settled down for a few hours of rest.

Birders call these spring flocks "waves," because of the way they roll across the landscape, unpredictable in their exact timing or landfall. (Unpredictable, that is, unless you visit one of the so-called migrant traps that are scattered across the continent, places where geography shoehorns the birds into small, wonderfully concentrated channels.) Catching the biggest waves of spring, when the world becomes a riot of color, movement, and song, is something that every birder hopes for—but it is also a goal that is more difficult to achieve with every passing year. Over the past three or four decades, scientists are increasingly certain, something has been mysteriously sapping the life out of this annual phenomenon, rendering it a ghost of its former glory—setting off a scramble to determine the cause, and sparking what may prove to be the largest, most inclusive conservation effort in history.

With the cold winds of autumn under their tails, songbirds hurry south in the fall, but their spring migration is a much more leisurely affair. Unless they are crossing a wide barrier like the Chihuahuan Desert, northbound passerines tend to move in relatively short hops once they're in North America, traveling only about thirty or forty miles a night, with frequent layovers for rest and food. The males are usually in the vanguard, arriving days or weeks ahead of the females, since it is the males that stake out territories. This is the final act in the yearly round of migration, the home stretch in a journey that began the previous autumn. The numbers are lower now, winnowed by disease and predators, accidents and bad weather. The survivors, blessed by the right genes and good luck, carry the future of their species north with them toward the breeding grounds.

In many parts of the continent, the pace of the spring migration is timed to coincide with the first appearance of new leaves on the trees, the "green line" that creeps steadily north across the landscape with the warm winds of April and May and rises each day higher and higher into the mountains. "The timing is hardly coincidental," writes ecologist John Terborgh, "because the larvae of several species of defoliating insects hatch out in synchrony with bud break to take advantage of tender, protein-rich foliage." The caterpillars, mostly inchworms, are themselves tender and protein-rich; unlike many insects, they are spineless and nontoxic, protected only by their overwhelming numbers, which the birds do their level best to reduce. In places like the upper Great Lakes, where the laid-back pace of spring migration still brings the birds well before the cold-numbed leaves open, the songbirds feed heavily on midges, minute insects that are barely visible to the naked eye but that emerge from the lakes at a time when caterpillars are unavailable.

The spring migration is a protracted affair, spanning three or four months and encompassing several major routes—overland through Mexico to the Southwest, across or around the Gulf, and across the Caribbean islands to Florida. By early March, the first black-throated gray warblers are crossing the Mexican border into Arizona, while the first hermit warblers may not show up there until a month later, eventually arriving on their Pacific Northwest breeding grounds in late April. Yellow warblers appear in Arizona in early March and in coastal California in mid-April, while a different subspecies, the most northerly breeding of all warblers, doesn't arrive on its Alaskan nesting grounds until late May or very early June.

The first sizable movement of neotropical songbirds across the Gulf of Mexico comes in mid-to-late March, the same time Caribbean migrants are island-hopping across the Antilles toward Florida. The passerine flights peak in the Southeastern states from late March to early April; at first the movements are dominated by warblers, but as the weeks go by, more and more orioles, buntings, thrushes, tanagers, and others show up in the mix.

Like ripples on a pond, riding the temperate south winds that prevail in spring, the successive waves of birds expand north, usually arriving in a predictable sequence, led by species that winter within the United States and followed weeks and months later by ranks of longer-distance migrants. For many years I've kept a record of first-arrival dates on an old, yellowed desk calendar, the spiral-bound pages now nearly filled with cramped, barely legible notations like "1st tree swallows '87" and "Heard first scarl. tan. '92." Those notes show that, year in and year out, the real harbingers of spring in the central Appalachians are turkey vultures, which reappear most often around the first week of February. They are followed about a week later by common grackles, and a few days after that by adult male red-winged blackbirds; the immature males and the females come several days thereafter. Robins, those celebrated "birds of spring," are comparative wimps, usually not making an appearance until the last week of February or early March—and then only if the snow has melted.

Many temperate migrants—such as waterfowl, eastern bluebirds, woodcock, and ring-billed gulls that wintered just a few hundred miles south—pass through here in March. As the days warm, more and more insect-eaters return, including tree swallows and eastern phoebes, both of which will eat last year's berries if the weather turns too cold for bugs. The big movement of songbirds begins locally in early April, first with hardy pine, palm, and yellow-rumped warblers that wintered along the Gulf, then blue-headed vireos, Louisiana waterthrushes, and brown thrashers, and finally a rapid crescendo from the tropics that climaxes the last week of April and the first week of May.

There may be no finer time of year in an Appalachian hardwood forest than daybreak on a mild, calm morning in early May. The oaks have a gauzy, buckskin color as their flowers and small, furry leaves break bud, a pointillistic collage mixed with the brighter greens of tuliptrees and the bronze of seed-bearing maples. The ovenbirds and wood thrushes, newly returned from the tropics and already claiming territories, fill the open, basilica spaces between the trees with song, while the night's fresh batch of migrants moves through the forest canopy searching for insects. Watching these spring waves always reminds me of the concourse of a large airport, where travelers from every walk of life and every corner of the world rub elbows. Beyond a need for trees, forest passerines pay relatively little attention to habitat at this time of year, especially when they have just finished a nocturnal flight and need to eat. I'll often see Baltimore orioles, which love deciduous forests, cheek-to-jowl with orange blackburnian warblers, which nest in conifers—two birds the color of a glowing campfire, setting the morning alight.

I started birding in the 1960s and got serious about it while I was in high school in the early '70s. I was largely a loner—birding was hardly the popular, accepted pastime it is today, especially among kids, and I learned to keep a low profile for fear of ridicule. When circumstances brought me into the field with a group of adult birders, though, I often detected an undercurrent of worry, especially during the spring. I was thrilled with the multicolored song-birds flooding the woods, but the older folks talked about the good old days, back when there were *real* waves, by God, not these anemic dribbles of birds. Grownups, I thought dismissively, paying them about as much attention as I did my great-grandfather when he said we hadn't had a *real* winter since 1936.

I didn't know it at the time but the discontent I was hearing echoed across much of North America. Birders griped that the spring waves weren't what they used to be, and worried that something bad was happening to the birds. Rachel Carson's warnings about the dangers of pesticides were still fresh in the environmental community's mind, and a lot of people wondered if the perni-cious effects of toxins like DDT were still being felt; maybe this was the start of a true silent spring. For the most part, though, many scientists put it all down to the same kind of selective recall that colored my great-grandfather's opinion of recent snowstorms.

Can a million birders be wrong? Sure. Nostalgia certainly plays a role; it's human nature to remember those shining moments when every branch held a brilliantly colored songbird, like living Christmas ornaments, and to forget the many times when you tromped around all morning and saw little but grackles and house sparrows.

Even if the impression is valid, and you really do see fewer birds through the course of a season, there could be many innocent explanations. Songbird migration is diffuse, largely hidden from sight by altitude and darkness, and subject to the whims of weather. As is the case along the Gulf Coast, at places like High Island or Fort Morgan, a good spring wave is often caused by a strong cold front meeting the northbound flights, grounding large numbers of birds. A spring that is dominated by mild, southerly winds, with few storms, will seem like a bust to humans, even though it is perfect for avian travel.

Or maybe there are a few big cold fronts that spawn terrific warbler waves, but they occur midweek, when most people are inside at work. That happened a few years ago here in the East. Since I work at home, setting my own schedule, I got to enjoy those mornings, and I considered that spring to be one of the best migrations in years. My friends who worked nine-to-five office jobs, on the other hand, were stuck with four or five drizzly weekends in a row, missed the big fallouts, and complained bitterly about how "poor" the spring flight had been.

These are all examples of what scientists call "observer bias"—one of the biggest drawbacks to using subjective, anecdotal evidence to gauge trends in migratory birds, and the reason why the complaints of birders were ignored (or at least downplayed) by many in the scientific community for so long. Yet I don't know an ornithologist who isn't also an avid birder, and they were growing uneasy, too. When biologists began looking for evidence that the declines were real, they found it.

The first solid indications surfaced in the mid-1970s, from long-term studies of breeding songbirds in the East, often involving large, protected tracts of forest close to urban centers like Washington, D.C. At these and other locations, the number and variety of forest-dwelling migrant songbirds seemed to be dropping sharply, even though there was relatively little habitat change. (In many cases, however, the surrounding land had become increasingly urban, isolating the sites.) At Rock Creek Park near Washington, which has been intensively studied since World War II, six species disappeared entirely, including Kentucky warblers and yellow-throated vireos, while others dropped to a fraction of their former abundance. In the 1940s at Rock Creek, observers recorded an average of more than forty pairs of red-eyed vireos on each census; by the 1970s it was down to ten pairs, and fewer than six a decade later. Acadian flycatchers, ovenbirds, and wood thrushes suffered similar reversals.

Studies started popping up in scientific journals, based on a variety of research—migration counts, banding studies, breeding-ground censuses, work on the wintering grounds. Over almost twenty years, researchers with the Manomet Bird Observatory (now the Manomet Center for Conservation Sciences) on Cape Cod had banded more than 90,000 birds and found significant declines in eleven species during that time. In Puerto Rico, biologists documented the erosion of wintering warbler populations in undisturbed dry tropical forest; northern parulas and prairie warblers, which were among the most common species when the study started in 1973, were completely gone by the end of the 1980s, even though the habitat was still intact.

Not all the evidence was so unambiguous, however. The largest data set available to scientists was the North American Breeding Bird Survey, through which every summer since 1965 volunteers have surveyed thousands of 25-mile routes across the continent, counting the number of singing, territorial male birds. The BBS was the brainchild of Chandler Robbins, a lean, weathered man with a thatch of steel-gray hair who has spent fifty-one years as a federal wildlife research biologist but is perhaps best known as the lead author of a popular field guide to the birds. In the 1960s, the peak of the national controversy over pesticides, Robbins recognized the need for a way to track changes in continental bird populations—a need that was not fully appreciated by his employers, who put an official reprimand in his personnel file for launching the survey without permission.

In 1986, Robbins and two coauthors published an analysis of the first fifteen years of the BBS, from 1965 to 1979. In it they examined population trends for 230 North American birds, from herons to sparrows. Confounding the expectations of many, they found that during that period most migratory species were stable or increasing, including thrushes that winter in the tropics, vireos, and warblers—the very groups that birders and other researchers were most concerned about. Among certain groups, like flycatchers, they found sharp decreases in some species and equally marked increases in others.

Important as it was, that initial study was preliminary. Three years later, Robbins and a different team of collaborators—John Sauer of the Patuxent Wildlife Research Center, Russell Greenberg of the National Zoo, and Sam Droege of the federal Office of Migratory Bird Management—released the results of a much more comprehensive analysis, this time including BBS data through 1987 that hadn't been available for the first report.

As scientific papers go, this one, published in the October 1989 *Proceedings of the National Academy of Sciences*, wasn't much to look at—just five pages long, a piker in a world where journals may devote four times that much space to a single, arcane detail of bird biology—but it packed quite a wallop. "Using data from the North American Breeding Bird Survey, we determined that most neotropical migrant bird species that breed in forests of the eastern United States and Canada have recently (1978–87) declined in abundance after a period of stable or increasing populations," the authors began. The first study hadn't been a mistake; populations of many neotropical migrants had indeed risen during the late 1960s and early '70s, perhaps as a result of the banning of hard pesticides like DDT. But after 1978 they nosedived, especially in the eastern two-thirds of the continent.

Interestingly, resident birds, or those that winter within the temperate zone, were mostly holding their own or increasing, while twenty neotropical migrants experienced significant declines, especially those, like wood thrushes, that nested and wintered in mature forests. Wilson's warblers were slumping by 6.5 percent a year, yellow-billed cuckoos by 5 percent, wood thrushes by 4, and Baltimore orioles by almost 3 percent. Only four species that migrate to the tropics had significantly increased, the study found.

It seemed clear that the long-rumored declines were real. What wasn't certain was their cause—or, more likely, causes. The first reports of disappearing songbirds, from the 1970s, were attributed to changes on the breeding grounds, particularly ongoing fragmentation of the once-contiguous northern and eastern forests, which for centuries had been chopped into smaller and smaller pieces by logging, farming, cities, roads, and other human activities.

There is little question that North America's forests, especially those in the Eastern and Midwestern United States, are tattered; nor is there any doubt that fragmentation brings a laundry list of problems for breeding birds, including increased pressure from nest predators like raccoons, blue jays, and crows, which prowl the edges of woodlands while avoiding deep, unbroken forest. The smaller the fragment, the larger the proportion of its area hit by this "edge effect" and the tinier the island of safe, deep-forest habitat at its core; cut the woods into small enough pieces and the safe zone vanishes entirely. Fragments are also haunted by brown-headed cowbirds, parasites that lay their eggs in the nests of other songbirds, where the chicks monopolize food and parental care.

But part of the reason so many researchers initially blamed the declines on breeding-ground troubles may have been a bias I've mentioned before—the almost subconscious tendency to think of migrants as northern birds that take a brief, relatively unimportant vacation in the tropics. In the 1970s, very little was known about the winter ecology of migrant songbirds, and the pressures that might affect them in the tropics were scarcely understood. In fact, at a Smithsonian-sponsored symposium on neotropical migrants in 1966, there was an almost universal sense that habitat destruction in the tropics would *benefit* northern migrants, many of which were known to winter in disturbed habitat.

Chan Robbins and his colleagues, in their second BBS paper, noted one difficulty in trying to figure out whether breeding- or wintering-ground pressures were more important—most birds use the same general type of habitat in both winter and summer. There's no easy way of knowing whether a warbler is more affected by fragmentation of northern forests or by destruction of tropical woodlands. So they looked at the relatively few species that commute between different habitats.

What they found was telling. "Species that winter in a more mature habitat than that in which they breed"—in other words, birds that nest in scrub but winter in deep forest—"show negative slope changes, whereas those that winter in earlier successional habitats generally show positive slope changes . . . We suggest that this analysis represents the strongest evidence to date that tropical deforestation is contributing to declines in migratory bird populations." In fact, they argued, "tropical deforestation is having a more direct impact on neotropical migrants than is the loss and fragmentation of forest habitats in North America." Other biologists concurred, and even those who believed that breeding-ground effects were the most important limiting factor on migrant populations acknowledged that the continuing loss of tropical forest might soon push that factor into the fore.

Robbins and his colleagues ended their paper by cautioning that forest losses in the tropics and the temperate zone were not competing theories to explain songbird declines; instead, they should be viewed as twin engines driving bird populations lower, and would in all probability get worse. But more and more attention was being turned to Latin America and the Caribbean. Also in 1989, John Terborgh, then a biology professor at Princeton, published a book, *Where Have All the Birds Gone?*, which expertly summed up the worrisome state of migrant passerines and laid out the case for a tropical connection. Changes on the breeding grounds certainly had an impact on bird populations, Terborgh acknowledged. "Still, the uncanny consistency with which the missing or diminished species are tropical migrants has raised the possibility of another explanation—loss of wintering habitat through tropical deforestation."

It was hard not to agree; during the 1980s, tropical deforestation exploded into the public's consciousness, fueled by staggering statistics—half the world's rain forest was already gone, and the rest was being destroyed at a rate of 50 acres a minute, an area the size of Ohio burned or cut each year. Whether the greatest pressures on migratory birds thus far came in the north or in the south, it was clear to everyone that the continuing loss of tropical forest would be a calamity for some migratory birds.

It was in this atmosphere of impending crisis, in December 1989, that more than 300 scientists from across the Americas gathered at Woods Hole, Massachusetts, for a symposium on neotropical migrants, sponsored by the Manomet Bird Observatory. In a very real sense, the four-day event—the largest ever devoted to the topic—marked a turning point in the study and conservation of migratory land birds. It produced at least some common ground between members of the two major camps in the decline debate—those who favored breeding-ground pressures as the primary cause and those who believed the problem lay in tropical deforestation. For the first time, there seemed to be a consensus that songbirds really were in decline. Most important, it gave those present a sense of unified purpose and direction, and highlighted a number of avenues for future research and conservation efforts.

But the single event that, more than any other, propelled the Manomet conference onto the front pages—and with it the subject of songbird declines in general—was a paper presented on the final day by Sid Gauthreaux. Gauthreaux, you'll recall, is the Clemson professor who had been using weather radar to study trans-Gulf migration since the 1960s. Now he used that long-running data set to put the problem into frightening perspective. By analyzing thirty-three years' worth of archived radar images, Gauthreaux found that the frequency of springtime migrant waves across the Gulf had declined by almost 50 percent. Few people, even ornithologists, had suspected the situation was that bad.

Gauthreaux could not say that the total *number* of birds had declined by that much—the relatively crude radar images weren't detailed enough for that kind of analysis—but that was the logical implication, and that's how his findings were interpreted by many. The report caused a sensation at the conference and was widely reported in the general press. Like a flock of warblers crossing the Gulf, songbird declines were on the national radar scope for the first time.

I've sat at my computer and, thanks to the Internet, watched brightly colored Doppler radar images from places a thousand miles to the south. I've seen the nocturnal flocks blossom across the screen in clouds of orange and red, as the multitudes take wing at sunset, advancing across the continent, knowing that they will reach me in a week or two and be transformed from abstract blips to flesh-and-feather animals.

Aloft in the night air, migrant songbirds have the freedom of angels, but if daybreak finds them over hostile territory, the need to rest and refuel often packs them by the hundreds of thousands into whatever small patches of habitat they can find. It can happen in an oasis like Dugout Wells, or in narrow ribbons of greenery along desert rivers like the Rio Grande in Texas or the San Pedro in Arizona. It may happen when the flocks must cross large metropolitan areas, where city parks and wooded cemeteries act as asylum in a desert of gray concrete; New York City's Central Park attracts hundreds of species each spring, as well as a large, enthusiastic following of birders. Even where the landscape is better suited to forest songbirds, in the eastern United States and Canada, the travelers are funneled into peninsulas that jut out into large bodies of water, what birders call migrant traps. The peninsulas may either serve as jumping-off places for birds, if they point toward the direction of travel, or as first landfall after a long water crossing.

Lake Erie, which forms a 250-mile barrier across the main flow of the spring flight, has two such peninsulas along its northern shore in Ontario, Point Pelee in the west and Long Point in the east. As with the Gulf crossing some weeks earlier, the birds are easily able to make the 50-mile flight over Lake Erie—but not if they encounter bad weather along the way, especially if the storm hits after midnight, when many of them are past the point of no return. Then the cold north winds drive them onto the first land they find, producing astounding fallouts with almost obscene numbers of birds, many of which are so tired and hungry that they hop around the open beach and dunes, snatching up midges and caddisflies among the legs of birders.

Of the two points, Pelee is far and away the more famous; the national park that occupies the narrow spit may reel under the shoes of almost 100,000 birders in an average May, and during the peak of the season, which falls around the second or third week of the month, it is impossible to get a motel room or campsite for a very long way in any direction. So jammed and popular is the park that officials have had to strictly control access, just to keep the vegetation from being trampled; but still the hordes come, from Canada, from the States, from abroad, hoping to be at Pelee on THE day of the spring when weather and migration and landscape link hands to produce the stuff of legends.

What gets lost in the stampede to Pelee is the fact that Long Point is just as good for spring birding, and maybe (say it quietly, for this is heresy), maybe a little better. Unlike Pelee, which hangs straight out into Lake Erie like a short, sharp dagger, Long Point hooks languidly almost due east, parallel to shore, casting a wider net into the flow of birds. It encompasses a fine bay and shelters expansive marshes along its fourteen miles of forest and dunes, the perfect haven for a weary bird.

You could argue that my timing, when I visited Long Point, was either very good or very bad. It was the third week of May, prime time for songbird migration; that was very good, and the days leading up to it had been a long sequence of mild south winds, which spurred the birds north. But it was also Victoria Day weekend, Canada's summer-launching equivalent of Memorial Day, and the roads along the north shore of Lake Erie were jammed with vacationers heading to the point. That was very, very bad, for while Long Point Provincial Park is not as famous among birders as Pelee, it is hugely popular with folks who like to swim, boat, fish, roast hot dogs over campfires, and sit in the wan, late-spring sun and tan their goosebumps.

I snagged one of the last campsites, in a sparse grove of cottonwoods just yards from the lake itself, where the surf hissed like sleet on dry leaves. As I set up my tent, congratulating myself on some stunning good luck—the CAMPGROUND FULL sign had gone up moments after I pulled in—I realized that the quality of the light was changing. To the north and west, the sky had turned a dark, cold Prussian blue as an enormous line of storms swept down on us. Yet the sun, just barely above the clouds, was still beaming on the point, making the water and the pale sand beneath it shine with an almost ivory glow—an eerie reversal of light and dark, as though I were looking at the world upside down.

I turned on my weather radio, picking up a forecast from across the lake in Pennsylvania. "This is a special weather statement from the National Weather Service office in Erie," the tinny voice said through static. "A severe storm warning has been issued for the Lake Erie shore, effective until 10 P.M. These storms may produce strong, damaging winds, deadly lightning, and large hail. Winds in excess of 75 miles per hour and 1¼-inch hail have been associated with these storms as they have crossed the upper Great Lakes. People in the path of these storms are advised to seek shelter inside strong buildings."

Somehow I didn't think they meant nylon tents. With the wind rising, I did what little I could to prepare—cross-staked the corners in the loose sand

and hauled every piece of gear that I had out of my car and into the tent for ballast. Then I tied a few ropes from the tent to the bumper of my car, so that if worse came to worst, I, my shelter, and all my worldly goods wouldn't go rolling away into Lake Erie. The squall line was right overhead as I finished, heavy mammatiform clouds hanging like breasts along its leading edge; to the west, beyond them, everything had gone a flat gray behind sheets of rain. Moments later the storm broke.

It was a long night. The rain and wind were vicious, and although we were spared the hail, thunder rolled across the point for hours in the most continuous barrage of sound I've ever heard; the lightning that spawned it flickered and flashed constantly, a psychotic light show, much of it striking so near that the white burst and crashing explosion were simultaneous. I spent much of the time crouched on a folded foam sleeping pad, on the theory (unproven, but comforting in the breach) that the rubber mat would offer some protection from a ground surge should the lightning hit a nearby tree.

Dawn was a cold, murky, drizzling mess. Snapped branches and small trees were scattered everywhere in the campground, and a number of tents had collapsed entirely. Rainwater stood in deep puddles wherever there was a low spot, with yellowlegs and solitary sandpipers wading through them, feeding on the bugs and worms flushed to the top.

I drove out of the park and into the small neighboring community of Long Point Beach, turning down Old Cut Boulevard. Like Boy Scout Woods at High Island in Texas, the Long Point Bird Observatory (LPBO) maintains a field station right in the middle of this small hamlet of summer cottages, and it, like the rest of Long Point, was pulsing with birds.

That line of storms had collided with a major northbound flight of songbirds, wringing them from the sky with the rain. The most common were two kinds of warblers, magnolias and yellow-rumpeds—"maggies" and "butterbutts" in the verbal shorthand of birders—but the scribbled list I kept in the back of a rain-wrinkled notebook ran for several columns and nearly a hundred species. There were Nashville warblers and ovenbirds, Lincoln's sparrows and golden-winged warblers, veeries and wood thrushes, tree swallows and whitethroats and purple martins and—you get the idea.

The LPBO has a small visitors' center and banding station at Old Cut, and as with the Sargents' site in Alabama, birders were welcome to creep along the trails and net lanes that meandered through the spruce forests and along the marshes that edge the inner bay. I had planned to introduce myself and chat about banding, but the crew was so completely overwhelmed with birds that I simply stood with the other onlookers and watched without interrupting. The banders worked in pairs, sitting in a small room behind a waist-high counter like a little store—retail banding, I thought with a smile. Every five or ten minutes, a volunteer tromped in wearing muddy rubber boots, leaving another armful of small, squirming drawstring bags made of brightly flowered fabric.

I've never seen such smooth, clockwork precision in a banding operation; checking my watch, I found that the average time to band and process each bird was a slick-as-oil 45 seconds. "Magnolia warbler, ASY male, band number 34"—the fellow crimped the tiny silver ring in place—"fat two, wing 61." Then after a few notations about molt, the bird went into a cardboard tube and onto a digital scale to be quickly weighed, and the open end of the tube was tipped into a nifty little release chute in the front wall, like the night-depository door at a bank in reverse. The freed bird flashed past the window, but the team was already working on the next, one of hundreds they banded that morning.

As I've said, bird migration is a diffuse phenomenon, only rarely concentrated by weather or geography into spectacles like Long Point. It is hard to get a good grasp on its scope, much less any trends in the populations of the birds themselves. Old techniques like moon-watching and more recent advances like Doppler radar give only the broadest of outlines, with no details about which species, and how many individuals of each, are aloft. There are new technologies on the horizon, however, that may eliminate those shortcomings. Bill Evans, a research associate at the Cornell University Lab of Ornithology, is developing an automated system that records the flight calls of nocturnal migrants, then processes them through a computer that analyzes, identifies, and tallies each call by its unique acoustic signature. The technology is still rudimentary, and it is hindered by a lack of such basic information as what the night calls of many songbirds sound like, or even whether all songbirds vocalize when they're flying—vireos and some flycatchers apparently don't, for instance. But Evans, along with others like Sid Gauthreaux, believes the new technique holds incredible promise for monitoring bird migration. Evans foresees a network of automated stations scattered across the continent, each one recording and counting the birds that pass within earshot of its microphones, with the pooled information from all of them painting an astoundingly detailed picture of where, when, and how many birds are flying.

In the meantime, though, long-term banding studies like the one at Long Point—projects that take place in the same place, with the same degree of effort, at the same time each year—provide one of the few reliable barometers of how migratory birds are faring. I watched as the bander released another bird, then extracted a rose-breasted grosbeak from a bag, a triangle of striking pink on the bird's breast and under its wings, its heavy beak the color of old bone; instead of biting a chunk out of the man's fingers (which were, I noticed, scabbed with old wounds), the bird patiently endured the quick procedure. At Long Point, where more than 400,000 birds have been banded since 1961, captures of rose-breasted grosbeaks have dropped by 2.5 percent a year, northern waterthrushes and eastern towhees by 4 percent, and wood thrushes by 6 percent—a frighteningly steep rate when plotted over the course of several decades. Here as elsewhere, neotropical migrants have shown the worst decreases, although some declining species, like towhees, brown thrashers, and white-throated sparrows, winter largely within the United States, where they may be suffering from habitat loss.

The fact that many of the most seriously reduced species are long-distance migrants is significant, biologists think. The act of traveling is an inherently dangerous one, fraught with storms and famines, predators and exhaustion. The longer the migration route, the greater the likelihood that critical links in the chain will be destroyed—forests burned, grasslands overgrazed, wetlands drained, habitat poisoned with chemicals or converted to strip malls. The longer the journey, the greater the possibility that human interference of one sort or another will claim a bird's life.

When I was in college, I worked as a volunteer in the natural history lab, preparing study skin specimens for the university's collection. Many of them were songbirds, which we picked up beneath several local radio antennas, metal needles poking hundreds of feet into the air. Flying through the darkness, often less than a thousand feet from the ground, flocks of passerines slammed full-tilt into them, with disastrous results. Some days we'd find dozens and dozens of birds at each site, other days only one or two, but almost always there were some casualties, the leftovers that hadn't already been scavenged by raccoons and foxes. We thought it was a shame that the birds had died, of course, but mostly this was a convenient way to obtain specimens. After all, there were only a few.

On a cumulative scale, though, the carnage from broadcast antennas is sobering. The Federal Communications Commission estimates that there are 75,000 towers in the United States tall enough to interfere with aviation, and on foggy or cloudy nights even those that are shorter pose a risk to birds. Though most are less than 400 feet tall, some antennas stretch more than 2,000 feet into the air, indiscriminate killers nearly half a mile high.

Studies at individual towers in New York, Florida, and elsewhere show that each may kill from several hundred to several thousand birds in a migration season; one tower in Eau Claire, Wisconsin, claimed more than 2,000 birds of thirty-seven species in a single night. Over five rainy nights in the fall of 1977, a cluster of towers in Elmira, New York, killed 3,862 birds of forty-eight species, including twenty-four kinds of warblers. Another set of towers in Buffalo has claimed more than 20,000 birds since the 1960s. What is remarkable about these figures isn't the number of birds killed but that someone bothered checking regularly enough to document them. Bill Evans at Cornell has recently made curbing tower kills a personal crusade, and a map he has compiled of tower sites in New York alone gives one pause: the map is speckled with color, showing roughly five hundred towers between 200 and 500 feet in height and another three dozen more than 500 feet tall. Louisiana, right on the trans-Gulf route, has more than fourteen hundred towers, including forty-three over 800 feet high. It is hard to imagine how anything could pass through such a picket line unscathed, much less cross an entire continent so encumbered.

In the 1970s, long before the recent boom in their construction, a Fish and Wildlife Service report estimated that towers taller than 200 feet killed at least 1.2 million birds in the United States each year. The figure, which was conservative even then, is undoubtedly far, far higher today, especially since so many of the tallest, most dangerous towers are along the Gulf Coast and southern Plains, right in the main flight path of the spring migration. Evans and other activists guess that 2 million to 4 million birds are killed in the eastern United States alone, and they admit the true number could be much higher.

As if that isn't bad enough, the move toward digital television will require that more than a thousand new towers be built, each at least 1,000 feet tall. Add to that burgeoning microwave, cell phone, and personal radio systems, and the landscape is bristling with antennas like a porcupine, a forest of deadly metal that seines migrants from the air. New towers, mostly two or three hundred feet high, are being added at a rate of five thousand a year, five times the rate of a decade ago. Now, hoping to expedite the switch to digital TV, the FCC wants to exempt the construction of those new 1,000-foot towers from federal, state, and local environmental review—a move that has sparked joint opposition from all four major ornithological groups and conservation groups like the National Audubon Society.

Towers themselves are dangerous enough, but when you add lights, the situation gets considerably worse. In bad weather, lights disorient night-flying songbirds, which rely on subtle glimmers from the stars and the shadowy tapestry of the landscape for navigation. Under low clouds or in fog, lights overwhelm these cues, forcing the birds to circle lighted towers and buildings like moths around a flame, eventually ramming themselves to death or collapsing from exhaustion. Blinking lights in poor weather conditions are worst of all, as was tragically demonstrated in the winter of 1998. A huge flock of Lapland longspurs, confused by fog, snow, and the strobe lights on a 400-foot tower in Kansas, kept circling the antenna, colliding with it, its guy wires, each other, even the ground, until as many as 10,000 of them lay dead.

Lights lure millions of migrants to their deaths. In the nineteenth and early twentieth centuries, the worst offenders were coastal lighthouses, but after aviation became commonplace, airport ceilometers—which aimed a powerful beam of light into the sky to determine the cloud height—took over as the leading killer. Ceilometers are now rarely used (those that are feature a fixed beacon, which is less disorienting), and today the ubiquitous office highrise—glass-fronted and usually brightly lit long after everyone has left for the night—levies the greatest toll.

The estimates of how many songbirds die in building collisions each year run into the millions, a morbid kind of guesswork that probably only hints at the magnitude of the problem. In Canada, the World Wildlife Fund and an organization known as FLAP, the Fatal Light Awareness Program, have been encouraging building owners and tenants in Toronto to turn off their lights at night during the peak migratory periods. Many large, corporate landowners have shown a surprising willingness to cooperate, and the program is being duplicated elsewhere.

At night, the valley where I live is spangled with the lights of farmhouses and a few brighter, dusk-to-dawn lamps in barnyards. We can still see the stars here, the faint ones that have been washed from the night in cities, and our skyline is still the black silhouette of the ridges instead of glowing buildings. But when I look to the mountain several miles to the south, I can see vertical lines of slow-blinking crimson strobes on several high towers that have sprouted there in the past decade. East and west along the ridges, more clusters have gone up recently, always on the highest spots, like traps waiting to be sprung. Whenever the clouds and fog hang low over the ridge in spring, masking the metal spires, I wonder how the flocks of migrants are faring, coming into those disorienting fields of red that appear and disappear, like nothing for which evolution has prepared the birds. Some blunder past and escape; others are captured by those fatal globes of light. The towers no longer seem like such a great place to collect scientific specimens. On such still, damp nights I sit outside, listening for the quiet calls of birds from the silken darkness, and wish them well.

Bird migration covers almost every square mile of the entire hemisphere, crossing ecological zones and national boundaries, jurisdictions and dominions. Protecting those migrants, therefore, is an enormously complicated undertaking.

Before they adjourned, the scientists who were gathered at the Manomet symposium in 1989 voted on a final resolution. "WHEREAS: Neotropical migratory landbirds are a shared international resource and therefore require a major conservation initiative that transcends international boundaries," it began, running through a bunch of other whereases that summed up the concerns and objectives. It ended: "It is critical that the scientific community in North, Central, and South America form cooperative efforts in the conservation of this resource, and that the nations of the Western Hemisphere recognize the value of migrant landbirds and the habitats on which they depend."

Less than a year later, Partners in Flight/*Aves de las Americas* was born. As its bilingual name suggests, PIF was meant to be the "major conservation initiative that transcends international boundaries" envisioned by the Manomet conference. Partners was hatched in 1990 by the National Fish and Wildlife Foundation, a nonprofit organization created by Congress to foster public-private ventures in conservation. The idea wasn't to create another level of bureaucracy—for its first five years, Partners in Flight didn't even have an office or a staff of its own—but to forge a coalition of scientists, government agencies, land managers, private landowners, conservation groups, foundations, and individuals, all working in a coordinated fashion to preserve migratory birds. Partners in Flight has, with justification, been called the most ambitious conservation effort in history, tackling one of the most complicated conservation problems we've yet encountered.

Today, PIF encompasses more than sixteen government agencies, forty NGOs, sixty state and provincial wildlife departments, and many other participants from academia and industry, operating through regional working groups to bolster communication and cooperation.

They have agreed on an overarching bird conservation strategy, usually referred to as the "Flight Plan," stressing commonsense approaches and based on the best available science. Partners in Flight participants believe that the

time to save a bird is while it is still common and that it is better to focus on
the conservation of habitat for many species, instead of reacting only on a
case-by-case basis as individual species become rare.

The annual WatchList, a PIF project discussed previously, was one of the
first attempts to scientifically rank the birds of North America by the degree
of risk they face—threats like habitat loss, population declines, limited geographic range, or long migration routes. The 107 species on the current list
are ranked in descending order of urgency, from those considered at grave,
immediate peril—like the black-capped petrel and the bristle-thighed
curlew—to species like the bobolink and chuck-will's-widow that show worrisome trends but are still relatively widespread and common. The Watch-
List, although it lacks the force of law, serves as an early warning system, a
way of targeting research and conservation work where it is needed the most.

Even though Partners has an international reach, it's a grassroots undertaking, anachronistic in these top-heavy days, and it works. I've sat in on PIF
sessions in states like Nevada, where to a lot of people "conservation" still
carries a whiff of Eastern elitism, watching as birders, ornithologists, ranch
owners, and federal agency personnel—by no means natural allies—worked
together to find ways to safeguard the migrants in their region, and looked
beyond the horizon to the other end of the migratory loop. Partners in Flight
is part of the new wave in bird conservation—international partnerships that
place birds in their global context, forging links with conservationists in a
species' breeding, wintering, and migratory range. Habitat conservation is
seen as a key to success; better to save the links in the migratory chain,
instead of waiting until a bird is critically endangered, then fighting an
expensive, often futile, last-ditch effort to save it.

The undertakings range from the all-inclusive, like the PIF Flight Plan, to
more specialized endeavors like the North American Waterfowl Management Plan, which involves people in and out of government from Canada,
the United States, and Mexico. The Nature Conservancy, which has preserved or helps to manage more than 84 million acres of land in twenty-four
countries, now zeroes in on migratory bird habitat through its "Wings of the
Americas" strategy. And taking a hint from Europe, conservationists have
launched the Important Bird Areas (IBA) program, which, like the Ramsar
Convention and the Western Hemisphere Shorebird Reserve Network, identifies critical breeding, wintering, or stopover habitat for birds of all kinds.

Modeled on a highly successful program pioneered by BirdLife International, the IBA system works on two levels. First, sites of statewide importance

are identified by volunteers working through the National Audubon Society and are evaluated by regional ornithologists. Those designated as state IBAs are then reviewed by experts at the American Bird Conservancy, and locations of national or global importance are specified. Similar efforts are under way in Canada, Mexico, Panama, and elsewhere in the hemisphere. As with WHSRN or the Ramsar Convention, the IBA designation doesn't, of itself, provide protection, but it is often a catalyst for study and preservation. In Europe and the Middle East, where the IBA program began in the 1980s, more than 600 sites encompassing 16 million acres have received recognition and some degree of protection.

One of the most important goals of Partners in Flight was to facilitate cooperation between the United States and Canada, which have strong programs for migratory bird research and conservation, and counterpart agencies and institutions in Latin America, where both basic research and effective management are grossly underfunded. Latin American biologists often lack the bare essentials for fieldwork—binoculars, spotting scopes, banding equipment, up-to-date reference books, and the like. Parks and refuges, while protected by statute, frequently go begging for enforcement and management personnel, leaving them open to many forms of destruction.

This Latin American connection was seen as especially vital, in light of the widespread suspicion that songbirds were declining because of trouble on the wintering grounds. But not everyone was convinced that the worst threats lay in the tropics. The pendulum of scientific opinion, which first blamed breeding-ground effects in the 1970s, then swung toward a tropical explanation in the 1980s, started back the other way. As PIF and other bird conservation programs gained impetus through the early 1990s, as more and more research was published on the intricacies of migratory bird ecology, prominent ornithologists and ecologists talked increasingly about the hazards facing neotrops on their northern breeding grounds.

To a layperson, this ebb and flow of scientific opinion, the flipflop of theory over time, the occasionally fierce and public disagreements between viewpoints, is as mystifying as it is frustrating—just ask anyone trying to learn whether or not eggs are bad for you, or how much dietary fat constitutes a health risk. This isn't a sign of science's weakness, though, but rather its strength, always testing assumptions, making challenges, drawing new conclusions based on the latest evidence. It may err too far in one direction, then too far in the other, correcting as it goes, but every step eventually brings us all a little closer to the truth.

Here's the truth about migratory songbirds, as we know it today: Migratory birds as a whole are being squeezed at every stage of their life cycle—summer, winter, and in transit. Some are holding their own, but some are clearly in trouble, and the trends cut both ways across a whole host of taxonomic and ecological lines. There does seem to be a link between the kinds of habitat used in summer and winter and whether or not the bird is declining. But teasing out the details, finding the subtle patterns amid a confusing, hemispheric phenomenon, isn't easy.

Natural systems are like those hollow Russian dolls, layer nested within layer within layer, always hiding a new riddle beneath the last one. Figuring out how they work is a herculean task, one with which humans have been grappling for a relatively short time. If you want clear-cut problems and neat solutions, try geometry. This is ecology, and it doesn't get any more complex and messier than this. But we're learning.

By the third week of May, the biggest waves have passed me by, here in the gentle hills of Pennsylvania, rushing the boreal songbirds north toward the spruce forests of Canada and New England. But the advancing waves leave behind a company of migrants in the greening oak forests—scarlet tanagers and Acadian flycatchers, blue-gray gnatcatchers and Swainson's thrushes, hummingbirds and redstarts, yellow-billed cuckoos and blue-headed vireos, and a hundred more, each adding its own aria to the morning chorus.

I took a long, daybreak walk this morning, past small meadows and up the side of the mountain, through hardwoods and hemlocks. To my ears, the woods sounded as rich with birdsong and as full of bird life as they did when I was a boy. But I know better than to trust a set of hazy memories. In fact, it turns out that I cannot trust even the immediate reality of a song-filled morning, for in much of North America the beautiful music masks a chilling fact.

Enormous areas that look like good songbird habitat and that thrum with melody on a May morning are little more than black holes, swallowing up the fruitless production of breeding season after breeding season. Scientists studying migrants on the breeding ground have uncovered an army of dangers that we, in our heedless manipulation of the natural world, are making worse.

Now it's time for bed, as I set the alarm clock for two hours before daybreak. In the early-morning cool tomorrow, I'll join some friends who are searching, one slow day of research after another, for a pattern amid the chaos and an answer to the question: What is really happening in the woods?

15

Trouble in the Woods

In late spring, there isn't a second to lose. Long before sunrise, when the only sign of dawn is the smudged absence of stars on the eastern horizon, male birds are already singing. The wood thrushes always seem to start first, their clear notes scrolling up and down like improvisations, looping back on themselves, then ringing out in lucent peals. When the thrush stops, it feels as though the forest is holding its breath.

The air is damp, charged with oxygen, heavy with the smells of wet tree bark and old leaves underfoot. I move among the pale trunks of the oaks, feeling the chilly dew from the shin-high blueberry bushes soaking through my pants legs. The sky brightens quickly, and other birds are singing: vireos, a towhee, a black-throated green warbler, a couple of juncos. Soon the songs lose their individuality as sunrise nears and more and more birds join in, their music merging into a pleasant babel with the rhythm of a fast-flowing stream.

A blue-headed vireo moves through the branches of an oak, the fresh green leaves matching its back, the color of its head a mirror to the fading darkness in the west. Its eyes are surrounded by white spectacles that perch on its face like a pince-nez. It finds a caterpillar, swallows it, hops a few inches, and finds another. This one it carries, limp in its beak, to a neighboring tree, where another vireo that looks just the same greets it with trembling wings, like a baby. The first is a male, carrying a gift of courtship, which the female accepts; he fluffs his feathers, showing his yellow flanks, bobs his head once or twice, and sings, a solo against the sweet murmur of the chorus.

After months of travel spanning half the world, the birds in this forest are finally home, most of them back in precisely the same swatch of Appalachian mountainside where they'd nested the year before. For the next three months, they will scarcely stray from a territory that may be less than a quarter acre. It seems so anticlimactic. Watching birds cross the planet's surface with the grace of clouds, it is easy to forget, as I do from time to time, that however awe-inspiring migration may be, however remarkable the feats or heroic the journey, it is not an end in itself. Migration is the exterior drama, bold and eye-catching, but secondary to what happens here on the breeding grounds. By surviving the rigors of a life on the wing, birds earn another chance to pass on their genes. Nothing else matters. Without progeny, even the most spectacular migration is a dead end.

So the journey concludes where it began, in the woods where the birds will court and mate and raise their chicks. But if the habitat that a thrush or a vireo choses for its territory cannot support it and its offspring—if instead of a safe home it picks one beset by predators or with marginal food supplies—then the thousands of miles of travel have been wasted. And if enough birds are forced to make that kind of choice because there's nowhere better to go, then the species itself begins a slide toward oblivion.

Is that the case today? It may be hard to tell, especially if you read the newspapers.

On June 10, 1997, *The New York Times* and *The Christian Science Monitor* both ran prominent stories on migratory bird conservation. The *Times* story, which occupied most of the front page of the day's science section, was "Something to Sing About: Songbirds Aren't in Decline." The *Monitor* piece, on the other hand, was titled "Requiem for the Songbird: Perilous Decline Puzzles Scientists."

Contradiction among reporters isn't new, but what made this situation so remarkable is that not only did the stories run on the same day but they were based, in part, on the *same* information.

"For almost two decades, it has been conventional wisdom that America's migratory songbirds are seriously declining in numbers and headed for trouble," the *Times* piece began. "Now a new analysis of 30 years of data from the North American Breeding Bird Survey has found that on the whole, continent-wide numbers of the birds have remained remarkably stable."

The *Monitor*, however, wrote: "Plenty of scientific evidence exists to support the gravity of the situation . . . Using the BBS and other monitoring data, conservationists have placed 35 of 157 North American neotropical migrants on a 'watch list' of birds at risk. Another revealing study, looking at high-resolution radar images of the nocturnal clouds of migratory birds along the Gulf Coast, showed a 50 percent decline in the size of flocks during the past two decades."

How could two newspapers draw such wildly different conclusions from essentially the same material? And more important, which article was right?

In truth, the two stories weren't as far apart as they might seem. For one thing, the cheerful *Times* headline did not accurately reflect its own story, ignoring the point (made by the writer within the first few sentences) that the BBS data showed catastrophic declines in grassland birds, in certain forest species like wood thrushes and cerulean warblers, and in entire forest-bird communities in places like the southern Appalachians and the Adirondacks.

Nevertheless, the *Times* article set off a brouhaha about songbirds. Conservative commentators used it to rail against the doom-and-gloom attitudes of environmentalists, while many scientists and birders reacted ferociously to what they saw as the newspaper's Pollyanna coverage. Some also blasted the journal article on which the *Times* piece was based, a detailed overview of continental bird populations written by a scientist named Scott K. Robinson.

Robinson, an ornithologist with the Illinois Natural History Survey and the University of Illinois, has strong credentials in bird research. Since the mid-1980s he has overseen a series of ambitious studies into the breeding success of migrant songbirds in North America, and he has decades of experience studying tropical birds in Central and South America. The article he wrote, "The Case of the Missing Songbirds," was published in *Consequences*, a small, federally supported journal available primarily over the Internet.

In his paper, reviewed in advance by several noted ornithologists but written without the footnotes and long list of references customary in scientific articles, Robinson synthesized much of the work published in recent years on migratory bird declines and their possible causes, including a number of analyses of BBS data.

Most researchers, Robinson said, agree on several major points—that some forest-dwelling species, such as olive-sided flycatchers and eastern wood-pewees, are declining rapidly, as are some species like prairie warblers and yellow-breasted chats, which need newly reverted scrubland for breeding. Virtually everyone agrees that grassland birds are in serious jeopardy, he said, and that in places like the Great Smoky Mountains, "many or even most" forest songbirds are in trouble.

But, said Robinson, "When viewed as a whole, and when averaged over large geographical areas, Neotropical migrant populations have remained generally stable. This seems especially true in the western half of the continent."

Some forest species, like ovenbirds and hooded warblers, "which might be expected to be declining because of lower reproduction and the loss of Neotropical winter habitat, have remarkably stable populations," he wrote.

In fact, Robinson's conclusions were neither revolutionary nor terribly controversial, and were it not for that somewhat misleading headline on the splashy *New York Times* article, they probably would have attracted little attention. Many ornithologists, in fact, were coming to realize that the situation involving migratory songbirds was far more complicated, and far less cut-and-dried, than it had seemed in the late 1980s and early '90s, after the first major warnings from Chan Robbins, Sid Gauthreaux, and others.

There were even some, albeit a tiny minority, who doubted whether the declines were real at all. Frances C. James, who teaches ecology at Florida State University, and several other researchers working with her have for years been outspoken in their belief that there is no compelling evidence for broad songbird declines. At the 1989 Manomet conference, James raised a cautionary note amid the general feeling of crisis and urgency. Although she acknowledged that some species, like the cerulean, prairie, and Canada warblers, have experienced severe population drops, James and her coworkers have analyzed BBS data for wood warblers as a whole and concluded that populations were stable everywhere except in the Appalachians and adjacent eastern foothills. The "decline" shown by Chan Robbins's BBS study in 1989, she thinks, reflects a natural sag after the unusually high populations in the 1970s caused by insect outbreaks that provided lots of bird food.

Today, ornithologists and ecologists find themselves confronting a much more complex puzzle than they had ever imagined. In the past, scientists tended to view migrant songbirds, especially those that travel to the tropics, as a monolithic block—reacting in similar ways to similar pressures throughout their ranges. But they're finding out that every species, often each regional population of each species, reacts differently to the unique set of circumstances they face on the breeding and wintering grounds, and along their particular migratory routes in between. Generalizations are usually wrong and, from a management perspective, often dangerous.

"So, are Neotropical migratory birds really declining?" asked the editors of a special section in the journal *Ecology*. "The answer would go something like this: Some species are declining in some parts of their geographic ranges and increasing in others, other species are increasing in most of their geographic ranges, yet others are decreasing throughout their ranges ... We give this answer realizing that as complex as it sounds, it is still probably simplistic relative to what's really happening. Yet it illustrates the kind of dilemma that ecologists studying continental ecosystems are likely to be faced with."

The difficulty, writes one team of ornithologists who have wrestled with this issue, is finding a balance between two extremes—on the one hand, panicking over what may be natural, short-term population declines and, on the other, refusing to acknowledge that anything is wrong unless the data meet impossible standards of certainty, by which time it may be too late—what they call the "Chicken Little Syndrome" and the "Tobacco Lobby Syndrome."

So are neotropical migrants really declining, justifying all the anguish and frantic activity of recent decades, or did wildlife biologists fall into a Chicken Little trap, misinterpreting a lot of ambiguous data and scaring the bejesus out of all of us who love birds? Or, as a lot of people are starting to think, is the truth somewhere in between?

"Now, let's see if I can stir up some trouble," Laurie Goodrich said, holding a small tape recorder over her head. She thumbed the switch, and the loud, driving song of an ovenbird filled the air.

She found trouble almost immediately, in the form of an indignant male that flashed past her head, nearly parting her long brown hair, and landed just a few yards away in a small sapling, belting out a song that was, if anything, louder than the tape. I fumbled with my binoculars, trying not to lose the pack and a heavy golf bag slung over my shoulder, and checked the bird's pink legs for bands. There were none.

"Okay, let's set up right here," Laurie said, dropping the equipment she was carrying as the ovenbird kept singing and singing, trying to intimidate a rival he'd heard but not seen. We drove steel rods into the ground and quickly set up two mist nets using poles from the golf bag, forming a right angle through the forest. Near the inside corner of the nets we placed the tape player and, perched on a tall stake, a wooden decoy of an ovenbird, painted with a spotted breast and rusty crown. The whole process took just ten minutes, and as I moved the remaining gear out of the area, Laurie again turned on the tape.

Settling down thirty yards away, we watched through a corona of blackflies buzzing around our faces. The taped song brought the male to the nets almost immediately, flitting back and forth past what he must have thought was an unusually stoic intruder. Nearby, we could hear the sharp *chips* of his mate, scolding the decoy from a safe distance. Finally, the male ovenbird made a wrong turn and hit one of the mist nets, and Laurie was sprinting for him before I could even unlimber myself and get to my feet.

We were working at Hawk Mountain Sanctuary, not far from my home and just over the hill from where I'd been watching the courting vireos a few hours before. My good friend Laurie is senior naturalist there, and every spring since 1988, she and the other researchers have been netting and color-banding ovenbirds as part of an ongoing study that examines how woodland-dwelling songbirds are reacting to the modern landscape, with its fragmented forests and escalating development. Some of the netting is done here, on the wide top of the Kittatinny Ridge, in the midst of a nearly contiguous forest of more than 50,000 acres, while other study sites include smaller forests of several hundred acres and tiny woodlots adrift in farmland. The Hawk Mountain study is one of only a handful to look in such detail, and over such a long period of time, at how songbirds handle the pressures of life in late-twentieth-century America.

Our "understanding of migratory birds' population dynamics remains at a primitive stage relative to the scope of the problem," caution Tom Sherry and Richard Holmes, who have been studying warblers in New Hampshire for thirty years and in Jamaica for fifteen.

For example, we do not know for any single Neotropical migrant bird species how mortality varies throughout the annual cycle. Nor do we know typical dispersal distances for individuals of most populations, nor where individuals within a particular breeding population spend the winter and vice versa. We cannot assess rigorously with the available information the relative importance of wintering or breeding areas for any single migratory species. Students of migratory bird populations are thus much like Rudyard Kipling's blind men, each of whom tries to understand the same elephant by examining a piece of the beast. Understanding the whole migratory phenomenon is essential for effective conservation and management.

Laurie caged the ovenbird gently in her hand, its head protruding between her index and middle finger and its feet gripping her pinky. Opening a tackle box, I flipped through a notebook and looked up the color combination to be used on this bird—SBWR, meaning silver (a numbered metal band), and a blue plastic ring on the left leg, white and red plastic bands on the right. Within minutes Laurie had banded the bird, sealing the plastic with carefully applied drops of acetone, taken a few measurements, and slipped the small bird into a sock so she could weigh it. I couldn't help noticing that the sock in question was petite and white, with a bow of pale blue ribbon on the ankle— the kind my sisters used to wear to church when they were eight.

"Nice sock," I said, handing Laurie the spring scale.

"I figured I could spare that one from my wardrobe," she replied blandly, the bagged ovenbird dangling from a clip at the end of the scale while she read off its weight. Then she slid the bird out of its wrapping and released it. Within seconds he was singing again.

We moved the rig a few hundred yards, into the territory of another unbanded male, and settled down again once the tape started playing. This male was either more cautious or less belligerent than the first, and we had a chance for a quiet talk before he was finally caught. I asked Laurie what she thought of the dust-up over Robinson's article, and the larger question of whether or not songbirds were in decline.

"Well," she said after a long pause, "I think some birds are doing okay— ovenbirds, for instance, seem to be increasing in some parts of their range, certainly in Pennsylvania. But I don't think it would take a lot to push them over the edge into the declining category. Just one more little pressure might do it, like loss of wintering habitat. I think the general message has been correct. Some birds are very clearly declining, like wood thrushes and cerulean warblers. But you don't want to be alarmist.

"We need to monitor birds in different ways than we've been doing. The Breeding Bird Survey is all right for vocal species, but it's a roadside survey, so you're sampling fragmented habitat right off the bat. We need to set up a lot more long-term monitoring plots, and we need reproductive data. The BBS just records presence or absence, but you need an additional layer of data if you want to know how the birds are really doing."

The fact is, a singing bird doesn't really tell you that much. An unmated male with a lousy territory, without a prayer of contributing to the gene pool, will sing just as loudly as one with a terrific piece of real estate, a fertile mate, and a bunch of chicks in the nest. In fact, the unmated male may sing more persistently each day, and longer into the breeding season, than a mated male, making itself more likely to be counted on a BBS survey route. And depending on local circumstances, this factor can make a big difference. A scientist studying ovenbirds in Missouri found that three-quarters of the males he heard singing were unmated, creating the illusion of reproductive health, when in fact the population was in dire straits.

Looking for that "additional layer of data" is what Hawk Mountain and other teams of researchers have been doing in recent years, but it isn't easy. Hawk Mountain's study is a good example of the toil involved: every new male ovenbird within the eleven study sites has to be caught and marked with a combination of color bands each spring. That's time-consuming enough, but then each of the dozens of ovenbird territories must be checked every four days throughout the breeding season, from the end of April to the middle of July, to see if the males are tending any chicks, which stay with their parents for a month after leaving the nest. That's the only way the researchers can determine which pairs were successful and which were not—only then can they hope to understand the variables like forest size and vegetation makeup that may influence nesting success.

As Laurie said, ovenbirds and a number of other neotrops seem to be doing pretty well. The most recent BBS data show that scarlet tanagers and fifteen or sixteen other species are stable or increasing—at least *overall*. The tanagers, for example, are increasing in parts of the Great Lakes and lower Ohio Valley, but have decreased sharply in much of the Southeast coastal plain.

Assuming the BBS trends are correct and, as Robinson and others assert, the continent's breeding songbirds are holding steady overall, it's hard to reconcile that fact with the disastrous trend Sid Gauthreaux reported from his radar studies, which suggest a 50 percent decline in trans-Gulf migration since the 1960s, or with migration banding studies from places like Long Point that show similar long-term decreases. One explanation may be "floaters," as biologists call the large pool of unmated birds, often low-status young of the previous year, which drift around the margins of life like wallflowers at a dance—what one biologist dubbed the "underworld" of birds.

Nature almost always produces more individuals than there is space for them to occupy, and by some estimates, floaters may comprise 30 to 50 percent of the total population of some songbird species. They cannot find a territory of their own, and male floaters usually do not sing, because doing so would earn them a whack upside the head from the resident male on whose turf they are skulking. But floaters are ready to pounce on an opening if a territorial bird is killed or incapacitated, sometimes within minutes of its demise.

In 1950, in an infamously cold-blooded experiment, two biologists in Maine armed with a 16-gauge shotgun and a federal collecting permit set out to shoot as many of the songbirds as they could in their forty-acre study site, part of a study on the impact of insect-eating birds on forest pests. In six days they'd killed 156 birds—vireos, thrushes, warblers—and for the next three weeks they methodically shot any new bird unlucky enough to claim a territory in their plot. By the time they finally quit, they'd killed 528 adult birds— three and a half times the number originally found there. The study ended, not because the supply of floaters had run out, but because the breeding season was drawing to a close and, with it, the hormonal urge to claim a territory.

So floaters might, for a time, mask the true dimensions of a population decline, at least from breeding surveys like the BBS. Let's say, just for the sake of argument, that the BBS is wrong and scarlet tanagers are actually becoming rarer in most of their range. The overall number of tanagers could drop by 10 percent, 20 percent, even 30 percent over the years—but because there are still extra floaters rushing in to fill the void, most of the breeding territories are occupied each spring. Volunteers running their BBS routes still hear the raspy, robinlike phrases of the tanager's song in most of the places they had heard them before, and despite the fact that the total population is way down, everything seems to be fine.

If this sort of thing could really happen in nature—and not everyone is convinced that it could—then the diminishing species will eventually hit a threshold, below which there are no more floaters, no more reserves, no more buffer—suddenly nothing to mask the fact that the species is in deep trouble and sinking fast. Nor is there any way of knowing where such a threshold might lie, or how close some birds might be to it. Even a species in clear jeopardy, like the wood thrush, still numbers in the millions of individuals, and for especially widespread songbirds like blackpoll warblers, the figure may be in the hundreds of millions. A slow, steady erosion might go on for decades before it became detectable through breeding surveys.

Another problem for researchers is trying to distinguish between human-induced changes and natural cycles, which affect virtually all living things to one degree or another. There are, for example, four kinds of warblers that breed primarily in Canada and feed heavily on a moth caterpillar known as spruce budworm. The budworm periodically explodes in outbreaks that defoliate vast areas of conifer forest, and these four species of birds—the blackpoll, bay-breasted, Cape May, and Tennessee warblers—specialize in finding and exploiting the caterpillar infestations.

When spruce budworms reach a peak of their boom-and-bust cycle, the warblers flood into the area to take advantage of the easy pickings; during one outbreak in Ontario, biologists counted nearly 600 pairs of bay-breasted warblers per square mile. Reproduction soars (when food is abundant, these species lay larger clutches of eggs than most warblers), and if the outbreak is large enough, the overall population of the warblers climbs rapidly. That happened in the 1970s, but by the 1980s the budworm cycle had crashed, perhaps short-circuited by pesticide spraying, and the warblers began to decline, too.

Because we humans don't notice changes that take place slowly, incrementally, we tend to underestimate their cumulative effects, even within the span of one or two lifetimes. One of my favorite walks in late spring follows the flat top of one of the local ridges, a worn footpath that parallels an old stone wall for the first quarter mile through the oaks and white pines. There are quite literally hundreds of miles of these stone walls winding through the woods in my home county, green with lichen and half-buried beneath humus and leaves. Many, like the one along my mountaintop trail, are in places that I would consider marginal for agriculture, and they are a constant reminder of how utterly the landscape here—and elsewhere in the East—has changed in the past seventy to one hundred years.

For all the talk about habitat loss, eastern North America actually has much more land suitable for woodland birds than it did a century ago, when farming and logging reduced the forest cover by more than half its original size. I've seen photographs of these mountains in the late 1800s, when they were utterly stripped of trees, both to make way for crops and to provide mine timbers for the coal fields just to our north. Since then, much of the East has reverted to forest, especially in New England and the Appalachians, where many farms were abandoned in the first half of the century.

A hundred years ago there was an imbalance between breeding and wintering habitat for forest-dwelling migrants—a relative scarcity of breeding habitat in the north, and an overabundance (compared to the population of birds) of wintering habitat in Central America and the Caribbean, where deforestation was still limited. Then the eastern forests recovered, balancing the two ends of the habitat seesaw—but not for long. The destruction of tropical forests accelerated after World War II, and especially in recent decades, at a time when forest cover has slowly started receding in North America again, and what's left is increasingly diced into smaller pieces. That probably means that many songbirds are heading for an ecological wall.

The fragmentation of once-intact forest is a major source of concern for conservationists. The study of how animals and plants are distributed in isolated habitats is a discipline known as island biogeography. It grew out of pioneering studies in the 1960s by Edward O. Wilson at Harvard and Robert

MacArthur at Princeton, who found that there is a predictable link between the size of an island and the diversity of species found on it; smaller islands have fewer species, and the risk of extinction (from environmental hazards or inbreeding, among other causes) is much greater on them.

MacArthur and Wilson originally applied their principles only to true islands, but ecologists have long since realized that the same rules hold for figurative islands as well, isolated fragments of habitat like forest patches surrounded by suburban sprawl or pieces of native grassland in a sea of crops. Biologists have found that forest-interior songbirds, a host of species like black-and-white warblers that habitually nest far from the edge of the woods, react most strongly to patches, disappearing from those that have been sepa-rated from other forests.

Small islands of forest lose bird species, and overall numbers of individuals, at a worrisome rate. But what is it about fragmentation, biologists wondered, that had such a deleterious effect on birds? It wasn't that forest-interior birds had a natural aversion to the edges of woodlands, to open space itself—stud-ies in very large tracts of forest showed that many species maintained terri-tories that ran right up the margins of lakes, fields, or other neighboring habitats. It wasn't voluntary avoidance. There had to be something else at work.

In the early 1980s, a young graduate student at Princeton named David Wilcove, working under the tutelage of John Terborgh, tackled the question for his doctoral thesis. Wilcove's working hypothesis was that edges are bad for nesting songbirds because they are good for nest predators, an idea that had been floated by many biologists but never rigorously tested. For one thing, it is difficult to find enough wild-bird nests to make for a statistically significant sample in a research project. So Wilcove adapted a technique developed a few years before in Wisconsin, which substituted artificial nests stocked with domestic quail eggs. His eleven study sites ranged from tiny sub-urban woodlots in Maryland to Great Smoky Mountains National Park, at half a million acres the largest undisturbed forest left in the East.

Half the dummy nests were placed on the ground, and half were wired into trees. Wilcove found a striking contrast in predation rates, depending on the size of the forest in which the nests were placed. While only 2 percent of the nests in the Smokies were destroyed by predators, the failure rate jumped to almost half in rural woodlots that were thirty acres or smaller. In suburban forests of similar size, the predation rate skyrocketed to 70 percent or more. (Wilcove himself acknowledged that his technique, while innovative, was not perfect. The dummy nests were of necessity not concealed as well as a real bird's nest would be, and they tended to be clumped, so that a successful raccoon, combing the nearby woods for another windfall, was likely to stumble on more. Nor could Wilcove do anything about his scent trail, which might easily lead a sharp-nosed predator from one experimental nest to another.)

But even if his results were a bit inflated, Wilcove's research showed that mammalian predators like raccoons, house cats, opossums, chipmunks, and skunks were exacting a terrific toll on songbird nests, as were avian predators like blue jays and crows. All are animals that thrive in disturbed habitat, in close association with people—feeding at our garbage cans and bird feeders, raiding our gardens, able to adapt easily to the mix of brush, small woodlots, and back yards that now dominate the East and Midwest. Such creatures exist in very much lower numbers in unbroken forest, where conditions are not nearly so amenable to them and where larger predators like coyotes and bobcats tend to suppress their numbers. Wilcove's research is considered one of the most important studies into the effects of forest fragmentation on birds, and his findings have been confirmed by a number of other biologists, using such refinements as automatic cameras to record the identity of the nest thieves.

What Wilcove and others have also found is that small forest fragments aren't bad for all birds. Many kinds of songbirds fare quite well in them— species like titmice, chickadees, woodpeckers, song sparrows, cardinals, and robins. The reasons probably include nesting strategies; many of the successful breeders lay their eggs in tree cavities, safe from most predators. (Wilcove created artificial nest holes in some of his hardest-hit study sites, stocked them with quail eggs, and confirmed that nest robbing is rare even in such situations.)

Neotropical migrants, on the other hand, tend to build simple, open-cup nests much more vulnerable to attack, and they are much more likely to build their nests on or near the ground. Most of the thriving species are also either full-time residents or short-distance migrants, which gives them a head start each breeding season. Unlike neotrops, which have time for just one nesting attempt per year, with no second chance if the first nest fails, these birds can produce two, even three sets of chicks, easily offsetting any predation. Many, like robins, are also hefty enough to drive off jays or grackles, something a small vireo or gnatcatcher cannot.

While most studies into nest predation have focused on wild animals, conservationists are increasingly fed up with one major, preventable source of bird mortality—free-running house cats. John Coleman and Stanley Temple, two researchers at the University of Wisconsin, recently conducted a four-year study of cat predation and concluded that domestic cats kill roughly 39 million birds per year in rural areas of that state alone, and hundreds of millions nationwide—not including the tens of millions more killed by cats in urban and suburban areas. Those figures may seem impossibly high, but they're actually quite conservative; Coleman and Temple found that the number of free-ranging cats (as opposed to house pets) in rural Wisconsin was 60 million, and estimated that they may actually kill as many as 217 million birds a year.

All this has led the American Bird Conservancy and other conservation groups to launch a national "Cats Indoors!" campaign to convince pet owners to keep Tabby in the house. It's a hard sell; while few people countenance letting dogs run loose to kill deer and rabbits, they see no problem with allowing their cats to wander free.

One major problem for forest-nesting birds isn't a predator at all, but a plant-eater. All but exterminated by forest destruction and uncontrolled shooting in the nineteenth century, white-tailed deer quickly recovered in the twentieth, becoming, by the 1930s, the most common large mammal in North America. Today, densities of more than twenty or thirty per square mile are normal in many regions. Wildlife managers are only now coming to grips with the profound effect that so many deer have on their environment.

Whitetails are browsers, feeding on woody twigs and seedlings, as well as on softer plants like wildflowers and grasses. After nearly seven decades of high deer populations, many forests in the East, Midwest, and Great Lakes states have been drastically altered. Tree reproduction is low or nonexistent, and the shrub layer found in healthy, mature forests is often absent or severely reduced, while overall plant diversity is limited. Ecologists refer to such woods as "fern parks," blanketed by endless expanses of inedible fern—the only thing that can grow within a whitetail's ravenous reach. Nest sites for mid-canopy birds like least flycatchers and yellow-billed cuckoos are absent, and so are the birds. Studies in the Allegheny National Forest of Pennsylvania show that, at deer densities above twenty per square mile, both bird abundance and species richness begin to decline sharply.

The same fragmentary, overbrowsed conditions that promote nest predators also benefit another creature of paramount importance to nesting songbirds, a species as close to universally vilified as a wild bird is likely to be. Last summer, while working in the garden one evening, I heard a commotion in the trees a short distance away—the frantic cheeping of a young bird begging for food. A chipping sparrow, its beak overflowing with green caterpillars, had landed next to what was clearly not its own chick. The baby was covered in dingy brown feathers quite unlike the smooth pearl and rust of the sparrow, and it was more than twice the adult's size, like a Sumo wrestler squatting beside its handler. Yet when the chick opened its mouth again, wings fluttering, and leaned forward with agitated calls, the sparrow stuffed the food down its maw and flew off for more.

On the surface, there was a Laurel and Hardy air of absurdity to the pair, but there was nothing funny about it. I was watching a hijacking of sorts, the piracy of parental instinct by a young brown-headed cowbird, the only habitually parasitic bird in North America.

Cowbirds are icterids, members of the same family as orioles, blackbirds, and meadowlarks. The males are a glossy purple, with dirt-brown heads, while the females are plain grayish-brown; both are little bigger than cardinals, and they would be wholly unremarkable birds were it not for their unusual breeding habits. Cowbirds are brood parasites, laying their eggs in the nests of other birds, leaving their chicks to be raised by foster parents. Because cowbirds were originally found largely on the mid- and shortgrass prairies of the West, it has always been assumed they evolved this technique so they could follow the bison (and, during the Pleistocene, camels and horses), feeding on seeds and insects in the trampled wake of the migratory herds.

It was no great leap from wild bison to domestic cows, and the felling of the eastern forest—which had once formed a substantial barrier to the open-country cowbirds—made enormous new areas available for the species. Although rare in the Eastern states before 1800, "cow-buntings" (as Audubon called them) quickly became common, and by the end of the nineteenth century they were among the most widespread farmland birds in the Midwest and the East. Through the twentieth century their range expanded through the West as well, into areas like the Sierra Nevada that hadn't originally supported large herds of grazing animals; in the East, their populations soared as a result of abundant food on farms (especially rice farming in the South, where many winter) and at back-yard bird feeders.

A female cowbird is an egg factory, able to produce two to five per week for a seasonal total of up to forty—the most eggs known to be laid by any wild bird in the world, and a trait that has led the cowbird to be dubbed "the passerine chicken." The female constantly scouts for nests under construction, sneaking back periodically to check for the presence of the host's first egg. (If she lays her own egg first, there is a greater chance the host will desert the nest.) Finding an untended nest with one or more eggs, she slips in, tosses or eats the existing egg, and lays her own, heavily speckled and often not looking at all like the host's.

Some birds, long exposed to the ways of cowbirds, have evolved defenses against this parasitism. Most such "ejector" species, as they are known, including Baltimore orioles and gray catbirds, simply heave the cowbird egg overboard, but yellow warblers are famous for their tendency to bury them— and their own clutch, if necessary—beneath a new nest floor and start again. One persistent female warbler, hit over and over again by cowbirds, ended up building six such layers. Blue-gray gnatcatchers, at least in rare circumstances, may go one step further—completely dismantling a nest that has been parasitized and using the material to build a new one elsewhere.

Other species, known as "acceptors," do not seem to recognize the cowbird egg as foreign; they may wind up with several more in their nest before all is said and done. Cowbirds are indiscriminate about choosing hosts, unlike European cuckoos, which have evolved a sophisticated form of mimicry in which the eggs of cuckoos from different regions closely match the color and pattern of the most common host. Cowbird eggs, on the other hand, have been found in the nests of more than 200 species of North American birds, including such unsuitable locations as those of blue-winged teal and ferruginous hawks. Cavity nesters are usually safe, although I once watched a cowbird try mightily to perch at the entrance to a birdbox, pressing its vent against the hole to dump an egg inside. It might have succeeded, in fact, had the two bluebirds nesting in the box not strafed the cowbird repeatedly, finally driving her off.

Mostly, though, cowbirds pick the open-cup nests of small passerines—and here's where the trouble begins. Most eastern songbird species had little or no experience with cowbirds until the last century or two—an eyeblink of time from an evolutionary standpoint. Relatively few of them are ejectors—most blindly accept the cowbird eggs as their own, incubating them without a qualm. The cowbird egg usually hatches a day or so earlier than any surviving host eggs, giving the chick a head start which it never relinquishes. Bigger and more aggressive than many songbirds, it elbows its foster-siblings aside, and while it does not actively kill them, it often starves them out by monopolizing the food the parents bring in.

Out of all the eggs a female cowbird broadcasts among the hapless nests of her neighborhood, only 2 or 3 percent will survive to become adult birds. But that's really a pretty good rate. On average, a female will produce 2.4 adult cowbirds in her lifetime—1.2 pairs, more than enough to replace her and her mate, and enough to ensure that the cowbird population will double every eight years. Today there are an estimated 50 million brown-headed cowbirds in North America, with the species pushing into new territory in places like the Yukon. Not only are they causing problems among forest songbirds, but the overwhelming numbers of cowbirds are also stressing prairie species that evolved in tandem with them. Scientists working on the northern Plains suspect that fragmentation of grasslands into small pieces, with lots of perches like fenceposts from which cowbirds can scan for nests, have tipped the balance in favor of the parasites.

Although they tend to avoid deep woods, a cowbird doesn't need much of an excuse to go bushwhacking; they'll follow roads, power lines, logging cuts, even grassy paths into what had been unbroken forest, ferreting out the nests of vireos, warblers, flycatchers, and other species with little or no experience with nest parasitism. Some years ago I joined a graduate student working at Hawk Mountain who was studying the impact of cowbird parasitism on wood thrushes. Jeff and I were working in the middle of a large mountain forest, but what from the valley appeared to be a solid canopy of green covering tens of thousands of acres was actually webbed with dirt roads and pocked here and there with weedy clearings, most no bigger than a city yard.

But that was big enough. Whenever Jeff found a thrush nest, a mud-and-grass cup like a robin's with long tendrils of dead leaves and shredded bark dangling beneath it, he would screw together several lengths of aluminum pole, tipped with a round mirror on an angled neck, and raise it above the nest to peer inside while the female scolded us angrily. Time after time, day after day, the nests he found had already been discovered by cowbirds, which had left their brown-spotted eggs among the pale blue of the thrush's. Those dirt tracks and tiny clearings, insignificant as they seemed, were a chink in the armor of the forest that the cowbirds were quick to exploit.

Nowhere has the cowbird invasion been as great a calamity as in the Midwest. In the Shawnee National Forest of extreme southern Illinois, Scott Robinson and his associates at the state Natural History Survey have been studying the impact of cowbirds on migrant songbirds, and it hasn't been pretty. A mosaic of federal forest and private farmland, the Shawnee is a cowbird's paradise and a songbird's hell; cowbird populations are extremely high, and parasitism rates for species like wood thrushes approach 90 percent. Overall, six out of ten songbird nests in the national forest contain one or more cowbird chicks. Farther north in Illinois, where the woods are scattered in clusters of even tinier islands, the rates are, if anything, worse; Robinson found that thrush nests contained, on average, four times as many cowbird eggs as thrush eggs. He found one wood thrush nest with twelve cowbird eggs, and none of the bird's own eggs at all.

In other parts of the United States, researchers have also found a convincing link between the incidence of parasitism and forest fragmentation. In Wisconsin, biologists checked songbird nests within a hundred yards of the edges of woodlots, and found that 65 percent were victimized by cowbirds. But among nests more than 300 yards into the woods, the rate dropped to fewer than 20 percent. There also seem to be regional differences in cowbird populations. In the Ozarks of southern Missouri, one of the largest stretches of intact hardwood forest left in the region, cowbirds remain relatively scarce, and parasitism is therefore low, barely a tenth the rate in Illinois. As far as cowbirds are concerned, much of the East lies somewhere between the extremes of the Ozarks and the Shawnee, and unlike in the Midwest, BBS data show that cowbird numbers there are declining slightly.

Robinson and his colleagues believe that songbirds in the Midwest are a good example of something known as "source-sink population dynamics." Remember, just because there are lots of birds singing in the woods, there's no guarantee that the population is healthy; the only way to be sure is to look at reproductive success. Robinson's team did that the hardest but most accurate way possible—by deploying an army of college students in the steamy, buggy woods to painstakingly locate songbird nests, then track their successes and failures.

And failures, they found, far outweighed successes. The Shawnee and other fragmented forests in Illinois are a reproductive trap for nesting songbirds, attracting breeders but wiping out any eggs and young they produce, so that there aren't enough new birds to compensate for natural mortality among adults. Besides cowbird parasitism, nest predation rates in small woodlots approached 80 percent. Such areas are a net loss, a "population sink" in ecological jargon. The only reason most of these populations do not quickly collapse (as some within Robinson's study area have done) is that they constantly attract new recruits from outside the region, where reproductive circumstances are more favorable. Those productive areas are known as sources, and the give-and-take between sources and sinks can stretch across huge regions. In one of Robinson's study areas, 85 percent of the adult birds disappeared between breeding seasons, a phenomenally high turnover rate that could be sustained only by immigrants.

More recently, other researchers have followed Robinson's lead, examining cowbird and parasitism rates in huge, intact forests and in highly fragmented habitat. Their findings suggest that large forests like the Ozarks, south-central Indiana, and northern Wisconsin are sources, producing a surplus of young birds each season that disperse to sink areas like Illinois; other likely sources elsewhere in the continent include the expansive forests of New England, eastern and central Canada, the Great Lakes states, and parts of the Appalachians.

As if the brown-headed cowbird wasn't bad enough, there are now two more kinds to worry about. Five cowbird species occur in Latin America, and in 1985 one of them, the shiny cowbird, was first discovered in the United States, in the Florida Keys. Similar in size and shape to the brown-headed but with a glossy purple head, the shiny cowbird is on a roll, having overrun the islands of the Caribbean in the past half century, driving at least one species, the yellow-shouldered blackbird of Puerto Rico, to the brink of extinction. Two years after landing in the Keys, the shiny cowbird spread to mainland Florida, then exploded across the Southeast, reaching as far as Oklahoma, Texas, and Alabama by 1990 and as far north as Virginia a few years later. Its numbers have remained low, but ornithologists are worried about its eventual impact on Southern songbirds. And while all this was going on, the bronzed cowbird—a larger, red-eyed species native to Texas and the Southwest—was expanding its range east and north. In some parts of the Gulf, all three cowbird species have been found in the same forest, a potential troika of doom for nesting songbirds.

Cowbirds elicit a visceral reaction from a lot of the biologists I know. Just recently I was talking to a friend who does some banding at a local sanctuary. "I caught a cowbird in the net the other day," Mark mentioned in passing, and when I gave him a questioning look, he shook his head, instantly understanding my meaning. "No, I had a group of visitors with me at the time, so I let it go. I decided not to play God."

Mark and I both know that, however satisfying it might be to wring a cowbird's neck, it would be wasted effort; killing a single cowbird would make not a whit of difference, given how many millions of them are out there. (Also, cowbirds are a native species, and thus protected by the same federal laws that safeguard other birds.) But there have been some occasions where conservationists, with official blessing, have been forced to "play God," attempting to control cowbirds to save critically imperiled birds. The best-known example involves the endangered Kirtland's warbler in Michigan, but several other birds suffer from the same problems—small populations, tiny breeding ranges, and too damned many cowbirds, including the least Bell's vireo and the willow flycatcher in California and the black-capped vireo in the southern Plains.

In all four cases, wildlife managers have tried to reduce the number of cowbirds within the threatened birds' breeding range; the standard approach is to build huge wood-frame-and-chicken-wire traps, bait them with grain and live cowbird decoys, and gas the multitudes of other cowbirds that crowd in. At Fort Hood in Texas, the military maintains more than fifty cowbird traps year-round—the reason the installation is home to one of the few thriving black-capped vireo populations. Balcones Canyonlands National Wildlife Refuge near Austin has been doing the same thing since 1992, with similar success. In Michigan, tens of thousands of cowbirds have been removed from the two-county core of the Kirtland's warbler's breeding range, plunging parasitism rates there from nearly 75 percent to barely 5, and effectively tripling productivity among this critically endangered migrant.

Unfortunately, what works on a limited scale is impossible across states or regions. Cowbird trapping is labor-intensive and expensive, and it simply isn't feasible to control them over large areas. Besides, they are the symptom, not the disease; instead of treating the fever—the cowbirds—it makes more sense to go after the virus of habitat fragmentation. One solution is to restore and consolidate tattered woodlands through reforestation, removing habitat for cowbirds and edge predators. That approach is already under way along the Cache River, a seasonally flooded cypress forest at the southern tip of Illinois, in the heart of the area studied by Robinson and his coworkers. The Nature Conservancy, with support from the U.S. Fish and Wildlife Service, Ducks Unlimited, the Illinois Department of Conservation, and other groups, has been acquiring and restoring forest land—38,000 acres so far, with a long-term goal of 60,000 acres.

It is also critical to preserve existing large, contiguous forests. Ornithologists have pointed out that most of the major "source" areas like the Ozarks have extensive national forest holdings at their core; management decisions, therefore, like how much timber to cut, where and how to cut it, will have a tremendous impact on the future of nesting birds not only within their boundaries but in distant "sink" areas now kept afloat by their emigrant songbirds.

From all the research, it is clear that turning contiguous forest into myriad tiny islands is bad for many songbirds. The pressing questions for land managers and conservationists, though, are: How small is too small? How big is big enough? What is the right mix of large core areas to smaller fragments? And how far into a forest does edge effect extend?

On this last question, different studies have come to different conclusions. One on ovenbirds suggested that edge effect reached as much as 300 yards into a forest; other researchers concluded it was less severe, no more than 50 yards. In all likelihood, the discrepancies stem from differences in the habitat, the makeup of the predator pool, and other local variables. And there are probably aspects of fragmentation that ecologists haven't even confronted yet. Recently, two biologists in Ontario, Dawn Burke and Erica Nols, sampled the abundance of invertebrates in forest fragments of various sizes and correlated it to the density and nesting success of ovenbirds. Scott Robinson—whose own research is famous for the numbing amount of fieldwork it requires—called the study "enormously labor-intensive," but the results were striking: insect prey was as much as thirty-six times more abundant in large forests than small fragments (perhaps because the smaller woodlots had less leaf litter and, being more exposed to sun and wind, a drier climate), and female ovenbirds carefully chose territories with the richest supply of food.

Even among closely related groups of birds, there may be big differences in how they react to fragmentation. For several years volunteers working with the Cornell Lab of Ornithology have been surveying thrushes, as part of the lab's Birds in Forested Landscapes project. Preliminary analysis showed that veeries were highly sensitive to fragmentation, while varied thrushes in the West were almost immune to it—provided the fragments had tall trees and a thick understory. Studies along the Cache River showed that narrow corridors of forest, linking larger tracts, were all but useless to species like Acadian flycatchers, which were hammered by predators and cowbirds in thin ribbons of woods. And not all songbirds are equally susceptible to the pressures of predation or parasitism. While the large, bulky nests of wood thrushes are fairly easy for cowbirds to find, worm-eating warblers build small, well-concealed ground nests, often covering them with dead leaves when they aren't around.

The upshot is that there is no recipe for an "ideal" forest fragment, except that size matters, and bigger is better. Chandler Robbins and two colleagues have estimated that a forest in the mid-Atlantic states must be at least 3,000 hectares—about 7,400 acres—to preserve all the most sensitive species. Not even a 2,500-acre plot in Missouri, scientists found, was big enough to prevent the loss of some species of nesting songbirds.

Cowbirds, deer, crows, raccoons, rat snakes, opossums—it's a rough world for a nesting bird. And now it's turning into a hot one, too.

A few days after I joined Laurie to band ovenbirds, the spring weather—which had been the perfect, temperate example of May in the central Appalachians, cool nights and pleasant days—took a hot turn. The thermometer climbed into the mid-90s every day for a week, with dripping humidity. The robin nesting in the red maple outside my bedroom window panted in the sultry midday sun, which poured through the half-grown leaves, and the bluebirds and catbirds lined up at the small garden pond for quick sips and cooling baths. Such late-spring heat waves are not unusual, but already the weather gurus were predicting a scorcher of a summer, with temperatures expected to be far above normal.

Actually, that itself has become the norm, with nine of the ten hottest years on record coming since 1987. To most climatologists, this fact is just one more in an array of evidence that points, more and more clearly, to the beginnings of widespread global climate change. Increasingly sophisticated computer models agree that by overloading the atmosphere with carbon dioxide, methane, and other "greenhouse gases," we have altered the climatic balance, propelling the earth into a period of draconian change.

That doesn't mean universal warming; some areas would become colder than they are now. The distribution of rainfall would change as well, with some areas becoming wetter, while others, especially in the middle of large continents, would be subject to great droughts. Storms like hurricanes would become more frequent and more severe, and even areas that receive the same amount of rain might see a shift in how it falls, in a few, gully-washing tempests instead of frequent, gentle showers. The details still aren't clear, but climate change promises drastic, traumatic differences for the earth's ecosystems.

By even the most conservative calculations, such adjustment in temperature and rainfall patterns would spark a radical shift in the timing of the seasons and the distribution of vegetation in North America, with the severity increasing the farther one goes from the equator. There is plenty of precedent for how climate change alters vegetation patterns; paleoecologists studying fossil pollen have rather precisely mapped the northward displacement of plant communities after the last ice age ended, and they've applied those lessons to predicting what a warmer world would look like.

Assuming a doubling of carbon-dioxide levels by the middle of the next century—a rate based on current trends—the average global temperature should rise several degrees Centigrade, shuffling the vegetation mosaic in

ways that would make the landscape nearly unrecognizable. The treeless tundra ecosystem would all but vanish, pushed out by boreal spruce forest edging
up from the south. Grassland would replace conifers over most of western
Canada and central Alaska, while in the East, the hardwood forest would
migrate north, occupying New England, Ontario, Quebec, and the Canadian
Maritimes. Most of the Appalachians and Southeast, now covered in solid
forest, would become a savannalike mix of grassland and pockets of trees.

And what of the birds? It's hard to imagine how the vast numbers of shorebirds and waterfowl that depend on the marshy tundra could survive in anything like their current numbers if that ecosystem virtually disappears. The
expansion of grasslands might be beneficial for some birds, although experts
caution that mid-continental droughts would create arid grasslands, very different from the rich tallgrass and midgrass prairies of the northern Plains. The
prairie pothole region, North America's "duck factory," already subject to
drought, would dry up permanently.

Forest songbirds would also face enormous challenges. Consider species
that breed today in the endless stands of conifers in Canada, Alaska, the
Rockies, and the northern Appalachians. The mountain forests would vanish
altogether, and the southern limit of the remaining spruce woods would move
from New England and the Great Lakes region to the fringe of the Arctic
Ocean. Most of the boreal habitat would be restricted to the Arctic islands of
Canada, very much reduced in extent from what it is today, and as much as
1,200 miles north of its current location—1,200 miles that a migrating songbird would have to add to each leg of its annual trip, additional fat it would
have to lay on, more stopover sites it would have to find along the way. A
journey that already pushes birds to the physical limit would become considerably harder, and perhaps impossible.

Climate change means a shift not only in physical distribution but in timing as well. Think of the red knots landing on the beaches of Delaware Bay
just as the horseshoe crabs are spawning, or the songbirds sweeping north
across the eastern hardwood forests just as the leaf-eating caterpillars are most
abundant. Think of shearwaters ranging thousands of miles along the Pacific
Rim, hitting each local bloom of fish and squid at peak abundance, and the
bar-tailed godwits in Beringia that hitchhike south to New Zealand on the
storm winds of early autumn. Timing is crucial to bird migration.

As ecologists have pointed out, climate changes that alter the growing season for plants would set off a cascade of severed connections among species
that depend, however tangentially, upon them. There is evidence that such a

transformation is already under way. In 1997, scientists from Boston University analyzing satellite data announced that they had documented a 12 percent increase in spring and summer plant growth across the Far North, and a lengthening of the growing season by an average of eight days in the spring and four days in the fall. That substantial change had occurred, the team said, over just ten years, from 1981 to 1991. A few months later, ornithologists in Great Britain announced that more than 74,000 nesting records showed that many birds in the United Kingdom (most of them nonmigratory residents) were laying their eggs an average of nine days sooner than in 1971. In Germany, ornithologists found that short-distance migrants in the fall were passing through the study area up to ten days later than they had in the 1970s, while in North America, a University of Michigan researcher found that spring migrants were arriving in the northern Great Lakes states up to three full weeks earlier than they had in 1960—more proof that the Northern Hemisphere is warming up faster in spring and staying warmer later in the fall than it had earlier in the century.

These studies show that migrants possess at least a little flexibility in the face of weather changes, but no one is sure how migratory birds will react to major shifts in global climate and ecosystems. Certainly they have coped before—just 18,000 years ago, the northern half of the continent was gripped in ice, and tundra extended south as far as the mid-Atlantic states and Iowa. There is speculation, you'll recall, that some of the more impressive long-distance migrations we see today, like the continent-hopping flight of western blackpoll warblers, evolved as a way of leapfrogging over the ice sheets that blanketed most of Canada, taking advantage of the prevailing winds generated by the glaciers.

After the world entered the current period of interglacial warming, the great bands of vegetation—tundra, boreal forest, hardwoods, southern pines, and grassland—flowed north in the wake of the ice, each one in succession as the climate warmed. For example, over the course of about 7,000 years, the white-spruce zone traveled north from New Jersey and Illinois to Labrador and central Ontario, carrying its dependent bird species along with it, stretching their migratory routes farther and farther each season.

But 7,000 years is a long time to cover a thousand miles—a rate of just a few hundred yards per year, one that plant seeds can cover easily on the wind or in the stomachs of hungry birds. One of the biggest worries about human-induced climate change is the projected timetable; what took the earth thousands of years to accomplish after the ice age, the reasoning goes, will take us only a century or so. No natural system, like tree dispersal or migration, can hope to keep up with such a cataclysmic rate of change, and scientists worry that ecological communities will crumble—plant species will be unable to move north quickly enough to escape the warmth and will tumble toward extinction, taking their dependents with them.

But the latest information coming out of studies in Greenland, where the deep ice cap preserves a record of past temperatures, indicates that the climate is even testier than we thought. By drilling into the glaciers, removing plugs of ancient ice, and studying the composition of the annual layers, like rings in a tree, scientists can infer the temperature when the ice formed; they also look at varying amounts of trapped dust, which is more common in colder, drier periods in the Arctic and less so in warm, moist periods.

What they have found suggests that the climate bends, in a sense, until it reaches a threshold of environmental pressure—and then it snaps, changing global temperatures suddenly and significantly. The switch from frigid conditions to global warmth at the end of the last ice age, once thought to have taken hundreds or thousands of years, now appears to have occurred with stunning speed, in two quick bursts, perhaps lasting as little as a decade each. Other scientists studying the Greenland ice say significant warming happened in even less time—just one to three years. If there are thresholds that trigger calamitous change, no one knows how close the planet might be to one. In the words of one prominent geochemist, "The climate system is an angry beast, and we are poking it with sticks."

There is both despair and hope in these findings. It was bad enough believing that birds would have to cope with a fundamental reordering of the natural world over the course of the next century; to think that wild animals and plants may have to do so in a matter of a few decades, even a few years, is extraordinarily disheartening. On the other hand, it is encouraging to discover that they may have been through something like this before. We don't know what species might have been lost along the way (after all, the end of the last ice age saw the extinction of more than fifty of North America's largest mammals, including mammoths and mastodons, whose bones fossilize much easier than birds'), but hundreds of species of migratory birds obviously survived those changes. Perhaps, the tiny core of optimism inside a modern birder pipes up, perhaps they'll muddle through this, too.

And were this still a pristine, unaltered planet, they might. But we've used up so much of the space and the resources that migratory birds may not have the room they need to maneuver through the coming crisis.

"We face the end of migrations in our lifetime," predicts a gloomy J. P. Myers, the biologist who helped create the International Shorebird Survey and who did much to draw attention to the need to preserve stopover habitat.

Not next year, not in the next 10 years, but in our lifetime. Migration, as we know it, will fade away, its threads unraveled, tattered and diminished beyond recognition . . . The migration events that now occur across the hemisphere each fall and spring—hundreds of thousands of southbound shorebirds in the Bay of Fundy each August, one-half million Sandhill Cranes on the Platte River in March, spring warbler rains in coastal Texas, autumn warbler fallout on Point Reyes, a million Sooty Shearwaters offshore of Pismo Beach each August, to name only a few—these spectacles will become events of the past—avian buffalo reduced to ecosystem irrelevancies if not eliminated altogether.

How likely are we to lose these marvels? Since Myers wrote that warning in 1991, one species from his list has already faltered—the once-staggering number of sooty shearwaters that summer off the coast of California has dropped by more than 90 percent in the last decade, a loss of potentially 4 million birds. Researchers at the Scripps Institution of Oceanography in San Diego believe the decline is linked to a 70 percent slide in zooplankton in the region, in turn caused by a sharp rise in surface water temperatures since the 1950s.

Remember Bicknell's thrush, the species recognized only in 1995, which breeds in eastern mountains and winters in the Caribbean? This bird could be the poster child for what migrants may face in the century ahead. It nests only in stunted forests of spruce and balsam fir, from high altitudes in New York and New England to coastal areas of Nova Scotia—an ecosystem that is already scattered and marooned, an archipelago in a sea of hardwood, limiting the species to perhaps no more than 5,000 birds. Nudge the global temperature up a few degrees and the spruces have nowhere to go; they've already climbed as far as they can into the mountains, and the thrush may lose its only home. Of more immediate concern, the spruces and firs have been dying, both from air pollution like acid deposition and from insect outbreaks probably exacerbated by the toxins, which weaken the trees' defenses. And the mountains where the thrushes nest are under growing assault by ski resorts, construction of windmills for electrical generation, and communication towers.

Bicknell's thrush apparently migrates along a narrow corridor, down the East Coast through Florida and out over the Antilles, putting it at risk from hurricanes, which are predicted to grow stronger and more frequent with global warming. One monster storm, colliding with the peak of the fall migration, could all but snuff out the species in a single stroke. And assuming they reach the Caribbean safely, they must find mature tropical forest in the mountains, a habitat that is greatly diminished and becoming more threadbare by the day. Without heroic conservation action, its winter refuge may disappear in the coming decades, and with it any chance of saving the thrush.

I am not as pessimistic as Myers, but I'm not sure why. We've stacked the deck against migrant birds, made life barely livable for many of them, pushed others right up to the edge of survival. We're poised to sever the last global links that allow many of them to traverse the planet. We're going to lose some, without a doubt, and when I'm an old man I'll probably be talking in hushed, awed tones to a younger generation about wonders I saw way back when, in the twentieth century, the way withered old men used to reminisce about passenger pigeons and Carolina parakeets in the nineteenth.

So my optimism isn't rooted in logic. It is a fragile emotion, much bruised by reality—a slender slip of a thing, but still standing. Besides, there's no future in pessimism. It may be nothing more than wishful thinking, but I suspect the migrants are a little more flexible than we give them credit for, a tad more supple in the face of humanity's juggernaut. We have started to recognize the problems and search for solutions, stitching together international coalitions to husband these precious creatures and the land on which they depend. Here, at the last possible moment, we have awakened to what we stand to lose—poised on the brink, but still, perhaps, with time to draw away from the edge.

Afterword

And now I just want to put aside all the worry and heartache about what the future may hold for migratory birds, the fretting and pondering and crystal-ball gazing. I want to forget it all for a short time and give myself over to this one, perfect May morning, along a stream in the mountains of northern Pennsylvania.

It was cold at daybreak, somewhere in the middle thirties, judging by the pinch it gave my nose and ears, but the sun is well up now, its light sliding majestically down the sides of the steep mountains, igniting the glorious green of new leaves. This narrow valley, among the tall, widely spaced trees, is a palette of wildflowers—golden ragwort everywhere in yellow masses, the last of the white foamflowers, pockets of violets, spring beauties, toothwort. The stream, swollen with two days of rain, runs deep and inviting, and the fly rod is leaning beside me against the piebald trunk of a fallen sycamore, but I'll leave the trout for later.

Instead, I sit with my eyes closed, enjoying the warmth on my face and the bird songs that pierce the constant sibilance of the water. I tick them off in my mind: A yellow warbler insisting *Sweet-sweet-sweet-I'm-so-sweet!* A blue-headed vireo wheedling a self-involved monologue. A scarlet tanager—no, I realize, a rose-breasted grosbeak; I always get those two mixed up in spring, when my ears are rusty. A song sparrow's three piping notes, the rest of the melody lost to the ruckus of the creek. The gobble of a distant turkey; the drumming of an even more distant grouse, not a sound at all, really, just a low, faint tapping I pick up through my bones. And a redstart, somewhere very close, his notes coming out in a tumbling rush.

The waves of migration have passed through here, but they roll on to the north, and in my mind's eye I try to map them out: the fingers of passerines reaching through the Great Lakes and threading the valleys of the Cascades and the Rockies and the northern Appalachians; ducks and shorebirds pushing up the prairies of central Canada, into the muskeg country around Great Slave and Great Bear and all the other spruce-rimmed lakes where caribou and wolves roam; flocks of Sabine's gulls and Arctic terns, jaegers and storm-petrels moving up the Pacific coast, the silhouettes of whales dark in the water below them.

I think of Alaska, of millions of sandpipers pouring through the Copper River delta, of blackpoll warblers fresh from South America, converging on the state even as the snow lingers in the shadows of the forests, of spectacled eiders deserting the rotted pack ice of the Bering Strait to head back for shore. The black brant would be streaming out of Izembek by now, leaving the snowy volcanoes and wide eelgrass pastures, funneling north through the strait to the coast of the Arctic Ocean.

And from the skies, exhausted and hungry, would have come the trans-Pacific shorebirds, the tattlers and turnstones and golden-plovers, back from the coral atolls and lonely islands. The bristle-thighed curlews would be home in Beringia now, wading through icy puddles of snowmelt, their brown and buff feathers matching the dead sedges and cottongrass. Back, too, would be the bar-tailed godwits, returned from New Zealand after 7,000 miles of labor. I can imagine the males, their heads and breasts the color of fresh cinnamon, flying in high, loud circles around their territories—calling constantly, swooping and gliding in complex displays, the polygonal pattern of the frost-heaved tundra wheeling below them. I see the brown females crouching in their small nests among the dwarf willows and grass tussocks on the higher ridges, hunching low over their eggs when the Arctic foxes pass by in the night.

All this I imagine, sitting in the Pennsylvania sun as a redstart sings. I open my eyes and he's right in front of me, in a low willow thicket that was half-flattened by the winter ice floes. He is no bigger than my thumb, all black except for the colorful patches on his wings, flanks, and tail—the same pink-orange color, it occurs to me, as the meat of the native brook trout that still live in the small headwater streams hereabouts, the same color as a monarch butterfly's wings, and the wild turks'-cap lilies that bloom here in summer. That symmetry feels proper, somehow, almost preordained.

The redstart is unaware of me; sitting motionless, I am simply another mis-shapen log to his eyes. He flits restlessly through the willows, fanning his tail, flicking his wings, zipping out to snatch midges from the air, his bill making a dry little *snap!* each time he does. Every thirty seconds or so, he stops, points his head skyward, and sings; I am so close I can see his small, pointed tongue and the tiny black feathers of his throat trembling. I can see the minute bris-tles around his mouth, the way his body feathers are layered like shingles, the scales on his delicate legs, the feather tracts that define his face. I can see the nostrils that pierce the needle of his beak, the way the sun reflects a white-hot speck in his obsidian eyes, the way the shafts of the black feathers mingle with the orange where they meet.

What I cannot see, no matter how closely I look, is what drives this small creature, barely heavier than air, to make the journeys that it must make. I may have seen this same redstart in an acacia forest in Jamaica, among the ruins of a Maya city in Belize, or in a half-dozen other places in the tropics. I can only imagine what has happened to it in its life—what near-brushes with predators it has escaped, what storms have tried to rake it from the sky, what females have taken it as a mate, what dynasties of redstarts it has founded. What thousands of miles have passed beneath its stubby wings, which seem so ill suited to the task but which have carried it back here again, to this mountain, this stream, this willow thicket. Its secrets are locked in that tiny packet of brain and muscle and instinct, a few feet away but separated from me by an immense, uncrossable distance. It knows, and I do not. And there seems to be a proper symmetry in that, too.

Notes and Bibliography

There is a rich literature of ornithology, for both scientific and general audiences, much of it having to do with migration. For broad overviews of the phenomenon, readers would do well to find *Bird Migration* by Thomas Alerstam (Cambridge University Press, 1993), *Bird Migration: A General Survey* by Peter Berthold (Oxford University Press, 1993), and *How Birds Migrate* by Paul Kerlinger (Mechanicsburg, PA: Stackpole, 1995). More specialized in their focus are *The Evolutionary Ecology of Animal Migration* by Robin Baker (New York: Holmes & Meier, 1978), *Neotropical Migratory Birds* by Richard DeGraaf and John Rappole (Ithaca, NY: Comstock, 1995), and Rappole's *The Ecology of Migrant Birds: A Neotropical Perspective* (Washington, D.C.: Smithsonian, 1995). Good books from a conservation perspective are *Where Have All the Birds Gone?* by John Terborgh (Princeton University Press, 1989), *Bring Back the Birds* by Russ Greenberg and Jamie Reaser (Mechanicsburg, PA: Stackpole, 1995), and the revised edition of Peter Matthiessen's classic *Wildlife in America* (New York: Viking, 1987).

Although specific citations and bibliographies for each chapter are given below, the following works were of use to me throughout the book: *Birds of the World: A Checklist*, fourth edition, by James Clements (Vista, CA: Ibis, 1991); the monumental *Handbook of Birds of the World* series (Barcelona, Spain: Lynx, 1992–97); *Ornithology* by Frank Gill (New York: W. H. Freeman, 1995); John Terres's *Audubon Society Encyclopedia of North American Birds* (New York: Knopf, 1980); and the ongoing *Birds of North America* series published by the Academy of Natural Sciences and American Ornithologist's Union.

1: Beringia

CITATIONS

11 "If a Blackpoll Warbler were burning gasoline": Timothy C. and Janet M. Williams, "An Oceanic Mass Migration of Land Birds," *Scientific American*, Vol. 239 (1978), 166–76.

17 "It is natural to suggest": Dean Amadon, "Migratory Birds of Relict Distribution: Some Inferences," *The Auk*, Vol. 70 (October 1953), 461–69.

21 "crooked, illogical, out-of-the-way routes": Joel Carl Welty, *The Life of Birds*, 2nd ed. (Philadelphia: W. B. Saunders, 1975), p. 473.

21 "Migration evolved": Paul Kerlinger, *How Birds Migrate* (Mechanicsburg, PA: Stackpole, 1995), p. 1.

23 Homer, Aristotle, Magnus, and Belon quoted in Jean Dorst, *The Migrations of Birds* (Boston: Houghton Mifflin Co., 1962).

25 "Winter quarters remain unfound": Josep del Hoyo, Andrew Elliott, and Jordi Sargatal, eds., *Handbook of Birds of the World*, Vol. 1 (Barcelona, Spain: Lynx Editions, 1992), p. 621.

BIBLIOGRAPHY

Allen, Arthur A., and Henry Kyllingstad. "The Eggs and Young of the Bristle-thighed Curlew." *The Auk*, Vol. 66 (October 1949), 343–50.

Anon. "Bridge over Northern Waters." *Science News*, Vol. 150 (July 20, 1996), 41.

Balogh, Gregory. "Spectacle on Ice." *Bird Watcher's Digest*, Vol. 19 (January/February 1997), 46–51.

———. "Secret Life of the Spectacled Eider." *National Wildlife*, Vol. 35 (April/May 1997), 36–39.

Dixon, James E. *Quest for the Origins of the First Americans*. Albuquerque, NM: University of New Mexico, 1993.

Dumond, Don E. *The Eskimos and Aleuts*, rev. ed. London: Thamas and Hudson, 1987.

Dunkel, Tom. "Eyeballing Eiders." *Audubon*, Vol. 99 (September/October 1997), 48–57.

Dunn, Jon, and Kimball Garrett. *A Field Guide to Warblers of North America*. Boston: Houghton Mifflin Co., 1997.

Ehrlich, Paul R., David S. Dobkin, Darryl Wheye, and Stuart L. Pimm. *The Birdwatcher's Handbook*. Oxford University Press, 1994.

Eliot, John C. "A Winter Spectacle of Eiders." *National Geographic*, Vol. 189 (January 1996), 137.

Gabrielson, Ira N., and Frederick C. Lincoln. *Birds of Alaska*. Harrisburg, PA: Wildlife Management Institute and Telegraph Press, 1959.

Handel, Colleen M., and Christian P. Dau. "Seasonal Occurrence of Migratory Whimbrels and Bristle-thighed Curlews on the Yukon-Kuskokwim Delta, Alaska." *The Condor*, Vol. 90 (November 1988), 782–90.

Johnson, Oscar W., Martin L. Morton, Phillip L. Bruner, and Patricia M. Johnson. "Fat Cyclicity, Predicted Migratory Flight Ranges, and Features of Wintering Behavior in Pacific Golden-plovers." *The Condor*, Vol. 91 (February 1989), 156–77.

Johnson, O. W., and P. G. Connors. "American Golden-plover (*Pluvialis dominica*) and Pacific Golden-plover (*Pluvialis fulva*)." *The Birds of North America*, No. 201–2, A. Poole and F. Gill, eds. Philadelphia: Academy of Natural Sciences and Washington, D.C.: American Ornithologists' Union, 1996.

Lanctot, R. B., R. E. Gill, Jr., T. L. Tibbitts, and C. M. Handel. "Brood Amalgamation in the Bristle-thighed Curlew, *Numenius tahitiensis*: Process and Function." *Ibis*, Vol. 137 (October 1995), 559–69.

Leopold, A. Starker, Ralph J. Gutiérrez, and Michael T. Bronson. *North American Game Birds and Mammals*. New York: Charles Scribner's Sons, 1981.

McCaffery, Brian J., and Robert E. Gill, Jr. "Anti-predator Strategies in Breeding Bristle-thighed Curlews." *American Birds*, Vol. 46 (Fall 1992), 378–83.

Marks, Jeffrey S. "Molt of Bristle-thighed Curlews in the Northwestern Hawaiian Islands." *The Auk*, Vol. 110 (July 1993), 573–87.

——, and Roland L. Redmond. "Migration of Bristle-thighed Curlews on Laysan Island: Timing, Behavior and Estimated Flight Range." *The Condor*, Vol. 96 (May 1994), 316–30.

——, Paul Hendricks, Roger B. Clapp, and Robert E. Gill, Jr. "Notes on Longevity and Flightlessness in Bristle-thighed Curlews." *The Auk*, Vol. 107 (October 1990), 779–81.

Murray, Bertram G., Jr. "A Critical Review of the Transoceanic Migration of the Blackpoll Warbler." *The Auk*, Vol. 106 (January 1989), 8–17.

Otis, D. R. "Migration—To the Moon, by the Stars." *Living Bird Quarterly*, Vol. 8 (Autumn 1989), 28–29.

Palmer, Ralph S., ed. *Handbook of North American Birds*. Vol. 3, New Haven, CT: Yale University Press, 1976.

Pielou, E. C. *After the Ice Age: The Return of Life to Glaciated North America*. University of Chicago Press, 1991.

Piersma, Theunis, and Robert E. Gill, Jr. "Guts Don't Fly: Small Digestive Organs in Obese Bar-tailed Godwits." *The Auk*, Vol. 115 (January 1998), 196–203.

Pratt, H. Douglas, Phillip L. Bruner, and Delwyn G. Berrett. *The Birds of Hawaii and the Tropical Pacific*. Princeton University Press, 1987.

Wilcove, David S. "A Tale of Two Curlews." *Living Bird*, Vol. 16 (Autumn 1997), 6–7.

Wilford, John Noble. "Excavation in Chile Pushes Back Date for Human Habitation of Americas." *The New York Times*, February 11, 1997.

Williams, Timothy, and Janet M. Williams. "Radar and Visual Observations of the Autumnal (Southward) Shorebird Migration over Guam." *The Auk*, Vol. 105 (July 1988), 460–66.

Zink, Robert M., Sievert Rohwer, Alexander V. Andreev, and Donna L. Dittmann. "Trans-Beringia Comparisons of Mitochondrial DNA Differentiation in Birds." *The Condor*, Vol. 97 (August 1995), 639–49.

2: A Far-flung Tapestry

CITATIONS

31 "has revealed movement patterns": Hugh Dingle, "Ecology and Evolution of Migration," in *Animal Migration, Orientation, and Navigation*, Sidney A. Gauthreaux, Jr., ed. (New York: Academic Press, 1980), p. 3.

32   "Migration is specialized behavior": Ibid., p. 4.

32   "regular seasonal movement": Peter Berthold, *Bird Migration: A General Survey*, Oxford University Press, 1993, p. 31.

33   "The phrases 'to winter' and 'wintering ground'": Neal G. Smith, "The Path Between North America and Limbo: The 'Wintering Grounds' Syndrome and Future Research on Migratory Raptors," in *Proceedings of Hawk Migration Conference IV*, Michael Harwood, ed. (Hawk Migration Association of North America, 1985), p. 387.

40   "There was great excitement": Albert M. Day, *North American Waterfowl* (Harrisburg, PA: Stackpole and Heck, 1949), p. 63.

49   "We do not need to look": John Rappole, *The Ecology of Migrant Birds: A Neotropical Perspective* (Washington, D.C.: Smithsonian Institution Press, 1995), p. 103.

50   "If the change is simply destructive": Quoted in Carol Kaesuk Yoon, "Bird-watching Biologists See Evolution on the Wing," *The New York Times*, December 22, 1992.

BIBLIOGRAPHY

Anon. "The Alluring English Winter" (changing blackcap migration). *Discover*, Vol. 14 (April 1993), 13.

——. "Migrants Seek Out Berry, Berry Rich Food." *Science News*, Vol. 150 (October 19, 1996), 252.

Batt, Bruce. "The Grand Passage of 1995." *Ducks Unlimited Duck Tails*, Vol. 1 (April 1996), 3.

Berger, Cynthia. "Finding Beautiful Facts in Feathers." *National Wildlife* (August/September 1997), 14–22.

Berthold, Peter. "Endogenous Control as a Possible Basis for Varying Migratory Habits in Different Bird Populations." *Experientia*, Vol. 34 (1978), 1451.

——, and Andreas Helbig. "Changing Course." *Living Bird*, Vol. 13 (Summer 1994), 24–29.

———, A. J. Helbig, G. Mohr, and U. Querner, "Rapid Microevolution of Migratory Behavior in a Wild Bird Species." *Nature*, Vol. 360 (1992), 668–69.

——, W. Wiltschko, H. Miltenberger, and U. Querner. "Genetic Transmission of Migratory Behavior into a Nonmigratory Bird Population." *Experientia*, Vol. 46 (1990), 107–8.

Connor, Jack. *Season at the Point*. New York: Atlantic Monthly Press, 1991.

Dingle, Hugh. *Migration: The Biology of Life on the Move*. New York: Oxford University Press, 1996.

Durant, Mary, and Michael Harwood. *On the Road with John James Audubon*. New York: Dodd, Mead & Co., 1980.

Fisk, Erma J. *The Bird with the Silver Bracelet: An Essay on Birdbanding*. South Orleans, MA: Arey's Pond Press, 1986.

Gauthreaux, Sidney A., Jr. "Historical Perspectives. Bird Migration: Methodologies and Major Research Trajectories." *The Condor*, Vol. 98 (May 1996), 442–53.

Gustafson, M. E., and J. Hildenbrand. "A Brief History of Bird Banding," 1998, Bird Banding Laboratory Homepage, ver. 2/18/98, http://www.pwrc.usgs.gov/bbl/homepage/history.htm.

Hayes, Floyd E. "Definitions for Migrant Birds: What Is a Neotropical Migrant?" *The Auk*, Vol. 112 (April 1995), 521–22.

Hill, Geoffrey E., Robert S. Sargent, and Martha B. Sargent. "Recent Change in the Winter Distribution of Rufous Hummingbirds." *The Auk*, Vol. 115 (January 1998), 240–45.

Karasov, William H. "In the Belly of the Bird." *Natural History*, Vol. 102 (November 1993), 32–37.

Kerlinger, Paul. "Night Flight." *Natural History*, Vol. 104 (September 1995), 66–69.

——, and Frank R. Moore. "Atmospheric Structure and Avian Migration." *Current Ornithology*, Vol. 6, Dennis M. Power, ed. New York: Plenum Press, 1989.

Kricher, John C. *Eastern Forests*. Boston: Houghton Mifflin Co., 1988.

Line, Les. "Migrating Songbirds Switch Their Diets for Long Trips South." *The New York Times*, November 18, 1997.

Miller, Ken. "Birds Head South in Record Numbers, Jamming Radar at 3 Midwest Airports." Gannett News Service, November 10, 1995.

Newfield, Nancy L. "Piecing Together the Hummingbird Puzzle." *Living Bird*, Vol. 15 (Spring 1996), 16–21.

Teale, Edwin Way. *Autumn Across America*. New York: Dodd, Mead & Co., 1956.

3: The Way South

CITATIONS

60 "I had no need": Quoted in Robin Baker, *The Mystery of Migration* (New York: Viking Press, 1981), p. 13.

64 "It appears that the magnetic navigational sense": Quoted in Malcolm W. Browne, "Migrating Birds Steer by Sunlight and Stars," *The New York Times*, September 28, 1993.

66 "I tried to anticipate": Donald R. Griffin, *Bird Migration* (New York: Doubleday & Co., 1964), p. 103.

67 "A migratory bird follows": Thomas Alerstam, *Bird Migration* (Cambridge University Press, 1993), p. 379.

68 "There is increasing evidence": Peter Berthold, *Control of Bird Migration* (London: Chapman & Hall, 1996), p. 296.

74 "The availability of food": Paul Kerlinger, *How Birds Migrate* (Mechanicsburg, PA: Stackpole, 1995), p. 15.

77 "Our major point": Douglas J. Levey and F. Gary Stiles, "Evolutionary Precursors of Long-distance Migration: Resource Availability and Movement Patterns in Neotropical Landbirds," *The American Naturalist*, Vol. 140 (September 1992), 447–76.

78 "Of particular interest": Timothy C. Williams and Thompson Webb III, "Neotropical Bird Migration During the Ice Ages: Orientation and Ecology," *The Auk*, Vol. 113 (1996), 105–18.

BIBLIOGRAPHY

Adler, T. "It Takes Two Compasses to Fly Right." *Science News*, Vol. 150 (September 14, 1996), 164.

Baskin, Yvonne. "Creatures of Habit." *Living Bird,* Vol. 13 (Spring 1994), 20–25.

Berger, Cynthia. "A Loft-y Conclusion." *Living Bird,* Vol. 12 (Summer 1993), 5.

Conover, Adele. "Fly Away Home." *Smithsonian,* Vol. 29 (April 1998), 62–72.

Feduccia, Alan. *The Origin and Evolution of Birds.* New Haven, CT: Yale University Press, 1996.

Hochbaum, H. Albert. *Travels and Traditions of Waterfowl.* St. Paul, MN: University of Minnesota, 1955.

Houston, C. S., M. I. Houston, and H. M. Reeves. "The 19th-Century Trade in Swan Skins and Quills." *Blue Jay,* Vol. 55 (1997), 24–34.

Jerome, Richard. "Animals and Magnetism." *The Sciences.* (January/February 1993), 7.

Line, Les. "Pilots Train Cranes to Fly Away Home." *The New York Times,* December 9, 1997.

Long, Michael E. "Secrets of Animal Navigation." *National Geographic,* Vol. 179 (June 1991), 70–99.

Walcott, Charles. "The Mystery of Jersey Hill and Other Stories." *Birdscope,* Vol. 4 (Spring/Summer 1990), 1–3.

———. "The Pigeon Project." *Birdscope,* Vol. 1 (Summer 1987), 1–2.

Waldvogel, Jerry A. "Olfactory Orientation by Birds." *Current Ornithology,* Vol. 6, Dennis M. Power, ed. New York: Plenum Press, 1989.

Wallraff, H. G., J. Kiepenheuer, M. F. Neumann, and A. Streng. "Homing Experiments with Starlings Deprived of the Sense of Smell." *The Condor,* Vol. 97 (February 1995), 20–26.

Waterman, Talbot H. *Animal Navigation.* New York: Scientific American Books, 1989.

Wiltschko, W., and R. Wiltschko. "Migratory Orientation of European Robins Is Affected by the Wavelength of Light as Well as by a Magnetic Pulse." *Journal of Comparative Physiology,* Vol. 177 (1995), 363–69.

Yoon, Carol Kaesuk. "March of Glaciers Challenged as Engine of Bird Evolution." *The New York Times,* September 30, 1997.

4: Riding the Sea Wind

CITATIONS

84   "This level of ignorance": Charles D. Duncan, "The Migration of Red-necked Phalaropes," *Birding,* Vol. 28 (December 1996), 482–88.

95   "I do not think": Quoted in Robert Cushman Murphy, *Oceanic Birds of South America,* Vol. 2 (New York: Macmillan Publishing Co., 1936), p. 669.

95   "There were certainly": Ibid., p. 668.

98   "a more vivid impression": Ibid., p. 1025.

99   "Eggs are stolen": Ibid., p. 1019.

101   "The nights we spent": P. W. Reynolds, "Notes on the Birds of Cape Horn," *Ibis,* Vol. 13 (1935), 65–101.

101   "During the 'seventies": Murphy, 1936, Vol. 2, p. 664.

BIBLIOGRAPHY

Adler, T. "Certain Seabirds Drawn by the Smell of Food." *Science News*, Vol. 148 (August 26, 1995), 133.

Ainley, David. "Ashy Storm-petrel (*Oceanodroma homochroa*)." *The Birds of North America*, No. 185, A. Poole and F. Gill, eds. Philadelphia: Academy of Natural Sciences and Washington, D.C.: American Ornithologists' Union, 1995.

Bent, Arthur C. *Life Histories of North American Petrels and Pelicans and Their Allies*. Washington, D.C.: Smithsonian Institution, 1922.

Cushman, Robert Murphy. *Oceanic Birds of South America*, Vols. 1 & 2. New York: American Museum of Natural History, 1936.

Fisher, Mildred L. *The Albatross of Midway Island*. Carbondale, IL: Southern Illinois University Press, 1970.

Haley, Delphine, ed. *Seabirds of Eastern North Pacific and Arctic Waters*. Seattle: Pacific Search Press, 1984.

Harrison, Peter. *Seabirds: An Identification Guide*. Boston: Houghton Mifflin Co., 1983.

Nelson, Bryan. *Seabirds: Their Biology and Ecology*. New York: A&W Publishers Inc., 1979.

Nettleship, D. N., J. Burger, and M. Gochfeld, eds. *Seabirds on Islands: Threats, Case Studies and Action Plans*. Cambridge, England: BirdLife International, 1994.

Palmer, Ralph. S. *Handbook of North American Birds*, Vol. 1. New Haven, CT: Yale University Press, 1962.

Ridl, Julie. "Albatross: Bird of Mystery and Myth." *Living Bird*, Vol. 10 (Autumn 1991), 14–19.

Skira, Irynej. "A Muttonbird in the Hand." *Natural History*, Vol. 104 (August 1995), 24–35.

Sklepkovych, B. O., and W. A. Montevecchi. "World's Largest Known Nesting Colony of Leach's Storm-Petrels on Baccalieu Island, Newfoundland." *American Birds*, Vol. 43 (Spring 1989), 38–42.

Thurston, Harry. "The Little Fish That Feeds the North Atlantic." *Audubon*, Vol. 90 (January 1988), 52–71.

Whittow, G. Causey. "Black-footed Albatross (*Diomedea nigripes*)." *The Birds of North America*, No. 65, A. Poole and F. Gill, eds. Philadelphia: Academy of Natural Sciences and Washington, D.C.: American Ornithologists' Union, 1993.

———. "Laysan Albatross (*Diomedea immutabilis*)," in *The Birds of North America*, No. 66, A. Poole and F. Gill, eds. Philadelphia: Academy of Natural Sciences and Washington, D.C.: American Ornithologists' Union, 1993.

5: *Rivers of Hawks*

BIBLIOGRAPHY

Andrle, Robert F. "Raptors and Other North American Migrants in Mexico." *The Condor*, Vol. 70 (October 1968), 393–95.

Goldstein, David L., and Neal G. Smith. "Response to Kirkley" (fasting by migrant Swainson's hawks). *Journal of Raptor Research*, Vol. 25 (Fall 1991), 87–88.

Kirkley, John S. "Do Migrant Swainson's Hawks Fast en Route to Argentina?" *Journal of Raptor Research*, Vol. 25 (1991), 82–86.

Ruelas Inzunza, Ernesto. "*El Niño* Visits the River of Raptors." *Hawk Mountain News* (Spring 1998), 26–28.

——. "Mexico Region." *Journal of Hawk Migration Studies*, Vol. 17 (1992), 43–45.

Thiollay, J. M. "Spring Hawk Migration in Eastern Mexico." *Journal of Raptor Research*, Vol. 14 (1980), 12–14.

Tilly, Frederick C. "Hawk-watching's Little-Known Sites." *Birding*, Vol. 24 (February 1992), 10–17

——, Stephen W. Hoffman, and Cathryn R. Tilly. "Spring Hawk Migration in Southern Mexico, 1989." *Journal of Hawk Migration Studies*, Vol. 15 (1990), 21–29.

Weidensaul, Scott. "Secret of the Hawks." *International Wildlife*, Vol. 24 (November–December 1994), 46–49.

6: La Selva Maya

CITATIONS

129   "He was wealth incarnate": David Freidel, Linda Schele, and Joy Parker, *Maya Cosmos* (New York: William Morrow & Co., 1993), p. 211.

130   "There is no obvious": John Rappole, *The Ecology of Migratory Birds: A Neotropical Perspective* (Washington, D.C.: Smithsonian Institution Press, 1995), p. 53.

132   "a behavioral about-face": Steven Hilty, *Birds of Tropical America* (Shelburne, VT: Chapters Publishing Ltd., 1994), p. 226.

135   "almost universally espoused the view": Robert A. Askins, James F. Lynch, and Russell Greenberg, "Population Declines in Migratory Birds in Eastern North America," in *Current Ornithology*, Vol. 7, Dennis M. Power, ed. (New York: Plenum Press, 1990), p. 30.

136   "This idea of the adaptable migrant": Eugene Morton and Russell Greenberg, "The Outlook for Migratory Songbirds: 'Future Shock' for Birds," *American Birds*, Vol. 43 (Spring 1989), 178–83.

138   "Because flocks cross some boundary": George V.N. Powell, "Structure and Dynamics of Interspecific Flocks in a Neotropical Mid-elevation Forest," *The Auk*, Vol. 96 (April 1979), 375–90.

139   "The sentinel": Hilty, *Birds of Tropical America*, p. 104.

145   "The task for students": Daniel R. Petit, James F. Lynch, Richard L. Hutto, John G. Blake, and Robert R. Wade, "Habitat Use and Conservation in the Neotropics," in *Ecology and Management of Neotropical Migratory Birds*, Thomas E. Martin and Deborah M. Finch, eds. (New York: Oxford University Press, 1995), p. 160.

147   "The great importance": Douglas Stotz, John Fitzpatrick, Theodore Parker III, and Debra Moskovits, *Neotropical Birds: Ecology and Conservation* (University of Chicago Press, 1996), p. 69.

148   "I am not suggesting": John Terborgh, "Perspectives on the Conservation of
      Neotropical Migrant Landbirds," in *Ecology and Conservation of Neotropical Migrant
      Landbirds*, John M. Hagan III and David W. Johnston, eds. (Washington, D.C.:
      Smithsonian Institution Press, 1992), p. 9.
149   "It seems clear": David Whitacre, et al., "Migrant Songbirds, Habitat Change, and
      Conservation Prospects in Northern Petén, Guatemala: Some Initial Results," in
      Ecology and Conservation of Neotropical Migrant Landbirds, p. 344.
150   "Our research has shown": Russell Greenberg, "Birds in the Tropics: The Coffee
      Connection," *Birding*, Vol. 28 (December 1996), 472–81.
151   "As migrant habitat": John Terborgh, *Where Have All the Birds Gone?* (Princeton
      University Press, 1989), p. 144.
152   "Sipping a cup of coffee": Russell Greenberg, *Coffee, Conservation and Commerce in the
      Western Hemisphere*, Robert A. Rice and Justin R. Ward, eds. (Washington, D.C.: Nat-
      ural Resources Defense Council and Smithsonian Migratory Bird Center, 1996), p. 1.

BIBLIOGRAPHY

Adler, T. "Coffee Can Give Many Species a Boost." *Science News*, Vol. 150 (August 31,
   1996), 132.
Boyle, William J., Robert O. Paxton, and David A. Cutler. "Hudson-Delaware Region."
   American Birds, Vol. 41 (Summer 1987), 260–63.
Buskirk, William H., George V.N. Powell, James F. Wittenberger, Ruth E. Buskirk, and
   Thomas U. Powell. "Interspecific Bird Flocks in Tropical Highland Panama." *The Auk*,
   Vol. 89 (July 1972), 612–24.
Gradwohl, Judy, and Russell Greenberg. "The Formation of Antwren Flocks on Barro Col-
   orado Island, Panama." *The Auk*, Vol. 97 (April 1980), 385–95.
Greenberg, Russell. "Phenomenon, Comments and Notes" (shade coffee research). *Smith-
   sonian*, Vol. 25 (December 1994), 24–27.
Hartshorn, Gary S. "Forest Loss and Future Options in Central America." In *Ecology and
   Conservation of Neotropical Migrant Landbirds*, John M. Hagan III and David W. John-
   ston (Washington, D.C.: Smithsonian Institution Press, 1992).
Howell, Steve N.G., and Sophie Webb. *A Guide to Birds of Mexico and Northern Central
   America*. New York: Oxford University Press, 1995.
Karr, James R. "On the Relative Abundance of Migrants from the North Temperate Zone
   in Tropical Habitats." *The Wilson Bulletin*, Vol. 88 (September 1976), 433–58.
Kricher, John C. *A Neotropical Companion*. Princeton University Press, 1989.
Loiselle, Bette, and John G. Blake. "Temporal Variation in Birds and Fruits Along an Ele-
   vational Gradient in Costa Rica." *Ecology*, Vol. 72 (February 1991), 180–93.
Lynch, James F., Eugene S. Morton, and Martha E. Van der Voot. "Habitat Segregation
   Between the Sexes of Wintering Hooded Warblers (*Wilsonia citrina*)." *The Auk*, Vol.
   102 (October 1985), 714–21.
Morley, Sylvanus G., George W. Brainerd, and Robert J. Sharer. *The Ancient Maya*, 4th ed.
   Stanford University Press, 1983.

Morton, Eugene S. "Food and Migration Habits of the Eastern Kingbird." *The Auk*, Vol. 88 (October 1971), 925–26.

——, James F. Lynch, Kimberly Young, and Patricia Mehlhop. "Do Male Hooded Warblers Exclude Females from Nonbreeding Territories in Tropical Forest?" *The Auk*, Vol. 104 (January 1987), 133–35.

Ornat, Arturo Lopez, and Russell Greenberg. "Sexual Segregation by Habitat in Migratory Warblers in Quintana Roo, Mexico." *The Auk*, Vol. 107 (July 1990), 539–43.

Poulin, Brigitte, and Gaëtan Lefebvre. "Dietary Relationships of Migrant and Resident Birds from a Humid Forest in Central Panama." *The Auk*, Vol. 113 (April 1996), 277–87.

Rainforest Alliance. *The Conservation Coffee Campaign Organizer's Kit*. New York, 1997.

Remsen, J. V., Jr., and T. A. Parker III. "Seasonal Distribution of the Azure Gallinule (*Porphyrula flavirostris*) with Comments on Vagrancy in Rails and Gallinules." *The Wilson Bulletin*, Vol. 102 (September 1990), 380–99.

Smithsonian Migratory Bird Center. "Why Migratory Birds Are Crazy for Coffee." Washington, D.C., 1997.

——. *Shade Management Criteria for "Bird-Friendly™" Coffee*. Washington, D.C., 1998.

Stuart, Gene S., and George E. Stuart. *Lost Kingdoms of the Maya*. Washington, D.C.: National Geographic Society, 1993.

Tramer, Elliot J., and Thomas R. Kemp. "Diet-correlated Variations in Social Behavior of Wintering Tennessee Warblers." *The Auk*, Vol. 96 (January 1979), 186–87.

Winker, Kevin. "Habitat Selection in Woodland Nearctic-Neotropical Migrants on the Isthmus of Tehuantepec in Autumn Migration." *The Wilson Bulletin*, Vol. 107 (March 1995), 26–39.

Wunderle, Joseph M., Jr. "Sexual Habitat Segregation in Wintering Black-throated Blue Warblers in Puerto Rico." *Ecology and Conservation of Neotropical Migrant Landbirds*, John M. Hagan III and David W. Johnston, eds. (Washington, D.C.: Smithsonian Institution Press, 1992).

7: Hopping Dick and Betsy Kick-up

CITATIONS

160 "The once extensive forests": John Terborgh, "Perspectives on the Conservation of Neotropical Migrant Landbirds," in *Ecology and Conservation of Neotropical Migrant Landbirds*, John M. Hagan III and David W. Johnston, eds. (Washington, D.C.: Smithsonian Institution Press, 1992), p. 9.

163 "The Canes are ten to twelve feet": William Bartram, *The Travels of William Bartram* (New York: Macy-Masius Publishers, 1928), pp. 198, 328.

164 "It was this 'gross imbalance'": John Terborgh, *Where Have All the Birds Gone?* (Princeton University Press, 1989), p. 62.

165 "Today, one can fly": Harold F. Mayfield, "Kirtland's Warblers in Winter," *Birding*, Vol. 28 (February 1996), 34–39.

BIBLIOGRAPHY

Arendt, Wayne J., et al. "North American Migrant Landbirds in the Caribbean: A Summary." In *Ecology and Conservation of Neotropical Migrant Landbirds*, John M. Hagan III and David W. Johnston, eds. (Washington, D.C.: Smithsonian Institution Press, 1992).

Blom, Erik. "The Changing Seasons." *American Birds*, Vol. 41 (Summer 1987), 248–52.

Bond, James. *Birds of the West Indies*, 5th ed. Boston: Houghton Mifflin Co., 1993.

Conover, Adele. "A New World Comes to Life, Discovered in a Stalk of Bamboo." *Smithsonian* (October 1994), 120–29.

Downer, Audrey, and Robert Sutton. *Birds of Jamaica*. Cambridge University Press, 1990.

Eyre, L. Alan. "Jamaica's Crisis in Forestry and Watershed Management." *Jamaica Naturalist*, Vol. 1 (1991), 27–34, 44.

Faccio, Steve. "Cutting Through the Fog" (Bicknell's thrush). *Living Bird*, Vol. 14 (Winter 1995), 20–25.

Forsyth, Adrian. *Portraits of the Rainforest*. Ottawa, Ont.: Camden House, 1990.

Kapos, Valerie. "Dry Limestone Forests of Jamaica." *Seasonally Dry Tropical Forests*, Stephen H. Bullock, Harold A. Mooney, and Ernesto Medina, eds. Cambridge University Press, 1995.

Loveless, A. R., and G. F. Asprey. "The Dry Evergreen Formations of Jamaica." *Journal of Ecology*, Vol. 45 (1957), 799–822.

McLaren, Ian A. "Field Identification and Taxonomy of Bicknell's Thrush." *Birding*, Vol. 27 (October 1995), 358–66.

Marra, Peter P., Thomas W. Sherry, and Richard T. Holmes. "Territorial Exclusion by a Long-distance Migrant Warbler in Jamaica: A Removal Experiment with American Redstarts (*Setophaga ruticilla*)." *The Auk*, Vol. 110 (1993), 565–72.

Nowak, Ronald M. *Walker's Mammals of the World*, fifth ed., Vols. 1 and 2. Baltimore, MD: Johns Hopkins University Press, 1991.

Ouellet, Henri. "Bucknell's Thrush: Taxonomic Status and Distribution." *Wilson Bulletin*, Vol. 105 (December 1993), 544–72.

Parker, Theodore A. III, Douglas F. Stotz, and John W. Fitzpatrick. "Ecological and Distributional Databases." In *Neotropical Birds: Ecology and Conservation*, Douglas F. Stotz et al., University of Chicago Press, 1996.

Radabaugh, Bruce E. "Kirtland's Warbler and Its Bahama Wintering Grounds." *Wilson Bulletin*, Vol. 86 (December 1974), 374–83.

Remsen, J. V., Jr. "Was Bachman's Warbler a Bamboo Specialist?" *The Auk*, Vol. 103 (January 1986), 216–19.

Rimmer, Chris. "Island Connection." *Bird Conservation* (Winter 1996), 13.

——. "A Closer Look: Bicknell's Thrush." *Birding*, Vol. 28 (April 1996), 118–23.

Sherry, Thomas W., and Richard T. Holmes. "Winter Habitat Quality, Population Limitation and Conservation of Neotropical-Nearctic Migrant Birds." *Ecology*, Vol. 77 (January 1996), 36–48.

Stevenson, Henry M. "The Recent History of Bachman's Warbler." *Wilson Bulletin*, Vol. 84 (September 1972), 344–47.

Sykes, Paul W., Jr. "A Closer Look: Kirtland's Warbler." *Birding*, Vol. 29 (June 1997), 220–27.
Wauer, Roland H. *A Birder's West Indies*. Austin, TX: University of Texas Press, 1996.
Wunderle, Joseph M., Jr., and Robert B. Waide. "Distribution of Overwintering Nearctic Migrants in the Bahamas and Greater Antilles." *The Condor*, Vol. 95 (1993), 904–33.

8: Aguilucheros

CITATIONS

176 "The migrations of the Swainson's Hawk": Leslie Brown and Dean Amadon, *Eagles, Hawks and Falcons of the World*, Vols. 1 and 2, reprint ed. (Secaucus, NJ: Wellfleet Press, 1989), p. 586.
177 "in Argentina": Ibid.
177 "East of the [Panama] Canal Zone": Ralph S. Palmer, *Handbook of North American Birds*, Vol. 5 (New Haven, CT: Yale University Press, 1988), p. 64.
179 "Both grasshoppers and ground squirrels": Ibid., p. 61.
191 "Assuming that each hawk": Quoted in Arthur C. Bent, *Life Histories of North American Birds of Prey*, Vol. 1 (Washington, D.C.: Smithsonian Institution, 1937), p. 229.
193 "The white-rumped sandpiper": Peter Matthiessen, *Shorebirds of North America*, Gardner D. Stout, ed. (New York: Viking Press, 1967), p. 49.

BIBLIOGRAPHY

Bent, Arthur C. *Life Histories of North American Shorebirds*, Vol. 2. Washington, D.C.: Smithsonian Institution, 1927.
Blom, Eirik. "Eskimo Curlew Record in Error." *The Skimmer* (October 1995), 2.
England, A. Sidney, Marc J. Bechard, and C. Stuart Houston. "Swainson's Hawk (*Buteo swainsoni*)." *Birds of North America*, No. 265, A. Poole and F. Gill, eds. Philadelphia: Academy of Natural Sciences and Washington, D.C.: American Ornithologists' Union, 1997.
Goldstein, M. I., et al. "An Assessment of Mortality of Swainson's Hawks on Wintering Grounds in Argentina." *Journal of Raptor Research*, Vol. 30 (1996), 106–7.
Houston, Stuart C., and Josef K. Schmutz. "Declining Reproduction Among Swainson's Hawks in Prairie Canada." *Journal of Raptor Research*, Vol. 29 (1995), 198–201.
———. "Swainson's Hawk Banding in North America to 1992." *North American Bird Bander*, Vol. 29 (July–September 1995), 120–27.
Hudson, W. H. *Far Away and Long Ago*. New York: Dutton, 1918.
Jaramillo, Alvaro P. "Wintering Swainson's Hawks in Argentina: Food and Age Segregation." *The Condor*, Vol. 95 (1993), 475–79.
McKay, Barry Kent. "The Eskimo Curlew." *Bird Watcher's Digest* (March/April 1996), 22–33.
Narosky, T., and D. Yzurieta. *Birds of Argentina and Uruguay*, 2nd ed. Buenos Aires, Argentina: Vazquez Mazzini Editores, 1993.
Tate, James, Jr., and D. Jean Tate. "The Blue List for 1982." *American Birds*, Vol. 36 (March 1982), 126–35.

Tubaro, Pablo, Fabian Gabelli, and Diego Gallegas-Luque. "Pampas Meadowlark." *World Birdwatch*, Vol. 16 (September 1994).
Weidensaul, Scott. "Eye on a Hawk." *Philadelphia Inquirer*, March 10, 1997.
Woodbridge, Brian, Karen K. Finley, and Peter H. Bloom. "Reproductive Performance, Age Structure, and Natal Dispersal of Swainson's Hawks in the Butte Valley, California." *Journal of Raptor Research*, Vol. 29 (1995), 187–92.
——, and S. Trent Seager. "An Investigation of Swainson's Hawks in Argentina." *Journal of Raptor Research*, Vol. 29 (1995), 202–24.

9: *When Anywhere Is Better than Home*

CITATIONS

203 "A flock of finches": Kenn Kaufman, "The Changing Seasons: Winter 1991–1992," *American Birds*, Vol. 46 (Summer 1992), 224–27.
213 "dead elk": John K. Terres, *The Audubon Society Encyclopedia of North American Birds* (New York: Alfred A. Knopf, 1980), p. 144.
213 "It would be tedious": Bernd Heinrich, *Ravens in Winter* (New York: Summit Books, 1989), p. 49.

BIBLIOGRAPHY

Bennet, Doug, and Tim Tiner. *Up North Again*. Toronto, Ont.: McClelland & Stewart Inc., 1997.
Bent, A. C. *Life Histories of North American Jays, Crows and Titmice*. Washington, D.C.: Smithsonian Institution, 1946.
——. *Life Histories of North American Cardinals, Grosbeaks, Buntings, Towhees, Finches, Sparrows and Allies*, Vol. 1. Washington, D.C.: Smithsonian Institution, 1968.
Blom, Eirik A. T. "Evening Grosbreaks." *Bird Watcher's Digest*, Vol. 19 (January/February 1997), 28–37.
Bull, Evelyn L., and James R. Duncan. "Great Gray Owl (*Strix nebulosa*)." *The Birds of North America*, No. 41, A. Poole and F. Gill, eds. Philadelphia: Academy of Natural Sciences and Washington, D.C.: American Ornithologists' Union, 1993.
Chartier, Bonnie. *A Birder's Guide to Churchill*. Colorado Springs, CO: American Birding Association, 1994.
Godfrey, W. Earl. *The Birds of Canada*, rev. ed. Ottawa, Ont.: National Museum of Natural Sciences, 1986.
Hayward, G. D., and P. H. Hayward. "Boreal Owl (*Aegolius funereus*)," *The Birds of North America*, No. 63, A. Poole and F. Gill, eds. Philadelphia: Academy of Natural Sciences and Washington, D.C.: American Ornithologists' Union, 1993.
Johnsgard, Paul A. *North American Owls*. Washington, D.C.: Smithsonian Institution Press, 1988.
——. "Changing Seasons." *Field Notes*, Vol. 50 (Summer 1996), 136–39.
Lehman, Paul. "The Changing Seasons." *American Birds*, Vol. 38 (May/June 1984), 287–92.

Line, Les. "Staying the Winter." *National Wildlife* (February/March 1995), 52–59.

Marchand, Peter J. *Life in the Cold*. Hanover, NH: University Press of New England, 1987.

Newton, Ian. *Population Ecology of Raptors*. Vermilion, SD: Buteo Books, 1979.

Parmelee, David. "Snowy Owl (*Nyctea scandiaca*)," No. 10, A. Poole, P. Stettenheim, and F. Gill, eds. Philadelphia: Academy of Natural Sciences and Washington, D.C.: American Ornithologist's Union, 1992.

Taverner, Percy A., and George Miksch Sutton. "The Birds of Churchill, Manitoba." *Annals of the Carnegie Museum*, Vol. XXIII (May 1934).

Tessaglia, Diane L. "Project FeederWatch Annual Report 1996–97." *Birdscope*, Vol. 11 (Autumn 1997), 1–8.

Wells, Jeffrey V. "A Redpoll Perspective." *Birdscope*, Vol. 10 (Winter 1996), 4.

10: *Uneasy Neighbors*

CITATIONS

233　　"The scale of the problem": Kenneth F. Abraham and Robert L. Jefferies, "High Goose Populations: Causes, Impacts and Implications," in *Arctic Ecosystems in Peril: Report of the Arctic Goose Habitat Working Group*, Bruce D.J. Batt, ed. (Washington, D.C.: U.S. Fish and Wildlife Service, 1997), p. 56.

233　　"As long as the mid-continent population": Ibid., p. 46.

243　　"Upon initial examination": Jim Winton, "Summary of 1996–97 Fish Pathology Findings," Northwest Biological Science Center, USGS Biological Resources Division. http://www.r1.fws.gov/news/saltn.htm.

BIBLIOGRAPHY

Alexander, Laurence L. "Trouble with Loons." *The Living Bird Quarterly*, Vol. 4 (Spring 1985), 10–13.

Ankney, C. Davison. "An Embarrassment of Riches: Too Many Geese." *Journal of Wildlife Management*, Vol. 60 (April 1996), 217–23.

Anon. "U.S. to Close Poisoned Refuge." *Audubon*, Vol. 87 (July 1985), 108–9.

———. "Progress Report No. 8: Response of Greater Snow Geese (*Chen caerulescens atlantica*) to Hunting at Bombay Hook NWR and Related Wetland Changes." Smyrna, DE: U.S. Fish and Wildlife Service, June 6, 1996.

———. "Newcastle Disease Identified in Salton Sea Die-off." National Wildlife Health Center press release, Madison, WI, June 17, 1997.

———. "Disease Takes Its Toll on Waterfowl Populations." U.S. Geological Survey press release, Reston, VA, October 9, 1997.

Blom, Eirik. "Doggone Geese." *BWD Skimmer*, Vol. 5 (August 1995), 3.

———. "Chemical Deterrence Setback." *BWD Skimmer*, Vol. 5 (August 1995), 4.

Boyle, Robert H. "Life—or Death—for the Salton Sea?" *Smithsonian*, Vol. 27 (June 1996), 86–97.

Brauning, Daniel W., ed. *Atlas of Breeding Birds in Pennsylvania*. University of Pittsburgh Press, 1992.

Brown, Lauren. *Grasses*. Boston: Houghton Mifflin Co., 1979.

Carter, Lloyd. "What a Mess!" *National Wildlife*, Vol. 26 (October/November 1988), 42–44.

Chadwick, Douglas H. "Sanctuary: U.S. National Wildlife Refuges." *National Geographic*, Vol. 190 (October 1996), 2–35.

Commission of Environmental Cooperation. *CEC Secretariat Report on the Death of Migratory Birds at the Silva Reservoir (1994–95)*. Montreal, Quebec: Commission of Environmental Cooperation and Ottawa, Ont.: Prospectus Inc., October 1995.

Dunn, John P. "Pennsylvania's Canada Geese: Giant Success or Giant Dilemma?" *Pennsylvania Game News*, Vol. 63 (December 1992), 16–21.

Ganster, Paul. "Environmental Issues of the California–Baja California Border Region." *Border Environment Research Reports*, No. 1 (June 1996), Regional Studies of the Californias, San Diego State University, San Diego, CA.

Graham, Frank, Jr. "The Numbers Game Puts Birds at Risk." *Audubon*, Vol. 91 (January 1988), 18–27.

Horton, Gary A. *Carson River Chronology*. Carson City, NV: Nevada Division of Water Planning, 1996.

Johnsgard, Paul A. *A Guide to North American Waterfowl*. Bloomington, IN: Indiana University Press, 1979.

Kosak, Joe. *The Pennsylvania Game Commission: 1895–1995*. Harrisburg, PA: Pennsylvania Game Commission, 1995.

Laycock, George. "What Water for Stillwater?" *Audubon*, Vol. 90 (November 1988), 14–25.

Meanley, Brooke. *Waterfowl of the Chesapeake Bay Country*. Centreville, MD: Tidewater Publishers, 1982.

National Wildlife Health Center. "Newcastle Disease Identified in Salton Sea Die-off." http://www.emtc.nbs.gov/http_data/nwhc/news/salt-new.html (June 17, 1997).

Norris, Ruth. "Poisoned Refuges." *Audubon*, Vol. 88 (January 1986), 118–23.

Riley, Laura and William. *Guide to the National Wildlife Refuges*. New York: Anchor Press, 1979.

Rockwell, Robert, Kenneth Abraham, and Robert Jefferies. "Tundra Under Siege." *Natural History*, Vol. 105, No. 11 (November 1996), 20–21.

———. "The Best Laid Plans." *Living Bird*, Vol. 16 (Winter 1997), 16–23.

Steinhart, Peter. "California's Selenium Dilemma." *Audubon*, Vol. 87 (March 1985), 111–14.

Teal, John and Mildred Teal. *Life and Death of the Salt Marsh*. New York: Ballantine Books, 1969.

U.S. Bureau of Reclamation, Southern California Area Office Home Page. "Salton Sea Study." http://www.lc.usbr.gov/~scao/saltsea.html (July 13, 1997).

U.S. Fish and Wildlife Service. "Salton Sea National Wildlife Refuge Home Page." http://www.r1.fws.gov/news/saltn.htm, including the following documents: "Summary of 1996 Avian Botulism Event," "1997 Fish and Wildlife Mortality Events," "Wildlife Mortality Estimates 1987–1996, Salton Sea."

——. "Salton Sea National Wildlife Refuge (CA) Contaminant Studies." National Wildlife Refuge System Home Page, http://bluegoose.arw.r9.fws.gov/NWRSFiles/OtherIssues/EC-Studies3.html (May 5, 1997).

U.S. Geological Survey. "Disease Takes Its Toll on Waterfowl Populations." USGS Home Page. http://www.usgs.gov/public/press/public_affairs/press_release/pr357m.html (October 9, 1997).

Wilkinson, Todd. "It's the Pits." *Wildlife Conservation*, Vol. 100 (November–December 1997), 18.

11: *The Gulf Express*

CITATIONS

253 "We were visited": Quoted in George H. Lowery, Jr., "Evidence of Trans-Gulf Migration," *The Auk*, Vol. 63 (April 1946), 175–211.

254 "Though this theory": George C. Williams, "Do Birds Cross the Gulf of Mexico in Spring?" *The Auk*, Vol. 62 (April 1945), 98–111.

254 "Williams probably succeeded in establishing a reasonable doubt": Lowery, 1946.

255 "If one compares": Sidney A. Gauthreaux, Jr., "The Use of Weather Radar to Monitor Long-term Patterns of Trans-Gulf Migration in Spring," in *Ecology and Conservation of Neotropical Migrant Landbirds*, J. M. Hagan III and D. W. Johnston, eds. (Washington, D.C.: Smithsonian Institution Press, 1992), p. 99.

261 "Tailwinds reduce energy consumption": Paul Kerlinger, *How Birds Migrate* (Mechanicsburg, PA: Stackpole, 1995), p. 121.

272 Estimated value of insect-eating birds per square mile of forest land quoted from Scott K. Robinson, "The Case of the Missing Songbirds," *Consequences*, Vol. 3 (June 1997), http://www.gcrio.org/CONSEQUENCES/vol3no1/toc.html

BIBLIOGRAPHY

Anon. "Birdwatching's Economic Impact in Texas." *Partners in Flight*, Vol. 4 (Winter 1994), 16, National Fish and Wildlife Foundation.

——. "What's a Bird Worth?" *Bird Conservation* (Spring 1997), 6–8.

Brown, R. E., and J. G. Dickson. "Swainson's Warbler (*Limnothlypis swainsonii*)." *The Birds of North America*, No. 126, A. Poole and F. Gill, eds. Philadelphia: Academy of Natural Sciences and Washington, D.C.: American Ornithologists' Union, 1994.

Carter, M., G. Fenwick, C. Hunter, D. Pashley, D. Petit, J. Price, and J. Trapp. "For the Future" (WatchList 1996). *Field Notes*, Vol. 50 (Fall 1996), 238–40.

Caudill, James D. *1991 National Economic Impacts of Nonconsumptive Wildlife-Related Recreation*. Arlington, VA: U.S. Fish and Wildlife Service, 1997.

Curson, Jon, David Quinn, and David Beadle. *Warblers of the Americas*. Boston: Houghton Mifflin, 1994.

Gauthreaux, Sidney A., Jr. "A Radar and Direct Visual Study of Passerine Spring Migration in Southern Louisiana." *The Auk*, Vol. 88 (1971), 343–65.

Johnson, Ray. "Gulf Coast Bird Observatory Network." *Partners in Flight 1994 Annual Report*, Vol. 5 (1995), 20–21, National Fish and Wildlife Foundation.

Kerlinger, Paul. "The Economic Impact of Birding Ecotourists on Ten National Wildlife Refuges." *Winging It*, Vol. 9 (September 1995), 10–11.

——, R. H. Payne, and Ted Eubanks. "Not Just for the Birds." *Bird Conservation* (Spring 1997), 12–13.

Line, Les. "Silence of the Songbirds." *National Geographic*, Vol. 183 (June 1993), 68–91.

Moore, Frank R., S. A. Gauthreaux, P. Kerlinger, and T. R. Simons. "Stopover Habitat: Management Implications and Guidelines." In *Status and Management of Neotropical Migratory Birds*, D. M. Finch and P. W. Stangel, eds. Gen. Tech. Rep. RM-229, Fort Collins, CO: U.S. Department of Agriculture, Forest Service, 1993.

——, and Ted R. Simons, "Habitat Suitability and Stopover Ecology of Neotropical Landbird Migrants." In *Ecology and Conservation of Neotropical Migrant Landbirds*, J. M. Hagan III and D. W. Johnston, eds. Washington, D.C.: Smithsonian Institution Press, 1992.

Robbins, Chandler S., John W. Fitzpatrick, and Paul B. Hamel. "A Warbler in Trouble." *Ecology and Conservation of Neotropical Migrant Landbirds*, J. M. Hagan III and D. W. Johnston, eds. Washington, D.C.: Smithsonian Institution Press, 1992.

Romero, Andrew, and Peter Stangel. *1997 Directory of Birding Festivals*. Washington, D.C.: National Fish and Wildlife Foundation, 1996.

Ross, Gary Noel. "Butterfly Wrangling in Louisiana" (chenier conservation). *Natural History*, Vol. 104 (May 1995), 36–43.

Teisl, Mario F., and Robert Southwick. *The Economic Contributions of Bird and Waterfowl Recreation in the United States During 1991*. Washington, D.C.: International Association of Fish and Wildlife Agencies and U.S. Fish and Wildlife Service, 1995.

United States Department of the Interior. *1991 National Survey of Fishing, Hunting and Wildlife-Associated Recreation: State Overview*. Washington, D.C.: U.S. Fish and Wildlife Service, December 1992.

Vaughan, Ray. *Birder's Guide to Alabama and Mississippi*. Houston, TX: Gulf Publishing Co., 1994.

Weidenfeld, D. A., and M. G. Weidenfeld. "Large Kill of Neotropical Migrants by Tornado and Storm in Louisiana, April, 1993." *Journal of Field Ornithology*, Vol. 66 (Winter 1995), 70–77.

Weidner, David, and Paul Kerlinger. "Economics of Birding: A National Survey of Active Birders." *American Birds*, Vol. 44 (Summer 1990), 209–13.

Williams, George G. "Lowery on Trans-Gulf Migrations." *The Auk*, Vol. 64 (1947), 217–37.

Yoon, Carol Kaesuk. "More Than Decoration, Songbirds Are Essential to Forests' Health." *The New York Times*, November 8, 1994.

12: *Heartland*

CITATIONS

277   "Here the east-west human transportation corridor": Allan Jenkins, *The Platte River: An Atlas of the Big Bend Reach* (Kearney, NE: University of Nebraska, 1993).
280   "a tinkling of little bells": Aldo Leopold, "Marshland Elegy," in *A Sand County Almanac* (New York: Oxford University Press, 1947), p. 101.
280   "mean fundamental frequency": T. C. Tacha, S. A. Nesbitt, and P. A. Vohs, "Sandhill Crane," in *The Birds of North America*, No. 31 (Philadelphia: Academy of Natural Sciences and Washington, D.C.: and the American Ornithologists' Union, 1992), p. 8.
289   "In the spring mornings": Theodore Roosevelt, *Ranch Life and the Hunting Trail* (New York: Gramercy Books, 1995; facsimile ed. of original 1888 ed.), p. 38.
294   "Beside many of the reedy ponds": Ibid., p. 51.
298   "Populations have generally declined": J. R. Sauer, J. E. Hines, G. Gough, I. Thomas, and B. G. Peterjon, *The North American Breeding Bird Survey Results and Analysis*, Version 96.3 (Laurel, MD: Patuxent Wildlife Research Center, 1997).

BIBLIOGRAPHY

Berkey, Gordon, Richard Crawford, Steven Galipeau, Douglas H. Johnson, David Lambeth, and Randy Kreil. "A Review of Wildlife Management Practices in North Dakota." Unpublished report submitted to U.S. Fish and Wildlife Service, region 6, Denver, CO, Jamestown, ND: Northern Prairie Wildlife Research Center Home Page (July 16, 1997), 1993.

Berry, Charles R., Jr., and Dennis G. Buechler. *Wetlands in the Northern Great Plains*. Brookings, SD: U.S. Fish and Wildlife Service (U.S. Prairie Pothole Joint Venture) and South Dakota State University, 1993.

Bleed, Ann, and Charles Flowerday, eds. *An Atlas of the Sand Hills*. Resource Atlas No. 5A, Lincoln, NE: Institute of Agriculture and Natural Resources, University of Nebraska, 1990.

Bock, C. E., V. A. Saab, T. D. Rich, and D. S. Dobkin. "Effects of Livestock Grazing on Migratory Landbirds in Western North America." In *Status and Management of Neotropical Migratory Birds*, D. M. Finch and P. W. Stangel, eds. Gen. Tech. Rep. RM-229. Fort Collins, CO: U.S. Department of Agriculture, Forest Service, 1993.

DeBleiu, Jan. *Meant to Be Wild*. Golden, CO: Fulcrum, 1991.

Eliot, John L. "Military Eyes in the Sky Aided Sandhill Cranes." *National Geographic*, Vol. 184 (July 1993), 134.

Gollop, J. B., T. W. Barry, and E. H. Iverson. *Eskimo Curlew: A Vanishing Species?* Regina, Sask.: Saskatchewan Natural History Society, Spec. Pub. 17, 1986.

Gruchow, Paul. "The Ancient Faith of Cranes." *Audubon*, Vol. 91 (May 1989), 40–55.

Johnsgard, Paul A. *The Platte: Channels in Time*, Lincoln, NE: University of Nebraska Press, 1984.

Kantrud, H. A., R. R. Koford, D. H. Johnson, and M. D. Schwartz. "The Conservation Reserve Program: Good for Birds of Many Feathers." *North Dakota Outdoors*, Vol. 56 (August 1993), 14–17.

Knue, Joseph. NEBRASKAland Wildlife Viewing Guide. Helena, MT: Falcon Press, 1997.

Levey, Brooke. "What Is the Cooperative Agreement?" Voices for the Platte (Fall 1997), 1–4.

Lingle, Gary R. Birding Crane River: Nebraska's Platte. Grand Island, NE: Harrier Publishing, 1994.

Meine, Curt D., and George W. Archibald, eds. The Cranes: Status Survey and Conservation Action Plan. Gland, Switzerland, and Cambridge, England: International Union for the Conservation of Nature, 1996.

Mitchell, John G. "On the Seacoast of Nebraska." Audubon, Vol. 91 (May 1989), 56–77.

Pashley, David, and Barbara Pardo. "The Prairie Pothole Region: Comprehensive Ecosystem Conservation." Partners in Flight, Vol. 4 (Winter 1994), 1, 27–28.

Pielou, E. C. After the Ice Age: The Return of Life to Glaciated North America. University of Chicago Press, 1991.

Robbins, Chandler S., Danny Bystrak, and Paul H. Geissler. The Breeding Bird Survey: Its First Fifteen Years, 1965–1979. Resource Publication 157, Washington, D.C.: U.S. Fish and Wildlife Service, 1986.

Rodenhouse, N. L., L. B. Best, R. J. O'Connor, and E. K. Bollinger. "Effects of Temperate Agriculture on Neotropical Migrant Landbirds." Status and Management of Neotropical Migratory Birds, D. M. Finch and P. W. Stangel, eds. Gen. Tech. Rep. RM-229, Fort Collins, CO: U.S. Department of Agriculture, Forest Service, 1993.

Sidle, John G. "A Prairie River Roost." The Living Bird Quarterly, Vol. 8 (Spring 1989), 8–13.

——, and Craig A. Faanes. Platte River Ecosystems Resources and Management, with Emphasis on the Big Bend Reach in Nebraska. U.S. Fish and Wildlife Service, Grand Island, NE and Jamestown, ND: Northern Prairie Wildlife Research Center Home Page (July 16, 1997).

Trimble, Donald E. The Geologic Story of the Great Plains. U.S. Geological Survey Bulletin No. 1493, reprinted Medora, ND: Theodore Roosevelt Nature and History Association, 1990.

U.S. Fish and Wildlife Service. The Prairie Pothole Joint Venture: Cultivating Cooperation for Wildlife and Agriculture. Denver, CO: U.S. Fish and Wildlife Service, 1996.

13: Hopscotch

CITATIONS

312   "The crux of the conservation problem": Brian A. Harrington, "Red Knot," in Audubon Wildlife Report 1986, Roger L. Di Silvestro, ed. (New York: National Audubon Society, 1986), p. 880.

320   "The birds were far less numerous": Witmer Stone, Bird Studies at Old Cape May, Vol. 1 (Philadelphia, PA: Delaware Valley Ornithological Club and Academy of Natural Sciences, 1937), p. 364.

321   "While we realize": Ibid., p. 366.

BIBLIOGRAPHY

Anon. "WHSRN Sites Threatened." WHSRNews. Vol. 2.1 Manomet Center for Conservation Sciences. http://www.manomet.org/whsrnews5.htm (May 1998).

——. "Anatomy of a Sister Reserve." *Manomet Bird Observatory Newsletter* (September 1986), 1–4.

Barash, Leah. "Mass Appeal." *National Wildlife*, Vol. 31 (June–July 1993), 14–19.

Biderman, John O. "Food for Flight." *Audubon*, Vol. 85 (May 1983), 112–19.

Botton, Mark L., Robert E. Loveland, and Timothy R. Jacobsen. "Site Selection by Migratory Shorebirds in Delaware Bay, and Its Relationship to Beach Characteristics and Abundance of Horseshoe Crab (*Limulus polyphemus*) Eggs." *The Auk*, Vol. 111 (July 1994), 605–16.

Castro, Gonzalo, and J. P. Myers. "Shorebird Predation on Eggs of Horseshoe Crabs During Spring Stopover on Delaware Bay." *The Auk*, Vol. 110 (October 1993), 927–30.

Clark, Kathleen E., and Lawrence J. Niles. "Abundance and Distribution of Migrant Shorebirds on Delaware Bay." *The Condor*, Vol. 95 (1993), 694–705.

González, Patricia M., Theunis Piersma, and Yvonne Verkuil. "Food, Feeding, and Refueling of Red Knots During Northward Migration at San Antonio Oeste, Rio Negro, Argentina." *Journal of Field Ornithology*, Vol. 67 (Autumn 1996), 575–91.

Gosner, Kenneth L. *A Field Guide to the Atlantic Seashore*. Boston: Houghton Mifflin Co., 1978.

Graham, Frank, Jr. "Shorebirds: Images and Projections." *Audubon*, Vol. 86 (March 1984), 80–92.

Hall, William R., Jr. *The Horseshoe Crab*. Marine Advisory Service Bulletin, Newark, DE: University of Delaware, 1995.

Hands, Helen. "Kansas Travelers." *Natural History*, Vol. 107 (May 1998), 54–55.

Harrington, Brian A. "Twenty-five Years and Counting." *Natural History*, Vol. 107 (May 1998), 46.

——, and Charles Flowers. *The Flight of the Red Knot*. New York: W. W. Norton, 1996.

Kerlinger, Paul. "Showdown at Delaware Bay." *Natural History*, Vol. 107 (May 1998), 56–58.

Lanctot, Richard B. "A Closer Look: Buff-breasted Sandpiper." *Birding*, Vol. 27 (October 1995), 384–90.

Lyons, James E., and Susan M. Hand. "Fat Content and Stopover Ecology of Spring Migrant Semipalmated Sandpipers in South Carolina." *The Condor*, Vol. 97 (May 1995), 427–37.

Myers, J. P. "Conservation of Migrating Shorebirds: Staging Areas, Geographic Bottlenecks, and Regional Movements." *American Birds*, Vol. 37 (January–February 1983), 23–25.

New Jersey Division of Fish, Game, and Wildlife. "Delaware Bay Shorebird Staging Area Fact Sheet." Trenton, NJ, 1997.

Niles, Lawrence, and Kathleen Clark. *The Decline of the Delaware Bay Stopover for Migratory Shorebirds*. Trenton, NJ: New Jersey Division of Fish, Game, and Wildlife, 1997.

Norton, D. W., S. E. Senner, R. E. Gill, Jr., P. D. Martin, J. M. Wright, and A. K. Fukuyama. "Shorebirds and Herring Roe in Prince William Sound, Alaska." *American Birds*, Vol. 44 (Fall 1990), 367–71, 508.

Swan, Benjie Lynn, William R. Hall, Jr., and Carl N. Shuster, Jr., "*Limulus* Spawning Activity on Delaware Bay Shores on June 5, 1993." Delaware Estuary Program.

Thompson, Ida. *The Audubon Society Field Guide to North American Fossils*. New York: Alfred A. Knopf, 1982.

Thurston, Harry. "Fundy's Fecund Barrens." *Audubon*, Vol. 87 (September 1985), 88–103.

Weidensaul, Scott. "The Fate of the Horseshoe Crabs." *Philadelphia Inquirer*, May 29, 1995.

Western Hemisphere Shorebird Reserve Network. "WHSRN Site Profiles," http://www.manomet.org (January 1998).

14: *Catching the Wave*

CITATIONS

326 "The timing is hardly coincidental": Terborgh, *Where Have All the Birds Gone?* (Princeton University Press, 1989), p. 84.

330 "Using data from the North American Breeding Bird Survey": Chandler S. Robbins, John R. Sauer, Russell S. Greenberg, and Sam Droege, "Population Declines in North American Birds That Migrate to the Neotropics," *Proceedings of the National Academy of Science*, Vol. 86 (October 1989), 7658–62.

332 "Species that winter": Ibid.

332 "Still, the uncanny consistency with which the missing": Terborgh, p. 59.

341 "WHEREAS: Neotropical migratory landbirds": "Symposium Resolutions" in *Ecology and Conservation of Neotropical Migrant Landbirds*, J. M. Hagan III and D. W. Johnston, eds. (Washington, D.C.: Smithsonian Institution Press, 1992), pp. 597–98.

BIBLIOGRAPHY

Anon. "Elliot Coues Award, 1997: Chandler S. Robbins." *The Auk*, Vol. 115 (January 1998), 257–58.

———. "News from a Neotropical Conference." *Living Bird*, Vol. 3 (Summer 1990), 5.

Banks, R. C. *Human-Related Mortality of Birds in the United States*. U.S. Fish and Wildlife Service Special Science Report—Wildlife, no. 215, Washington, D.C., 1979.

Bonney, Rich, Susan Carlson, and Martha Fischer, eds. *Citizen's Guide to Migratory Bird Conservation*. Ithaca, NY: Cornell Laboratory of Ornithology and New York: National Audubon Society, 1995.

Butcher, Greg, and Rick Bonney. "Save Now or Pay Later." *Living Bird*, Vol. 11 (Summer 1992), 8–13.

Chiles, James R. "'We've Got Us Some Sky Today, Boy!'" *Smithsonian*, Vol. 28 (July 1997), 44–51.

Connor, Jack. "Cleared for Landing." *Nature Conservancy*, Vol. 45 (November–December 1995), 16–23.

Dunne, Pete. *The Feather Quest*. New York: Dutton, 1992.

Evans, Bill. "Deadly Towers." *Living Bird*, Vol. 17 (Spring 1998), 5.

Faaborg, John, and Wayne J. Arendt. "Long-Term Declines of Winter Resident Warblers in a Puerto Rican Dry Forest: Which Species Are in Trouble?" In *Ecology and Conservation of Neotropical Migrant Landbirds*, J. M. Hagan III and D. W. Johnston, eds. Washington, D.C.: Smithsonian Institution Press, 1992.

Gallagher, Tim. "A Little Night Music." *Living Bird*, Vol. 15 (Spring 1996), 10–14.

Galli, A. E., C. F. Leck, and R.T.T. Forman. "Avian Distribution Patterns in Forest Islands of Different Sizes in Central New Jersey." *The Auk*, Vol. 93 (April 1976), 356–64.

Graham, Frank, Jr. "Little Calamities." *Audubon*, Vol. 87 (September 1985), 24–27.

Hagan, John M. III, Trevor L. Lloyd-Evans, Jonathon L. Atwood, and D. Scott Wood. "Long-term Changes in Migratory Landbirds in the Northeastern United States: Evidence from Migration Capture Data." In *Ecology and Conservation of Neotropical Migrant Landbirds*, J. M. Hagan III and D. W. Johnston, eds. Washington, D.C.: Smithsonian Institution Press, 1992.

Morton, Eugene S. "What Do We Know about the Future of Migrant Landbirds?" *Ecology and Conservation of Neotropical Migrant Landbirds*, J. M. Hagan III and D. W. Johnston, eds. Washington, D.C.: Smithsonian Institution Press, 1992.

Robbins, Chandler S., Danny Bystrak, and Paul H. Geissler. *The Breeding Bird Survey: Its First Fifteen Years, 1965–1979.* Resource Publication 157. Washington, D.C.: U.S. Fish and Wildlife Service, 1986.

Sauer, John R., and Sam Droege. "Geographic Patterns in Population Trends in Neotropical Migrants in North America." In *Ecology and Conservation of Neotropical Migrant Landbirds*, J. M. Hagan III and D. W. Johnston, eds. Washington, D.C.: Smithsonian Institution Press, 1992.

Semmes, Anne W. "Another Neotropical Migrant?" *Living Bird*, Vol. 16 (Summer 1997), 35–36.

Wille, Chris. "Mystery of the Missing Migrants." *Audubon*, Vol. 92 (May 1990), 80–85.

15: *Trouble in the Woods*

CITATIONS

347   "When viewed as a whole": Scott K. Robinson, "The Case of the Missing Songbirds," *Consequences*, Vol. 3 (June 1997), http://www.gcrio.org/CONSQUENCES/vol3no1/toc.html.

348   "So, are Neotropical migratory birds really declining?": Brian A. Maurer and Marc-André Villard, "Continental Scale Ecology and Neotropical Migratory Birds: How to Detect Declines amid the Noise," *Ecology*, Vol. 77 (January 1996), 1–2.

350   "Our 'understanding of migratory birds'": Thomas W. Sherry and Richard T. Holmes, "Summer versus Winter Limitation of Populations: What Are the Issues and What Is the Evidence?" in *Ecology and Management of Neotropical Migratory Birds*, T. E. Martin and D. M. Finch, eds., 1995 (New York: Oxford University Press, 1995), p. 86.

369   "We face the end of migrations": J. P. Myers, "Nowhere to Run; Nowhere to Hide," *American Birds*, Vol. 45 (Fall 1991), 369–71.

BIBLIOGRAPHY

Adler, T. "Think Big to Save Birds on the Edge." *Science News* Vol. 147 (April 1, 1995), 198–99.

Anon. "A Catastrophe for Birds." *Bird Conservation* (Summer 1997), 10–11.

Askins, Robert A., James F. Lynch, and Russell Greenberg. "Population Declines in Migratory Birds in Eastern North America." *Current Ornithology*, Vol. 7, Dennis M. Power, ed. New York: Plenum Press, 1990.

Brawn, Jeffrey D., and Scott K. Robinson. "Source-sink Population Dynamics May Complicate the Interpretation of Long-term Census Data." *Ecology*, Vol. 77 (January 1996), 3–12.

Burke, Dawn, and Erica Nol. "Influence of Food Abundance, Nest-site Habitat, and Forest Fragmentation on Breeding Ovenbirds." *The Auk*, Vol. 115 (January 1998), 96–104.

Coleman, John S., Stanley A. Temple, and Scott R. Craven. *Cats and Wildlife: A Conservation Dilemma*. Madison, WI: U.S. Fish and Wildlife Service and University of Wisconsin-Extension, 1997.

deCalesta, David S. "Effect of White-tailed Deer on Songbirds Within Managed Forests in Pennsylvania." *Journal of Wildlife Management*, Vol. 58 (1994), 711–18.

Faaborg, J., M. Brittingham, T. Donovan, and J. Blake. "Habitat Fragmentation in the Temperate Zone." In *Ecology and Management of Neotropical Migratory Birds*, T. E. Martin and D. M. Finch, eds. New York: Oxford University Press, 1995.

Hames, Stefan. "BFL Analyzes Early Results." *Birdscope*, Vol. 12 (Spring 1998), 14, 16.

Hensley, M. Max, and James B. Cope. "Further Data on Removal and Repopulation of the Breeding Birds of a Spruce-Fir Forest Community." *The Auk*, Vol. 68 (October 1951), 483–93.

Line, Les. "Curse of the Cowbird." *National Wildlife*, Vol. 33 (December 1994), 40–45.

Monastersky, R. "Planet Posts Temperature Record for 1997." *Science News*, Vol. 153 (January 17, 1998), 38.

——. "Warmth in North Pushes Spring Forward." *Science News*, Vol. 151 (May 10, 1997), 290.

Peterjohn, Bruce G., John R. Sauer, and Chandler S. Robbins. "Population Trends from the North American Breeding Bird Survey," In *Ecology and Management of Neotropical Migratory Birds*, T. E. Martin and D. M. Finch, eds. New York: Oxford University Press, 1995.

Porneluzi, P., J. C. Bednarz, L. J. Goodrich, N. Zawada, and J. Hoover. "Reproductive Performance of Territorial Ovenbirds Occupying Forest Fragments and a Contiguous Forest in Pennsylvania." *Conservation Biology*, Vol. 7 (September 1993), 618–22.

Robinson, Scott K. "Nest Losses, Nest Gains." *Natural History*, Vol. 105 (July 1996), 40–47.

——. "Looking at the Big Picture." *Living Bird*, Vol. 15 (Spring 1996), 9.

——. "Population Dynamics of Breeding Neotropical Migrants in a Fragmented Illinois Landscape." In *Ecology and Conservation of Neotropical Migrant Landbirds*, J. M. Hagan III and D. W. Johnston, eds. Washington, D.C.: Smithsonian Institution Press, 1992.

——, S. I. Rothstein, M. C. Brittingham, L. J. Petit, and J. A. Grzybowski. "Ecology and Behavior of Cowbirds and Their Impact on Host Populations." In *Ecology and Management of Neotropical Migratory Birds*, T. E. Martin and D. M. Finch, eds. New York: Oxford University Press, 1995.

Scripps Institution of Oceanography. "Dramatic Drop in Zooplankton Reported off California Coast." University of California at San Diego, March 3, 1995.

Sexton, Charles W. "Cowbirds and Vireos." *Bird Conservation* (Summer 1997), 14–15.

Sherry, Thomas W., and Richard T. Holmes. "Population Fluctuations in a Long-distance Neotropical Migrant: Demographic Evidence for the Importance of Breeding Season Events in the American Redstart." In *Ecology and Conservation of Neotropical Migrant Landbirds*, J. M. Hagan III and D. W. Johnston, eds. Washington, D.C.: Smithsonian Institution Press, 1992.

Smith, P. William, and Alexander Sprunt IV. "The Shiny Cowbird Reaches the United States." *American Birds*, Vol. 41 (Fall 1987), 370–71.

Stevens, William K. "If the Climate Changes, It May Do So Fast, New Data Show," *The New York Times*, January 27, 1998.

Sullivan, Jerry. "Sinking in the Shawnee." *Living Bird*, Vol. 11 (Autumn 1992), 14–19.

Tramer, Elliot J. "Global Warming: An Imminent Threat to Birds?" *Living Bird*, Vol. 11 (Spring 1992), 8–12.

Waller, Donald M., and William S. Alverson. "The White-tailed Deer: A Keystone Herbivore." *Wildlife Society Bulletin*, Vol. 25 (Summer 1997), 217–26.

Wilcove, David. "Empty Skies." *The Nature Conservancy Magazine*, Vol. 40 (January 1990), 4–13.

——. "Nest Predation in Forest Tracts and the Decline of Migratory Songbirds." *Ecology*, Vol. 66 (1985), 1211–14.

——, and John W. Terborgh. "Patterns of Population Decline in Birds." *American Birds*, Vol. 38 (January 1984), 10–13.

Acknowledgments

This is where the author is supposed to say he couldn't have written the book without the people listed below. That isn't true—I could have written it anyway, but it wouldn't have been worth much. These folks were guides, mentors, and instructors to me; they exhibited the patience of Job and the enthusiasm of wide-eyed children, and most of them are tenaciously, ferociously dedicated to preserving migration in all its glories, despite frequently steep odds. Usually they had no clue who this guy was who was calling them, out of the blue, and asking to join them in their research. It is a credit to their generosity that they said yes (though usually after a long, thoughtful pause). Many of them also reviewed relevant parts of the manuscript; they deserve the credit for what I got right, and none of the blame for what I messed up.

For my time in Argentina, my thanks to Brian Woodbridge and Marc Bechard, Michael Goldstein, Sonia Beatriz Canavelli, José Hernán Sarasola, Juan José Maceda, Mike Kochert, and Scott Baker, for allowing me to join them in their work and to witness the spectacle of 10,000 Swainson's hawks surrounding us each evening like a living galaxy. And I can never adequately acknowledge, much less properly repay, Agustín Lanusse and his family at Estancia La Chanilao—Silvina, Agustina, Candé, Ayelén, Juan, and Pedro— for enveloping me in their warmth and extraordinary hospitality, and for doing so much to make the pampas safer for the *aguiluchos*.

In Mexico, my good friend Ernesto Ruelas Inzunza and his dedicated colleagues at *Rio de Rapaces* gave me the opportunity to realize the dream of a hawk lover's life. Thanks to him, and to Octavio Cruz-Carretero, María Liliana Vanda, Carlos Zavala Blas, Carlos Armenta Contreras, Sandra Luz Mesa, Flor Elena Galán Amaro, César Tejeda, Hector Eduardo Hemera, Juan Manuel García Rodriguez, Adriana Alvarez Andrade, Evodia Silva Rivera, Omar Gabriel Gordillo Souis, Rafael Villegas, Héctor López Pale, Jeanne Tinsman, John Haskell, and Christian Fregat.

In Jamaica, Leo R. Douglas was a gracious host and teacher, and I am grateful as well to Catherine Levy, Robert Sutton, and Ann Haynes-Sutton. Thomas W. Sherry at Tulane

University and Peter Vogel at the University of the West Indies were both generous with their time in helping me set up my visit.

For my travels in Alaska, I want to acknowledge Greg Siekaniec and Tracy Schafer at Izembek National Wildlife Refuge, Charlie Summerville in King Salmon, and my old friend Stanley E. Senner at the Exxon Valdez Trustee Council. I am especially grateful to Robert E. Gill, Jr., of the U.S. Geological Survey's Alaska Biological Science Center in Anchorage. Bob was a gold mine of information about curlews, godwits, and trans-Pacific shorebird migration, and when an injury forced him to cancel long-planned fieldwork just days before I was to join him, he made alternate arrangements that really saved my bacon.

I had such a good time with Bob and Martha Sargent and Duane and Donna Berger in Fort Morgan, Alabama, that I could hardly call it work. They and their crew soon felt like family, and I can't wait to get back to band with them again. In California, Debi Shearwater was an irrepressible companion, both on land and at sea. My thanks go as well to Frank Smith at Bombay Hook National Wildlife Refuge; David Prior in Halifax, Nova Scotia; David G. Campbell at Grinnell College in Iowa; Bill Finch and his colleagues at the Mobile (Ala.) *Register*; Jon McCraken, program manager at Long Point Bird Observatory, Ontario; Gordon Berkey in North Dakota; Chris Wille at the Tropical Conservation Newsbureau in Costa Rica; Bill Evans at Cornell University; Megan Hill with the National Fish and Wildlife Foundation in Guatemala; Ernesto Saqui at the Cockscomb Jaguar Reserve in Belize; Kevin DesRoberts at Ruby Lake National Wildlife Refuge in Nevada; Mike Mossman of the Wisconsin Department of Natural Resources; Robert S. Ridgely and David J. Agro of the Philadelphia Academy of Natural Sciences; and J. P. Myers of the W. Alton Jones Foundation. Charles Fergus, a fellow Pennsylvanian whose writing I greatly admire, was kind enough to look over early drafts and make many good suggestions.

At Hawk Mountain Sanctuary, thanks to my valued friend Laurie Goodrich, as well as to inventory biologist Mark Monroe and the rest of the staff. I have often used Hawk Mountain's fine library, but never more so than while writing this book, and I especially want to thank research director Keith Bildstein, who was always happy to supply source material from his own library to fill in gaps.

During the course of this long project I have enjoyed the attention of double sets of agents and editors, and all four deserve my deep thanks. Lizzie Grossman and Peter Matson are my once-and-future agents, and I owe Lizzie a particular debt for helping me shape this book in its early stages and finding it a good home. At Farrar, Straus and Giroux, I worked initially with Elisheva Urbas, whose advice was invaluable, and then with Ethan Nosowsky, who was a fine, patient collaborator through the rest of this long process. Thanks also to Lilly Golden and Carmel Huestis for their early encouragement and generosity of spirit.

No one can write a book without friends—friends to complain to, exult with, to pat your back or kick your butt. Some of mine have learned a great deal more about bird migration than they ever wanted to know and no doubt heaved a sigh of relief when I finished this project. But none of them started avoiding me or ducking my calls, at least that I noticed. A few of them joined me in the field from time to time, which made the long research trips away from home a bit easier to take. Many of them—Bruce and Alison, Rick and Margie, Jeanne, Sharon, Mike, Lynn and Steve, Dave and Ruth, and Sandy and Rick among them—were sounding boards for ideas and rough drafts, and without them I'd have wadded the whole thing up and tossed it in the trash long ago.

To Dave and Sylvia Young, a special thanks for friendship and support that spans the years. Finally, all my love to the four generations of my family, who provide the anchor that always pulls me back, no matter how far I wander.

Index

Able, Kenneth and Mary, 63–64
Academy of Natural Sciences of
    Philadelphia, 194–95
Acadian flycatcher. *See* flycatchers
agriculture: California irrigation water,
    241–42; and chemical contamina-
    tion, 179, 182, 184, 185, 190–91, 241,
    242; farmland *vs.* forest, 299–300;
    farmland *vs.* prairie, 295, 298; pampas
    farming, 174–75; and prairie potholes,
    295; sun-coffee farms, 151–52
Agro, David, 194
agrochemicals, 175, 241; *see also* chemical
    contamination
Alaska: Alagnak River, 10; Amak Island,
    18; birds in, 5–21; decline of specta-
    cled eider, 25, 26; Izembek National
    Wildlife Refuge, 5–10; during last ice
    age, 78; role of weather systems in
    bird migration, 79; Yukon River, 13
albatrosses: breeding grounds, 95–96, 97;
    flight techniques, 96–97; migration
    of, 83, 90, 95–96, 97; and *The Rime of
    the Ancient Mariner*, 89–90
    black-browed, 90, 97
    black-footed, 83, 95–96, 97
    Laysan, 65, 96
    short-tailed, 101
    yellow-nosed, 90

Alerstam, Thomas, 67
altitudinal migration, 21–22, 43, 146
Amadon, Dean, 17–18
Amak Island, Alaska, 18
American Bird Banding Association, 38
American Bird Conservancy, 185, 343, 357
American coots, 295
American golden-plover. *See* golden-
    plovers
anhingas, 117
animal migration, defined, 31–32; *see also*
    bird migration
anis, smooth-billed, 154
antbirds, 140
antennas, as bird risk, 338–39
Antilles, 161
ant swarms, 140–41
aplomado. *See* falcons
Appalachians: Kittatinny Ridge, 34, 35,
    47–48; migrant birds in forests,
    327–28
Aransas National Wildlife Refuge, Texas,
    284–86
Arctic Goose Habitat Working Group,
    231, 232, 234
Arctic tern. *See* terns
Arctic warbler. *See* warblers
Argentina: chimangos in, 186–87, 189;
    Estancia La Chanilao, 174–75, 181,

Argentina (*continued*)
    184, 186, 187–88; pampas farming,
    174–75; resident *vs.* migrant bird
    species, 192–93; rheas in, 195–96;
    spot-winged pigeons in, 186, 189;
    Swainson's hawk deaths, 175, 182–85;
    Tierra del Fuego, 313–14; use of
    insecticide monocrotophos, 182, 184,
    185, 190, 191; as wintering grounds,
    174, 193
Aristotle, 23, 24, 48, 84, 249
army ant swarms, 140–41
Atlantic flyway, 39, 92, 222
Atlantic States Marine Fisheries Commis-
    sion, 317
Audubon, John James, 38, 162, 195, 293
austral migrants, defined, 33
avian botulism, 244–45
avian cholera: at Fallon National Wildlife
    Refuge, Nevada, 242; at Salton Sea,
    California, 243
avocets: breeding grounds, 290; description
    of, 290; diet of, 290; habitat and
    resource requirements, 306–7; migra-
    tion of, 291

Bachman, Rev. John, 162
back-yard bird feeders, 50, 51, 52, 199,
    201, 203, 204, 209–10
Baker, Robin, 32
Balogh, Greg, 26
Baltimore orioles. *See* orioles
banding: and Bob and Martha Sargent,
    259–66; history of, 38; Long Point Bird
    Observatory, Ontario, 336–38; over-
    view, 38–41; sharp-shinned hawks,
    37–38; and tourists, 263–65; Veracruz
    station, Mexico, 119–20, 125
banks, offshore, 85, 86
bar-tailed godwit. *See* godwits
Bartlett, Jen and Des, 73
Bartram, William, 163
Bayard, Peter, 222
Bay of Fundy, Canada, 307, 310
BBS. *See* North American Breeding Bird
    Survey

becards, rose-throated, 59
Bechard, Marc, 184
bellbirds, three-wattled, 146
Belon, Pierre, 24
Bent, Arthur Cleveland, 194
Beringia, 6, 8, 13, 20, 25
Bering Sea, 5, 6, 20, 25–26
Berthold, Peter, 32, 48–49, 50, 68, 74
Bicknell's thrush. *See* thrushes
Big Bend National Park, Texas, 323
Bigi Pan hemispheric reserve, Suriname, 314
biological species concept, 167
Bird Banding Laboratory, Laurel, Mary-
    land, 38, 40
bird feeders, 50, 51, 52, 199, 201, 203, 204,
    209–10
birding and birders: as big business,
    270–71; and BirdSource project, 205;
    boom in, 202; *vs.* casual visitors, 269;
    characteristics of birders, 269–70; and
    fallouts, 251, 252; festivals, 272; and
    High Island, Texas, 268–70, 271; in
    Jamaica, 157; at Midwest Birding
    Symposium, 58–59; and sandhill
    cranes, 283–84; spirit of sharing
    among, 58–59; Texas sites, 270; and
    tourism, 271–72; and "waves" of
    spring flocks, 325–29
BirdLife International, 342
bird migration: adaptability of migrants,
    22–23, 26–27, 51, 52, 130, 135–44,
    156, 229–30; alternative to, 216–17;
    biggest raptor flights, 105, 106, 114,
    117; complexity of, 21; defined, 32;
    diet during, 112–13, 120; direction of
    flight, 21–22; elliptical, 292; failure to
    survive trip, 15; in first-year migrants,
    143; and global climate changes,
    366–70; historical understanding of,
    23–25; how routes evolve, 16–17, 19,
    20, 74–80, 130; impact of weather,
    47–48; as instinctive urge, 15, 16–17,
    20–21; intratropical, 33, 77, 145–46;
    during last ice age, 78; as learned
    behavior, 68–74; length of routes, 7,
    10–11, 16, 19, 173–74, 176, 193, 338;

nocturnal, 29–31, 35, 36, 61, 334; "northern home" theory *vs.* "southern home" theory, 75–76; numbers of birds, 21, 30–31, 44, 130; preparation for, 14, 16, 261; protecting migrants, 121, 312–14, 341–44; and ratio of daylight to darkness, 46–47; role of food supply, 6, 21, 46, 130, 200, 217; role of genetics, 48–51, 66, 67–68; role of wind, 79–80; as rule among bird species, 43; and seasons, 31, 32, 46–47; shortstopping phenomenon, 52–54, 203, 225; threats to habitat and resources, 26–27, 44, 46, 50, 130–31, 133, 344, 346; tools for studying, 25, 34, 35; trans-Gulf, 251–53; transpacific, 11, 14–18; vertical, 21–22, 43, 146; what it is, 31–32; what to call migrants, 32–34; and wing feathers, 19–20

birds: adaptation to changes in habitat, 22–23, 26–27, 51, 52, 130, 135–44, 156, 229–30; avian botulism, 244–45; avian cholera, 242, 243; capturing for banding, 36–38, 262–63; characteristics of wing feathers, 19–20; colic caeca in, 217–18; complete *vs.* partial migrants, 18, 19; counting, 108–11, 115–17, 123, 124; and electromagnetic fields, 62–64; extinctions, 161–62; floaters, 352–53; and hibernation, 23–24; homing ability, 60, 62, 64–65, 66; and insect control, 272–73, 353; Jamaican names for, 159; mixed flocks, 137–40; navigational abilities, 58, 60–66; nocturnal migration, 29–31, 35, 36, 61, 334; nonmigratory, 43, 212–13, 216–17, 235, 236–38, 284–85; protecting migrants, 121, 312–14, 341–44; relict species, 17; residents *vs.* migrants, 137, 139, 156–59, 172, 192–93; Spanish names for, 174; species splitters *vs.* species lumpers, 166–67; towers as risk, 338–39; *see also* bird migration; *names of specific birds*

Birds in Forested Landscapes project, 364
BirdSource, 204–5
bird watching. *See* bird feeders; birding and birders
blackbirds
  red-winged, 327
  yellow-headed, 294
  yellow-shouldered, 362
black-browed albatross. *See* albatrosses
blackcaps, 48–50
black-footed albatross. *See* albatrosses
blackpoll. *See* warblers
black-throated blue warbler. *See* warblers
Blas, Carlos Zavala, 108
bluebirds, eastern, 327
blue jay. *See* jays
blue-winged teal. *See* teals
blue-winged warbler. *See* warblers
bobolinks: declining numbers, 298, 299; description of, 288–89; magnetite in, 63; migration of, 44, 176; sounds of, 288
Bombay Hook National Wildlife Refuge, Delaware, 221, 222, 224–28, 229, 230–31, 234
Bono, Sonny, 245
boreal migrants, defined, 33
bottlenecks, hawk, 105, 106, 121
botulism, avian, 244–45
bownets, 35, 37
brant, 5, 8, 9
Breeding Bird Survey (BBS). *See* North American Breeding Bird Survey
Brewster, William, 163
bristle-thighed curlew. *See* curlews
broadcast antennas, as bird risk, 338–39
broad-winged hawk. *See* hawks
brown creepers, 138
Bruner, Lawrence, 292
Buller's shearwater. *See* shearwaters
Bullock's oriole. *See* orioles
buntings: migration of, 327
  indigo, 61–62, 256–57
  lark, 298
  snow, 18–19, 198, 214
Burke, Dawn, 364

Butte Valley National Grasslands, California, 177

Calakmul Biosphere Reserve, Mexico, 149
Canada geese. *See* geese
Canadian Wildlife Service, 38, 185, 194
Canavelli, Sonia Beatriz, 184
canebrakes, 163–64
Cape Churchill, Manitoba: and polar bears, 213, 214–15; as wintering grounds, 212–19; *see also* La Pérouse Bay, Manitoba
Cape May warbler. *See* warblers
captive birds, 60–62
Cardel, Mexico, 106, 107, 109–10, 112, 113–14
cardinals: capturing for banding, 262–63; as common bird-feeder species, 199; range expansion of, 20
Careys, defined, 89; *see also* storm-petrels
Caribbean, 153–72
Carson Sink, Nevada, 242
Cartier, Jacques, 223
Castro, Gonzalo, 310
cats: as danger to birds, 102, 357
Central America: deforestation in, 44–45, 133, 145; habitat preservation efforts, 147, 148–49; numbers of bird species, 130; as wintering grounds, 44, 130; *see also* Latin America
cerulean warbler. *See* warblers
chachalacas, 131
Chapman, Frank, 163
chats, yellow-breasted: in first-year migration, 143; stopping over in Texas desert, 324
chemical contamination: and Salton Sea, 241, 242; and Swainson's hawks, 179, 182, 184, 185, 190–91
cheniers, 267–68
Cheyenne Bottoms, Kansas, 313
chickadees: as common bird-feeder species, 199; in mixed flocks, 138
  boreal, 212, 215, 217
Chihuahuan Desert, Texas, 323, 324–25
chimangos, 186–87, 189

cholera, avian: at Fallon National Wildlife Refuge, Nevada, 242; at Salton Sea, California, 243
Ciba-Geigy, 185
climate changes, 365–70
coffee, 149–52
Coleridge, Samuel Taylor, 89
Columbus, Christopher, 154
complete migrants, defined, 18
Connor, Jack, 40
Conservation Reserve Program (CRP), 296, 297
Cooper's hawk. *See* hawks
Coppename estuary, Suriname, 307
Copper River Delta, Alaska, 307, 310–11
cordgrass, 221, 222, 227–29
cormorants, 83
Cornell Laboratory of Ornithology, 147, 203, 205, 364
corvids, 213; *see also* crows; jays; ravens
counting birds, 108–11, 115–17, 123, 124
cowbirds: controlling, 362–63; and forest fragmentation, 331, 357, 361; and nesting songbirds, 358–63; and wood thrushes, 148, 360–61, 364
cranes: in history, 23; migration of, 23, 36
  sandhill: description of, 276, 278–79; migration of, 73–74, 277, 279; Platte River, Nebraska, as staging ground, 277–81; population level, 277; sounds of, 279–80; and stopover ecology, 307
  whooping: in Canada, 285; as endangered species, 284–85; in Florida, 285; migration of, 73–74, 284, 285–86; nonmigratory flocks, 284, 285
crossbills: diet of, 216; in Vermont wintering grounds, 198
  red, 167, 202, 204
  white-winged: irruptions of, 204, 205
crows: migration of, 36
  hooded, 66
Cruz-Carretero, Octavio ("Tavo"), 113, 115, 121, 122–23
cuckoos
  Jamaican lizard-, 155, 172
  yellow-billed, 331

curlews
    bristle-thighed: as relict species, 17–18;
        transpacific migration of, 9, 11,
        13–15, 17–18
    Eskimo: diet of, 293; migration of, 176,
        194, 292; as "prairie pigeons," 293;
        rarity of, 193–94; shooting of, 293;
        type specimens, 195

Darwin, Charles, 60, 95
Dauphin Island, Alabama, 249–51, 256,
    259–66
deforestation: Gulf Coast, 27, 267–68; on
    Hispaniola, 169; in Jamaica, 160–61,
    172; tropical, 44–45, 133, 134,
    144–45, 332–33
Delaware Bay, 303, 304–6, 307, 308–10,
    312, 314–17
deserts, as migratory barrier, 324
Desiderata (boat), 81–82, 83, 84, 90–91
dickcissels: declining numbers, 298
dinoflagellates, 84
dinosaurs, 196
dippers, 23
DNA analysis, 42, 167, 168
Doppler radar systems, 42, 333–34
Douglas, Leo, 153, 154–55, 158, 161,
    169–71
doves: as hawk bait, 37; in history, 23
    Caribbean, 172
dowitchers, long-billed, 20
dragonflies, 122, 188
Droege, Sam, 330
ducks: banding, 39; diet of, 294; habitat
    loss, 296; migration of, 21, 31, 36,
    39–40, 68–69; pollution deaths, 243,
    244; population levels on prairies,
    295–97; Salton Sea as wintering
    grounds, 240; tagging of, 41; see also
    eiders, spectacled; mallards; teal
    pintails, 63
    redheads, 21
    ruddy, 294
    white-faced whistling-, 192
Ducks Unlimited, 297
Dugout Wells, Texas, 323, 324–25

dunlins, 7, 308, 311
Dust Bowl, 295, 300

eagles, 35, 41
eastern kingbird. See kingbirds
eatouts, 228–29
egrets, cattle, 20
eiders
    common, 218–19
    spectacled, 25–26
    Steller's, 8
electromagnetic fields, 62–64
elliptical migrations, 292
Emlen, John and Stephen, 61–62
Emlen funnel, 47, 61
emperor penguins, 217
endangered species, 185, 257–58, 283,
    284–85, 290
eruptions vs. irruptions, 199
Eskimo curlew. See curlews
Estancia La Chanilao, Argentina, 174–75,
    181, 184, 186, 187–88
Evans, Bill, 337, 339
evening grosbeak. See grosbeaks
Exxon Valdez, 310, 311

falconry, 122–123
falcons: capture of, 35; dangers to,
    120, 131
    aplomado, 121, 192, 193
    gyrfalcon, 212, 215
    peregrine, 105–6, 120, 121, 176
Fallon National Wildlife Refuge, Nevada,
    242
fallouts, songbird, 251, 252, 256–57, 334
farmland. See agriculture
feathers, bird wing, 19–20
Feduccia, Alan, 75
FeederWatch, 203–4
festivals, birding, 272
fieldfares, 20
finches: irruptions of, 199, 202, 204, 205;
    lack of irruptions, 202–3
    gray-crowned rosy, 22
    house, 209, 211
    northern, 199, 205

finches (*continued*)
  purple, 52, 198, 202, 204
  winter, 202–3
Finley, Karen, 180–83
fires, prairie, 300–1
Fitzpatrick, John W., 44, 147
flamingos, Chilean, 192
flats. *See* Leatherberry Flats, Delaware Bay;
  salt marshes
flickers: lumping of species, 167; migration
  of, 36
  gilded, 167
  red-shafted, 167
  yellow-shafted, 167
flightless birds, 195–96
floater birds, 352–53
flocks, mixed, 137–40
Fly Away Home (movie), 73
flycatchers: diet of, 77, 137
  Acadian, 136, 364
  fork-tailed, 189–90
  rufous-tailed, 155
  sad, 172
  scissor-tailed, 117, 120, 122
  willow, 363
  yellow-bellied, 136
  See also kingbirds
flyways, 39–40, 92, 222
Folsom, Jim, Jr., 266
food supply: bird preparation for migration,
  14, 16, 261; diets of migrating birds,
  112–13, 120; diets of tropical
  migrants, 77, 137; role in
  migration, 6, 21, 46, 130, 200, 217;
  weather sensitivity of, 6; *see also*
  insectivores
forests: breeding grounds *vs.* wintering
  grounds, 331–32, 354; coastal, 27,
  267–68; edge effect, 331, 364; fragmentation, 331, 354–64; hardwood,
  267, 268, 299, 354–64; tropical,
  133, 136, 146–47, 332–33; *see also*
  deforestation
Forsythe National Wildlife Refuge, New
  Jersey, 230

Fort Morgan, Alabama, 259–61, 262, 264,
  265–66
Franklin, Benjamin, 84
fulmars, northern, 90, 93

gadwalls, 293
gallinule, azure, 145–46
game birds, 271, 297; *see also* hunters
gannets, 66
Gaughan, Sharon, 153, 154–55, 158
Gauthreaux, Sidney A., Jr., 255–56, 333,
  337, 352
geese: banding, 39; in history, 23; hunting
  of, 53–54; migration of, 31, 36,
  39–40, 68–69; pollution deaths, 243;
  Salton Sea as wintering grounds, 240;
  tagging of, 41
  barnacle, 23
  blue, 223
  brant, 5, 8, 9
  Canada: attempts to control, 237–38;
    and Bill Lishman, 73; diet of,
    236–37; and golf courses, 235,
    236–37, 238; migration of, 35, 48,
    52–53, 214, 235–36, 238; nonmigra-
    tory, 235, 236–38; population
    increase, 236
  emperor, 5, 8
  snow: breeding ground problems,
    231–34; as cautionary tale, 222; as
    conservation success story, 221–22;
    and cordgrass eatout, 227–29; descrip-
    tion of, 221, 222–23; diet of, 225–26;
    greater *vs.* lesser, 223; as grubbers,
    228, 232; hunting of, 221, 223,
    230–31, 234; at La Pérouse Bay,
    Manitoba, 231–34; migration of,
    48, 225–26; population increase,
    222, 223, 225, 231–32; subspecies
    of, 223
genetics: role in bird migration, 48–51, 66,
  67–68
Gill, Robert G., Jr., 16, 79
glaciers, 287, 288
global climate changes, 365–70

gnatcatchers
  black-capped, 59
  blue-gray, 139–40
godwits
  bar-tailed, 8, 9, 11, 15–16, 79, 366
  Hudsonian, 8, 176, 212, 292
  marbled, 307
golden-plovers: length of migration, 176, 193
  American, 7–8, 9, 21, 176, 192, 291–92
  Pacific, 7, 9, 18
goldfinches, 35, 204
Goldstein, Michael I., 184–85, 186, 191–92
golf courses: and Canada geese, 235, 236–237, 238
Goodrich, Laurie, 349, 350–51
gooney birds, 96
goshawks: capture of, 35; diet of, 209; irruptions of, 206; migration of, 208–9
Gosse Bird Club, Jamaica, 157, 158
grackles, common, 327
Grand Island, Nebraska, 278, 283
grassland birds: characteristics of, 289–90; declining numbers, 297–301; and global climate changes, 366; varieties of, 288
grassquits, yellow-faced, 155
gray-cheeked thrush. See thrushes
gray jay. See jays
Grays Harbor, Washington, 314
Great Salt Lake, Utah, 314
Great Texas Coastal Birding Trail, 270
grebes, eared, 243, 245
Greeks, ancient, 23
Greenberg, Russell, 22, 76, 136, 150, 151, 330
greenhouse gases, 365
Greenland, 368
grosbeaks: capturing for banding, 262–63
  evening: breeding grounds, 200; description of, 199–200; diet of, 201; irruptions of, 201–2, 204; wintering grounds, 198

pine, 202, 204
  rose-breasted: declining numbers, 338; migration of, 31
grouse
  blue, 22
  spruce, 212
Gulf Express, 251
Gulf Islands National Seashore, 250, 266–67
Gulf of Mexico: barrier islands, 249–50; coast as migratory stopover, 251–53; Dauphin Island, Alabama, 249–51
gulls, 5, 83, 95; and horseshoe crabs, 308–9; pollution deaths, 243, 245; Salton Sea as wintering grounds, 240
  ring-billed, 327
  Sabine's, 97, 374
  western, 95
gyrfalcons, 212, 215

habitat: bird adaptation to changes, 22–23, 26–27, 51, 52, 130, 135–44, 229–30; vs. niche, 136–37, 139; preservation efforts, 50, 147, 148–52, 342; threats to birds, 26–27, 44, 46, 50, 130–31, 133, 344, 346
harriers, northern, 226, 300
Harrington, Brian A., 41, 312, 316
Haskell, John, 107, 116
Hastings, Nebraska, 284
Hawk Mountain Sanctuary, Pennsylvania, 34, 106, 107, 114, 124, 350–51, 352
hawks: capturing, 34, 35, 36–38, 39, 186; counting, 108–11, 115–17; dangers to, 119, 120–21; dependence on thermal updrafts, 108; diet during migration, 112–13, 120; diet of, 206, 207; flocking behavior, 118–19; kettling of, 108, 110–11, 115–17, 125; migration of, 34, 105–25; roosting sites, 119, 120; training, 122–23; Veracruz bottleneck, 105–25; in Vermont wintering grounds, 197–98

hawks (*continued*)
    broad-winged: counting, 109, 110–11, 115–16, 124; dangers to, 118; description of, 107; diet of, 112, 120; kettling of, 35–36, 115–16; Lake Erie migration corridor, 124; migration of, 105–6, 107–8, 112, 114, 115–16, 118, 124; roosting sites, 120
    Cooper's: dangers to, 120; diet of, 120; migration of, 52;
    Harris's, 123
    red-tailed: capture of, 35, 37; in Nebraska Sandhills, 275–76; tagging of, 41
    rough-legged: diet of, 207; irruptions of, 206, 207; migration of, 199; in Vermont wintering grounds, 197–98, 199
    sharp-shinned: banding, 37, 39; capture of, 37; dangers to, 120; diet of, 120; migration of, 52
    Swainson's: banding, 188; breeding grounds, 176–77; capturing, 186, 187–88; counting, 124; dangers to, 118; declining numbers, 177, 179; description of, 175, 176, 187; diet of, 112–13, 120, 177, 179, 182, 188–89; length of migration, 176; map of migratory route, 178; migration of, 105–6, 108, 118, 124, 176, 179–80; pampas deaths, 175, 182–85; Spanish name for, 174; study by Brian Woodbridge, 179–85; tracking of, 180, 188; wintering grounds, 177
Hawks Aloft Worldwide initiative, 124
HawkWatch International, 107
Hayes, Floyd, 33–34
Heinrich, Bernd, 213
Helmholtz coils, 62
herons: banding, 38; lumping of species, 167; migration of, 20
    capped, 146
    gray, 38
Hesperornis, 75
High Island, Texas, 268–70, 271
high-rise buildings, as bird risk, 340

Hill, Geoffrey, 263
Hochbaum, H. Albert, 68
Holmes, Richard, 158, 172, 350
Homer, 23
homing pigeon. *See* pigeons
honeycreepers, red-legged, 131
hooded warbler. *See* warblers
Horicon National Wildlife Refuge, Wisconsin, 52
horseshoe crabs: background of, 309–10; declining numbers, 314–17; and Delaware Bay, 304, 306, 308, 314–17; and fertilizer industry, 318, 319; and Lenape Indians, 317; migration of, 303, 304
house cat predation, 102, 357
Hubbard Brook Experimental Forest, New Hampshire, 158
Hudson, W. H., 181
Hudson Bay, Canada: snow goose damage to wetlands, 232–33
Hudsonian godwit. *See* godwits
hummingbirds: capturing for banding, 263; in Central America, 134; and hibernation, 24; in Jamaica, 156, 159; migration of, 21, 36, 51, 57; tagging of, 41; in tropics, 139
    broad-tailed, 57
    ruby-throated, 39, 52, 261
    rufous, 50–52
hunters: and geese, 53–54; and migratory birds, 271; and snow geese, 221, 223, 230–31, 234
hyperphagia, 47

ibises, white-faced: in Argentina, 192, 193
ice ages, 19, 20, 75, 76, 77–79, 287–88
Ichthyornis, 75
Imperial Valley, California, 242
Important Bird Area (IBA) program, 342–43
indigo bunting. *See* buntings
insectivores: and insect control, 272–73, 353; in mixed flocks, 139–40; vs. seed-eaters, 6; among tropical migrants, 137; young birds as, 272–73, 294

International Shorebird Surveys, 312
intratropical migration, 33, 77, 145–46
irruptions: absence of, 202–3; defined, 199; *vs.* eruptions, 199; northern finches, 199, 205; occurrences of, 200–5; and predation cycles, 206–7; purple finches, 202, 204; tracking, 203–5
Izembek National Wildlife Refuge, Alaska, 5–10

jacamars, 131
jaegers, 83
    long-tailed, 97
Jamaica, 153–62
James, Frances C., 348
jays: migration of, 36
    blue, 35, 199
    gray, 43, 212
    scrub, 167
    Steller's, 43
Jenkins, Allan, 277–78
juncos: lumping of species, 167
    dark-eyed, 22, 31, 48

Kahansink Indians, 222
Kaufman, Kenn, 203
Kentucky warbler. *See* warblers
Kerlinger, Paul, 74, 261
Kesterson National Wildlife Refuge, California, 241
kestrels: dangers to, 120, 121; diet during migration, 120; migration of, 116
kettling, 35–36, 108, 110–11, 115–17, 125
kingbirds
    eastern, 44, 129, 132, 139
    loggerhead, 154
    thick-billed, 59
    tropical, 189
kinglets, 138
Kirtland's warbler. *See* warblers
Kirtland's Warbler Festival, 272
kites: counting, 109, 110; dangers to, 120; diet during migration, 120; in history, 23
    hook-billed, 110
    Mississippi, 108, 109, 110, 120

Kittatinny Ridge, Pennsylvania, 34, 35, 47–48, 350
Klamath National Forest, California, 177
Kramer, Gustav, 60–61
Kumlein, Ludwig, 215

LaHave Bank, Nova Scotia, 85, 86
Lake Champlain basin: as migrant wintering grounds, 198
Lake Erie: area as hawk migration corridor, 124; Long Point Provincial Park, Ontario, 203, 334, 335–38; Midwest Birding Symposium, 58, 59; Point Pelee, Ontario, 334
Lanusse, Agustín, 182, 184, 186, 190
La Pérouse Bay, Manitoba, 231–34
larks: in history, 23
    horned, 198
Larned, Bill, 26
Latin America: Amazon basin, 133, 138; bird conservation programs, 148–49, 343; coffee in, 150–52; deforestation in, 44–45, 133, 134, 144–45, 332–33; habitat preservation efforts, 147, 148–52, 343; intratropical migration, 33, 77, 145–46; numbers of migrating bird species, 21, 44, 130; rain forests, 133, 136, 146–47, 332–33; Veracruz, Mexico, 105–25
Laysan albatross. *See* albatrosses
Leask, Michael J.M., 63
Leatherberry Flats, Delaware Bay: as example of salt marsh, 226–27; impact of cordgrass eatout, 228–29
Lefebvre, Gaëton, 137
lemmings, 206–7
Leopold, Aldo, 280
Levey, Douglas, 77
Levy, Catherine, 157
Lewis, Meriwether, 24
lights, as bird risk, 339–40
Linnaeus, Carolus, 24
Lishman, Bill, 73
lizard-cuckoo. *See* cuckoos
lodestones, 62

Long Point Provincial Park, Ontario, 334,
    335–36; bird observatory, 203,
    336–38
longspurs, Lapland: migration of, 18–19,
    214; in Vermont wintering grounds,
    198
loop migrations, 292
Lostwood National Wildlife Refuge,
    Missouri, 288–301
Lowery, George H., Jr., 254–55

MacArthur, Robert, 130, 355
magnetic fields, 62–64
magnolia warbler. See warblers
Magnus, Olaus, 24
mallards: banding, 39; migration of, 48, 69
Manomet Center for Conservation Sci-
    ences (Manomet Bird Observatory),
    Massachusetts, 41, 312, 330, 333, 341
maps: autumn migratory route of birds
    leaving Izembek National Wildlife
    Refuge, Alaska, 9; autumn migratory
    route of blackpoll warbler, 12; breed-
    ing and wintering ranges of magnolia
    warbler, 45; migratory route of Swain-
    son's hawk, 178; northbound migra-
    tory route of red knot, 305; original
    breeding range of trumpeter swan, 70;
    prairie pothole region, 289; Salton
    Sea, California, 239; submerged banks
    and basins of Nova Scotia, 86
Marismas Nacionales, Mexico, 314
market hunting: shorebirds, 319–20; and
    snow geese, 221, 223
masting, defined, 200
Maya Biosphere Reserve, Guatemala, 149
Mayan civilization, 129, 130, 133–34
Mayfield, Harold, 165
meadowlarks, 298
mergansers, red-breasted, 214
merlins, 120
Merriam, C. Hart, 166, 191
Mexico: habitat preservation efforts,
    148–49; Veracruz, 105–25; see also
    Latin America
Midwest Birding Symposium, 58–59

migrant traps, 325, 334
migration, defined, 31–32; see also bird
    migration
Migratory Bird Treaty Act of 1918, 69,
    224, 320
migratory routes: birds leaving Izembek
    National Wildlife Refuge, Alaska, in
    autumn, 9; blackpoll warbler in
    autumn, 12; distance examples,
    173–74; how they evolve, 16–17,
    19, 20, 74–80; northbound route
    of red knot, 305; protecting, 121,
    312–14, 341–44; Swainson's
    hawk, 178
Mississippi kite. See kites
mist nets, 34, 35, 37, 260, 262, 263–64
mixed flocks, 137–40
mockingbirds: calling with recorded bird
    sounds, 172; as Chihuahuan Desert
    residents, 325; range expansion of, 20
monocrotophos (insecticide), 182, 184,
    185, 190, 191
Monterey Bay, California, 93, 94, 95, 96,
    97–98, 99–101, 103
moon watching, 30–31, 255, 337
Mortensen, Hans, 38
Morton, Eugene S., 22, 136, 139, 143
Moskovits, Debra K., 44, 147
Mother Carey's chickens, 89; see also
    storm-petrels
motmots, 131
Murphy, Robert Cushman, 95, 98, 99, 101
muttonbirds. See shearwaters: short-
    tailed
Myers, J. P., 310, 369

National Audubon Society, 205, 343
National Survey on Recreation and the
    Environment, 271
national wildlife refuges: Aransas National
    Wildlife Refuge, Texas, 284–86;
    Bombay Hook National Wildlife
    Refuge, Delaware, 221, 222, 224–28,
    229, 230–31, 234; focus on waterfowl,
    297; Lostwood National Wildlife
    Refuge, North Dakota, 288–301; Red

Rock Lakes National Wildlife Refuge, Montana, 69–70, 71, 72; Salton Sea, California, 240, 242
natural selection, 76, 79
Nature Conservancy, 342
Nearctic migrants, defined, 33, 76
Nebraska Sandhills, 275
neotropical migrants: defined, 33–34; diet of, 77; intratropical, 33, 77, 145–46; population levels, 338, 347, 348–49; studying, 131, 331–32
New River, California, 242
Newton, Alfred, 60
Newton, Sir Isaac, 84
niches vs. habitats, 136–37, 139
night flying, 29–31, 35, 36, 61, 334
nightjar, 24
nocturnal migration, 29–31, 35, 36, 61, 334
noddies, defined, 89
Nols, Erica, 364
nonmigratory birds, 43, 212–13, 216–17, 235, 236–38, 284–85
North American Breeding Bird Survey, 147, 298, 330–31, 332, 346, 352; cerulean warbler status, 258; limitations of, 351; Swainson's warbler status, 259
North American Waterfowl Management Plan, 342
North Star, 62
Novartis, 185
Nova Scotia, 81, 84–85, 86
nutcracker, Clark's, 22, 43
nuthatches: in mixed flocks, 138 red-breasted, 198, 202

oil tankers, 311
oldsquaws, 214
Old Testament, 23
orangequits, 159
orioles: lumping of species, 166–67; migration of, 327; species pairs, 76
Baltimore: in Central America, 134–35; declining numbers, 331; lumped into northern oriole, 166–67; migration of, 33, 135, 328; paired with Bullock's oriole, 76; as tropical birds, 135
black-backed, 167
Bullock's, 76, 166–67
orchard, 22–23
oropendolas, 131
Ouellet, Henri, 168
ovenbird. See warblers
owls: diet of, 206, 207, 208; irruptions of, 206, 207–8
great gray, 208
snowy: description of, 206; diet of, 215; irruptions of, 206, 207; and lemmings, 206–8; migration of, 32, 207; wintering grounds, 198, 212
oystercatchers, 306–7

Pacific golden-plover. See golden-plovers
Pacific loons: migration of, 212; new species, 167
Pallas' sandgrouse: irruptions of, 203
palm warbler. See warblers
Pan-American Shorebird Program, 312
Parker, Ted, 44, 146, 147
parrots, 21, 131
partial migrants, defined, 19
Partners in Flight, 169, 257, 341–43
parula. See warblers
passenger pigeon. See pigeons
passerines. See songbirds
Patagonia Picnic Table Effect, 59
Patuxent Wildlife Research Center, Maryland, 298
pelagic birds: avoiding dehydration, 88; and birding, 100; dangers to, 101–3; defined, 83; examples of, 83
pelicans, 83; in history, 23; migration of, 23, 117, 122; pollution deaths, 242, 243, 244–45
white, 117, 122, 242, 243, 245
peregrine. See falcons
pesticides: and Swainson's hawks, 179, 182, 184, 185, 190–91
Peterson, Roger Tory, 280
petrels: Jamaican, 161; migration of, 83; species of, 100; see also storm-petrels

phalaropes: description of, 83; habitat and resource requirements, 306–7; migration of, 83–84; reversed sexual roles, 83, 291
    red-necked, 83
    Wilson's, 176, 290–91
philopatry, 231
phoebes, eastern, 327
photoperiods, 46–47
phylogenetic species concept, 167
PIF. *See* Partners in Flight
pigeons
    homing, 61, 62, 64–65
    passenger, 293; type specimens, 195
    spot-winged, 186, 189
pintails, 63
piping plover. *See* plovers
pishing, 155, 171
Plains. *See* prairies
plankton, 84–85
Platte River, Nebraska: Big Bend Reach, 277–81; as braided river, 277, 283; changes to, 281–83; as sandhill crane staging area, 277–81; and water issues, 281–83
plovers: habitat and resource requirements, 306–7; tagging of, 42
    mountain, 299
    piping, 283, 290
    Wilson's, 307
    See also golden-plovers
Point Pelee, Ontario, 334
polar bears, Cape Churchill, 213, 214–15
Polaris (North Star), 62
Poor Knights Island, New Zealand, 102–3
poorwill: and hibernation, 24; in history, 24
potholes. *See* prairie potholes
Poulin, Brigitte, 137
prairie fires, 300–1
prairie pigeons, defined, 293; *see also* curlews: Eskimo
prairie potholes, 288, 290; as food factories, 294; impact of weather, 295; loss of, 295; map of region, 289; preserving, 296; variety in chemical composition, 290

prairies: and bird migration, 288–301
predators: and lemmings, 206–7; and nesting songbirds, 331, 355–57; *see also* cowbirds; raptors
primary projection: defined, 19; in warblers, 20; in wheatears, 19
Prince William Sound, Alaska, 310, 311
Project FeederWatch, 203–4
prothonotary warbler. *See* warblers
ptarmigans
    white-tailed, 22, 43
    willow, 212, 215, 217–18
purple finch. *See* finches

quails, scaled, 324
quetzals, 146

radar, 42, 255–56, 333–34
radio antennas, as bird risk, 338–39
radio transmitters, 25, 42–43
rain forests, 133, 136, 146–47, 332–33
Ramsar Convention on Wetlands, 313
Rappole, John H., 49, 130
raptors: attitude toward, 107; biggest migratory flights, 105, 106, 114, 117, 124; capture of, 34, 35; dangers to, 120–21; dependence on thermal updrafts, 108; irruptions of, 206–8; migration of, 34, 35, 108; as protected species, 120; and stopover ecology, 307; in Vermont wintering grounds, 197–98; *see also* hawks
ravens: migration of, 36; as scavengers, 213, 214, 215; wintering grounds, 212–13; winter survival, 212–13, 216–17
recorded bird sounds, 155, 170, 172, 349
redheads, 21
red knots: diet of, 304; and horseshoe crabs, 304–6, 308; migration of, 193, 303, 304–6, 308, 316, 366; tracking, 41
redpolls
    common: diet of, 204; irruptions of, 202, 204

hoary: diet of, 216, 217; wintering
    grounds, 212, 215–16
Red Rock Lakes National Wildlife Refuge,
    Montana, 69–70, 71, 72
redstarts
    American: calling with recorded bird
        sounds, 155; description of, 143;
        gender segregation, 143; in Jamaica,
        159, 160; studying winter *vs.* summer
        ecology, 158; wintering grounds, 144
    European, 23
red-tailed hawk. *See* hawks
relict species, 17
Remsen, Van, 146, 163
return migration, 32
rheas, 195–96
Rimmer, Chris, 169
ringing. *See* banding
Rio Bravo Conservation Area, Belize, 149
Robbins, Chandler, 330, 332
robins: banding, 39; European, 63; in
    history, 23; migration of, 35, 36, 48,
    327
Robinson, Scott K., 347–48, 351, 364
Rock Creek Park, Washington, D.C., 329
rose-breasted grosbeak. *See* grosbeaks
rosy-finches: lumping of species, 167;
    migration of, 22
    gray-crowned, 22
rough-legged hawk. *See* hawks
Rowan, William K., 46
ruddy duck. *See* ducks
Ruelas Inzunza, Ernesto, 106–7, 110, 112,
    114, 119–20, 122–23, 124, 125

salt marshes: Bombay Hook National
    Wildlife Refuge, Delaware, 221, 222,
    224–28, 229, 230–31, 234; and cord-
    grass, 221, 222, 227–29; transforma-
    tion to mudflats, 229–30; *see also*
    Leatherberry Flats, Delaware Bay
Salton Sea, California: map, 239; national
    wildlife refuge, 240, 242; origins of,
    240; pollution problems, 241–46;
    salinity increase, 240–41; selenium in,
    242; as wintering grounds, 240

sanderlings, 306, 308, 316
sandgrouse, 203
sandhill crane. *See* cranes
sandpipers: habitat and resource require-
    ments, 306–7; length of migration,
    176, 193; tagging of, 42
    Baird's, 307
    buff-breasted, 293
    dunlins, 7, 308, 311
    least, 7
    pectoral, 20
    rock, 7
    semipalmated, 304, 306, 310, 316
    tattlers, 7
    upland, 33, 174, 193
    white-rumped, 193, 214, 292
    yellowlegs, 8, 193
    See also red knots
Sargent, Bob and Martha, 259–66
Sauer, Franz and Eleanor, 61
Sauer, John, 330
scarlet tanager. *See* tanagers
scissor-tailed flycatcher. *See* flycatchers
seabirds: length of migration, 79–80;
    migration of, 81–103; in Monterey
    Bay area, 93–103; off Nova Scotia,
    81–93; sense of smell, 64; and
    stopover ecology, 307; tubenoses, 88;
    See also names of specific birds
Seager, Trent, 180–83
seasons, 31, 32
seed-eating birds, 6, 52, 200–1, 217
selenium, 241, 242
semipalmated sandpiper. *See* sandpipers
Senner, Stanley E., 310, 311
shade-grown coffee, 150–51
sharp-shinned hawk. *See* hawks
Shearwater, Debra Love, 93–103
shearwaters: migration of, 8, 83, 366;
    species of, 100
    Buller's, 100, 102–3
    greater, 87–88
    Manx, 66
    short-tailed, 9, 10, 21, 102
    sooty, 91, 94–95, 101, 103
Sherry, Tom, 158, 172, 350

Shoemaker, Jesse, 278
shorebirds: banding, 41; around Cape
Churchill, Manitoba, 212; diversity of
breeding arrangements, 291; and
global climate changes, 366; habitat
and resource requirements, 306–7; and
horseshoe crabs, 304–6, 308, 309; at
Izembek National Wildlife Refuge,
Alaska, 5, 6–7; during last ice age, 78;
length of migration, 176, 193; market
hunting, 319–20; migration of, 83–84,
193, 292, 304–6; pollution deaths, 243,
244; protecting major staging sites,
312–14; Salton Sea as wintering
grounds, 240; shooting of, 293; and
stopover ecology, 307, 310–11, 312,
313, 321; tagging of, 41–42; transpacific migration, 11, 14–18; see also
names of specific birds
shortstopping, 52–54, 203, 225
short-tailed shearwater. See shearwaters
shrikes
loggerhead, 210
northern, 209–11
Shuster, Carl, 314, 318
silvereyes, 63
siskins, pine: irruptions of, 202, 203, 204;
migration of, 52; in Vermont wintering grounds, 198
skuas: description of, 98; diet of, 98–99;
migration of, 83, 99
south polar, 98, 99
skyscrapers, as bird risk, 340
Smith, Frank, 223–25, 226, 227, 228, 230,
231, 234–35
Smith, Neal, 33
snowberries, 300
snow bunting. See buntings
snow geese. See geese
Society for Caribbean Ornithology, 157
solar compass, 61
songbirds: adaptation to changes in habitat, 26–27, 135–44, 156; caged, 47,
121; declining numbers, 255–56,
298–99, 329–33, 338, 346–49; diet of,
267, 326; fallouts, 251, 252, 256–57,

334; and global climate changes, 366;
grassland species, 289–90, 297–301;
and hardwood forests, 267, 268, 299,
354–64; during last ice age, 78; length
of migration, 338; and migrant traps,
325, 334; migration of, 35, 44, 130,
326–27, 338, 366; most endangered
species, 257–59; nesting, 331,
355–57; nocturnal migration, 29, 30,
35, 36, 61, 334; and photoperiods, 47;
preparing for migration, 261; and
rain-forest preservation, 147, 148–52,
299; recording night calls, 337;
sounds of, 351–52; and stopover
ecology, 307, 324, 325, 369; threats
to, 27, 44, 46, 50, 130–31, 133,
256–57, 344, 350, 358–63; trans-
Gulf migration, 251–53, 261–62;
as tropical birds, 135; "waves" of
spring flocks, 325–29; wintering
grounds, 135–44; see also names of
specific birds
sooty shearwater. See shearwaters
South America. See Latin America
sparrows: declining numbers, 298, 338;
migration of, 8, 214
American tree, 214
black-throated, 324
fox, 8
golden-crowned, 8
grasshopper, 298
Henslow's, 298
sharp-tailed, 167
song, 199
white-crowned, 63, 66, 324
white-throated, 338
species: defined, 167; lumping vs. splitting,
166–67
species pairs, 76
spectacled eider. See eiders
sportsmen, 297; see also hunters
spot-winged pigeon. See pigeons
Sprague's pipits, 289
spruce budworms, 353–54
starlings: captive, 60–61; Dutch: navigational testing, 65

Steller's eider. *See* eiders
Stiles, Gary, 77
Stillwater National Wildlife Refuge, Nevada, 242
stopover ecology, 307, 310–11, 312, 313, 321, 325, 369
storks: in history, 23; as intratropical migrants, 146; tagging of, 41
jabiru, 146
maguari, 146
wood, 117, 122
storm-petrels: description of, 89, 91; extinct species, 102; migration of, 79–80, 83, 91–92, 93; species of, 91–92, 102
ashy, 93
fork-tailed, 39, 93
Leach's, 91–92
Wilson's, 79–80, 91–92
Stotz, Douglas, 44, 147
strip mines, reclaimed, as bird habitat, 300
sun-coffee farms, 151–52
superflights, 204–5
Swainson's hawk. *See* hawks
Swainson's thrush. *See* thrushes
Swainson's warbler. *See* warblers
swallows: in history, 23–24; migration of, 36, 176
bank, 176
barn, 174, 176
cliff, 176
tree, 35, 327
swans: migration of, 68, 69; tagging of, 42
mute, 237
trumpeter, 69–74
tundra, 69, 72
Swenk, Myron, 293
swifts: migration of, 36, 40, 122
chimney, 40, 122

tanagers: migration of, 327
scarlet: capturing for banding, 265; population level, 352, 353; sounds of, 353; wintering grounds, 136
scarlet-rumped, 134
summer, 139
tattlers, 7

teal: migration of, 69
blue-winged, 176, 293
green-winged, 32
Teale, Edwin Way, 36
temperate migrants, defined, 33
Tennessee warbler. *See* warblers
Terborgh, John, 148, 150, 151, 160, 164, 326, 332
terns
Arctic, 21, 39, 57, 92–93, 193
common, 176
least, 283
roseate, 176
Texas: Aransas National Wildlife Refuge, 284–86; Big Bend National Park, 323; birding in, 269, 270; Chihuahuan Desert, 323, 324–25; High Island, 268–70, 271
thermal updrafts: dependence of raptors on, 108; and flocking behavior, 118–19
thrashers: migration of, 327
brown: declining numbers, 338
thrushes: and army ant swarms, 140–41; diet of, 77, 137; migration of, 36, 327
Bicknell's: elevation to full species status, 168; and global climate changes, 369–70; interest in, 169; protection of, 168–69; wintering grounds, 151, 165–66
gray-cheeked: *vs.* Bicknell's thrushes, 168; migration of, 8, 59–60, 66, 212, 255; navigational abilities, 66–67
hermit: stopping over in Texas desert, 325
Swainson's, 9, 31, 143
white-chinned, 159
wood: capturing for banding, 265; and cowbirds, 148, 360–61, 364; declining numbers, 147–48, 331, 338, 351; migration of, 328; population level, 353; sounds of, 147, 345; tracking, 148; wintering grounds, 136
tidal marshes. *See* salt marshes; wetlands
Tierra del Fuego, Argentina, 313–14
time-and-direction programming, 67, 68

time-distance programming, 67, 68

Tinsman, Jeanne, 107, 108, 109, 110, 113, 116

titmice: as common bird-feeder species, 199; lumping of species, 167; in mixed flocks, 138

toucans: migration of, 21
    keel-billed, 134

towers, as bird risk, 338–39

towhees: declining numbers, 338; lumping of species, 167

trans-Gulf migration, 251–53

transmitters, 25, 42–43

tree swallow. See swallows

Tristan da Cunha islands, 87

tropical rain forests, 133, 136, 146–47, 332–33; and coffee, 149–52; preservation, 147, 148–52, 299

tropics. See neotropical migrants; tropical rain forests

trumpeter swan. See swans

tubenoses, 88

tundra swan. See swans

turkey vultures: counting, 110, 116, 123, 124; migration of, 108, 110, 123, 124, 327

turnstones, ruddy: description of, 308; migration of, 7, 9, 304, 306, 316

turtle-doves, 24

type specimens, 194–95

upland sandpiper. See sandpipers

U.S. Fish and Wildlife Service, 185, 224, 270

Vanda, María Liliana, 109, 110, 111, 113

Veracruz, Mexico, 105–25

Vermont: as migrant wintering grounds, 197–98

vertical migration, 21–22, 43, 146

vireos: calling with recorded bird sounds, 172; cowbird threat, 363; splitting of species, 167
    Bell's, 324–25
    black-capped, 363

black-whiskered, 172

blue-headed, 327, 345

Jamaican, 172

least Bell's, 363

white-eyed, 139, 262, 264

yellow-green, 271

Vogel, Peter, 158, 172

von Middendorff, Dr., 60, 62

wagtails, yellow, 20

Walcott, Charles, 62, 65

warblers: adaptability to environmental disturbance, 159–60; diet of, 77, 137, 353; gender segregation, 142–43; in Jamaica, 155, 156; migration of, 327; in mixed flocks, 139, 140; threats to, 26–27
    Arctic, 8, 9, 20
    Bachman's, 162–64
    bay-breasted, 353–54
    black-and-white, 140
    blackburnian, 328
    blackcap, 48–50
    blackpoll: diet of, 353, 354; migration of, 10–11, 12, 20, 21, 26–27, 367; population level, 353; threats to, 26–27; wing length, 20
    black-throated blue: calling with recorded bird sounds, 155; diet of, 170; gender segregation, 143; in Jamaica, 155, 170; studying winter vs. summer ecology, 158
    black-throated gray, 326
    black-throated green, 140
    blue-winged, 140, 264
    Canada, 46
    Cape May: diet of, 170, 353, 354; in Jamaica, 156, 170
    cerulean: as casualty of Gulf Coast fallout, 257; choice of habitat, 257–58; declining numbers, 351; migration of, 258; threats to, 257–58
    garden, 61, 64
    golden-winged, 140
    hermit, 32, 326

hooded: description of, 142; gender segregation, 142, 143; migration of, 142; wintering grounds, 142

Kentucky: and Gulf Coast fallout, 257; preparing for migration, 261; wintering grounds, 136

Kirtland's: breeding grounds, 164–65; cowbird threat, 165, 362, 363; wintering grounds, 164–65

magnolia: breeding range vs. wintering range, 44, 45; at Long Point Provincial Park, 336; migration of, 44

ovenbirds: calling with recorded bird sounds, 155, 172, 349; capturing, 349–51; in first-year migration, 143; at Hawk Mountain, Pennsylvania, 349–52; in Jamaica, 159; migration of, 328; sounds of, 349, 352

palm, 155, 156, 172

parula: calling with recorded bird sounds, 172; northern, declining numbers in Puerto Rico, 330

prairie, 172, 330

prothonotary, 249, 270

Swainson's: adaptability to environmental disturbance, 160; calling with recorded bird sounds, 155; choice of habitat, 258–59; declining numbers, 259; as endangered songbird, 258; and Gulf Coast fallout, 258; migration of, 259

Tennessee: description of, 141; diet of, 141, 353, 354; as forest birds, 268; migration of, 268; in mixed flocks, 141

waterthrushes
    Louisiana, 327
    northern, 338

Wilson's: declining numbers, 331; migration of, 8, 9, 143; stopping over in Texas desert, 325

worm-eating: adaptability to environmental disturbance, 159; calling with recorded bird sounds, 172; capturing for banding, 264; diet of, 250; in Jamaica, 159; migrant habitat, 151; migration of, 59; in mixed flocks, 140

yellow, 324, 325, 327

yellow-rumped: Audubon's warbler, 144; breeding grounds, 144; at Long Point Provincial Park, Ontario, 336, 337; myrtle warbler, 144; subspecies segregation, 144; wintering grounds, 144

yellowthroats, common: in Jamaica, 156; wintering grounds, 136

WatchList, 257, 342

water. See Platte River, Nebraska; Salton Sea, California

waterfowl: in American heartland, 276; diet of, 294; and global climate changes, 366; and migration as learned behavior, 68–74; migration of, 214, 245, 327; need for protection, 296, 297; nesting site preferences, 293–94; pollution deaths, 244, 245; population levels on prairies, 295–97; Salton Sea as wintering grounds, 240

waterthrushes. See warblers

waxwings, cedar, 35

weather, 47–48, 365; see also global climate changes

weather radar, 42, 255–56, 333–34

Webb, Thompson, III, 77

Weindler, Peter, 64

Western Hemisphere Shorebird Reserve Network, 242, 312–13, 314

wetlands: American heartland as major migratory corridor, 276; global significance list, 313; Leatherberry Flats, Delaware Bay, 226–27; loss of prairie potholes, 295; protecting, 296, 297, 312–14; and snow geese, 222–33; see also national wildlife refuges; salt marshes

wheatears, 8, 9, 19, 20

whimbrel, 13; see also curlews: bristle-thighed

white-crowned sparrow. See sparrows

white-eyed vireo. See vireos

white-faced whistling-duck. See ducks

white-rumped sandpiper. See sandpipers

white-tailed deer, 357

Whitman, Christine Todd, 316

whooping crane. *See* cranes
Wiedenfeld, David and Melissa, 256–57
Wilcove, David, 355
Williams, George G., 253–54, 255
Williams, Timothy C., 77
Wilson, Alexander, 320
Wilson, Edward O., 354–55
Wilson's warbler. *See* warblers
wind: role in bird migration, 79–80
wing feathers, 19–20
Wings Over the Platte Festival, 272, 284
Winker, Kevin, 143
Woodbridge, Brian, 177, 179–85
woodcocks, 327
woodpeckers: migration of, 36, 43; in
    mixed flocks, 138
    downy, 199

Jamaican, 159
Strickland's, 43
wood stork. *See* storks
wood thrush. *See* thrushes
wrens
    cactus, 324
    winter, 59
Wright, Angela, 145

yellowlegs, greater: in Argentina, 193;
    migration of, 8
yellow-rumped warbler. *See* warblers
yellowthroat, common. *See* warblers
yellow warbler. *See* warblers
Yukon River, Alaska, 13

Zugunruhe, 47, 48, 60